Folding for the Synapse

Andreas Wyttenbach • Vincent O'Connor
Editors

Folding for the Synapse

Editors
Andreas Wyttenbach
School of Biological Sciences
University of Southampton
Southampton
United Kingdom
aw3@soton.ac.uk

Vincent O'Connor
School of Biological Sciences
University of Southampton
Southampton
United Kingdom
voconno@soton.ac.uk

ISBN 978-1-4419-7060-2 e-ISBN 978-1-4419-7061-9
DOI 10.1007/978-1-4419-7061-9
Springer New York Dordrecht Heidelberg London

Library of Congress Control Number: 2010937760

Printed on acid-free paper

Springer is part of Springer Science+Business Media (www.springer.com)

Contents

1 Folding for the Synapse 1
Andreas Wyttenbach and Vincent O'Connor

Part I The Regulation of Protein (mis)folding, Aggregation and Degradation of Proteins by Molecular Chaperones

2 Protein Aggregation: Opposing Effects of Chaperones and Crowding 9
R. John Ellis

3 Molecular Chaperones as Facilitators of Protein Degradation 35
Jörg Höhfeld, Nikolaus Dick, and Verena Arndt

4 The Small Heat-Shock Proteins: Cellular Functions and Mutations Causing Neurodegeneration 49
C. d'Ydewalle, J. Krishnan, V. Timmerman, and L. Van Den Bosch

Part II The Making of the Synapse: Transport of Proteins, Vesicles and Organelles

5 Keeping it Together. Axonal Transport to the Synapse and the Effects of Molecular Chaperones in Health and Disease 81
Christopher Sinadinos and Amrit Mudher

6 Mechanisms of Neuronal Mitochondrial Transport 105
F. Anne Stephenson and Kieran Brickley

Part III Chaperone Modalities and Homeostatic Mechanisms in the Synaptic Compartment

7 **Molecular Chaperones in the Mammalian Brain: Regional Distribution, Cellular Compartmentalization and Synaptic Interactions** 123
Andreas Wyttenbach, Shmma Quraishe, Joanne Bailey, and Vincent O'Connor

8 **Cysteine-String Protein's Role at Synapses** 145
Konrad E. Zinsmaier and Mays Imad

9 **The Role of Protein SUMOylation in Neuronal Function** 177
Kevin A. Wilkinson and Jeremy M. Henley

10 **The Ubiquitin–Proteasome System in Synapses** 201
Suzanne Tydlacka, Shi-Hua Li, and Xiao-Jiang Li

Part IV Chronic Neurodegeneration Associated with Protein Misfolding and Synaptic Dysfunction

11 **VAPB Aggregates and Neurodegeneration** 215
P. Skehel

12 **Synaptic Dysfunction in Huntington's Disease** 233
Dervila Glynn and A. Jennifer Morton

13 **Synaptic Dysfunction in Parkinson's Disease: From Protein Misfolding to Functional Alterations** 257
Tiago Fleming Outeiro and Luísa Vaqueiro Lopes

14 **Synapses and Alzheimers's Disease: Effect of Immunotherapy?** 269
Nathan C. Denham, James A.R. Nicoll, and Delphine Boche

15 **Prion Protein Misfolding at the Synapse** 289
Zuzana Šišková, V. Hugh Perry, and Ayodeji A. Asuni

Index 313

Contributors

Delphine Boche
Division of Clinical Neurosciences, University of Southampton,
Southampton, UK
D.Boche@soton.ac.uk

R. John Ellis
Department of Biological Sciences, University of Warwick, Coventry,
CV4 7AL, UK
R.J.Ellis@warwick.ac.uk

Jeremy M. Henley
Department of Anatomy, MRC Centre for Synaptic Plasticity,
School of Biochemistry, Medical Sciences Building,
University of Bristol, University Walk, Bristol, BS8 1TD, UK
J.M.Henley@bristol.ac.uk

Jörg Höhfeld
Institut für Zellbiologie, Rheinische Friedrich-Wilhelms Universität Bonn,
Ulrich-Haberland-Str. 61a, D-53121 Bonn, Germany
hoehfeld@uni-bonn.de

Xiao-Jiang Li
Department of Human Genetics, Emory University School of Medicine,
Atlanta, GA, USA
Xli2@emory.edu

A. Jennifer Morton
Department of Pharmacology, University of Cambridge, Tennis Court Road,
Cambridge CB2 1PD, UK
ajm41@cam.ac.uk

Vincent O'Connor
Southampton Neuroscience Group, School of Biological Sciences,
Life Sciences Building (B85), University of Southampton,
Southampton SO17 1BJ, UK
voconno@soton.ac.uk

Tiago Fleming Outeiro
Cell and Molecular Neuroscience Unit, Instituto de Medicina Molecular, Av. Prof. Egas Moniz, 1649-028 Lisboa, Portugal
touteiro@gmail.com

Christopher Sinadinos
Southampton Neuroscience Group, School of Biological Sciences, University of Southampton, Highfield Campus, SO17 1BJ,UK
cas1@soton.ac.uk

Zuzana Šišková
School of Biological Sciences, Mailpoint 840 LD 80B, South Lab and Path Block, Southampton General Hospital, Southampton SO16 6XD, UK
Z.Siskova@soton.ac.uk

P. Skehel
Centre for Integrative Physiology, University of Edinburgh, Hugh Robson Building, George Square, Edinburgh EH8 9XD, UK
Paul.Skehel@ed.ac.uk

F. Anne Stephenson
School of Pharmacy, University of London, 29/39 Brunswick Square, London WC1N 1AX, UK
anne.stephenson@pharmacy.ac.uk

L. Van Den Bosch
Laboratory of Neurobiology, Department of Experimental Neurology, VIB and University of Leuven, Campus Gasthuisberg O&N2 PB 1022, Herestraat 49, Leuven B-3000, Belgium
ludo.vandenbosch@vib-kuleuven.be

Andreas Wyttenbach
Southampton Neuroscience Group, School of Biological Sciences, Life Sciences Building (B85), University of Southampton, Southampton SO17 1BJ, UK
aw3@soton.ac.uk

Konrad E. Zinsmaier
Department of Neuroscience, University of Arizona, Gould-Simpson Building 627, P.O. Box 210077, 1040 E. 4th Street, Tucson, AZ 85721–0077, USA
kez4@email.arizona.edu

Chapter 1
Folding for the Synapse

Andreas Wyttenbach and Vincent O'Connor

Abstract This book was *invited* after we had organized a symposium at the 2007 British Neuroscience Association (BNA) entitled "Synaptic origami: protein folding at the synapse." This poetic title was derived from our intention to drive a convergent discussion on protein folding pathways from insights on the emerging molecular neurobiology of synapse function and its involvement in major neurodegenerative brain diseases. This aim remains in its infancy but the original inception has encouraged Springer to facilitate the production of a book built around the concepts that fired the original symposia.

We believe that protein folding mechanisms that are classically studied in the context of protein biogenesis and degradation are inevitably going to figure as significant determinants in the regulation and maintenance of synaptic function. This seemed plausible on the basis of two distinct aspects of fields in which the editors have a particular background and/or interest. First, there was the detailed understanding that has continued to burgeon, highlighting how synaptic transmission is regulated by an intricate sequence of temporally and spatially precise protein–protein interaction cascades. There were already strong hints that the selective uses of classic protein chaperones known to regulate protein folding were important in the aspects of these protein interaction cascades.

The second input centers around the emerging understanding of neurodegenerative diseases pinpointing to a pivotal role for misfolded proteins in brain dysfunction. In view of the central role that the chaperones play in protein homeostasis, the significance of such activity during misfolding disease appears likely and there is indeed important research that has recently been undertaken to show this (Behrends et al. 2006; Fonte et al. 2008; Wacker et al. 2009; reviewed in Wyttenbach and Arrigo 2007). There is an increasing and continued indication that synaptic dysfunction occurs at an early point during the chronic neurodegeneration observed

A. Wyttenbach (✉) and V. O'Connor (✉)
Southampton Neuroscience Group, School of Biological Sciences, Life Science Building (B85), University of Southampton, Southampton SO17 1BJ, UK
e-mail: aw3@soton.ac.uk; voconno@soton.ac.uk

A. Wyttenbach and V. O'Connor (eds.), *Folding for the Synapse*,
DOI 10.1007/978-1-4419-7061-9_1,

during proteinopathies that ultimately end with widespread neuronal loss. There is clearly much to be learnt about how early the synaptic dysfunction is coupled to protein misfolding. However, because protein misfolding underpins diseases of the central nervous system (CNS), showing that an early synaptic dysfunction supports the notion, that investigation and discussion of synaptic origami will be fruitful.

Of course, the context for these musings is the synapse or more precisely the trillion or so synapses that ensure intercellular communication in the brain. The consequence of synaptic activity ensures the co-ordinated behavior of an organism. Specialized adherens junctions are at the core of synapses to allow molecular recognition between two cells through abutting sub-compartment specialization. These specializations are laid down during development and cause juxtaposition of a pre- and post-nerve via the pre- and postsynaptic compartments. Individual neuronal cell bodies will have several nerve processes along which many of these discrete synaptic compartments accumulate. What is clear from electron microscopy (EM) studies is that synapses are highly enriched in protein depositions. This is readily visualized in the EM by the accumulation of osmium stains at the points of contact between the pre- and post-nerve. Embedded in these protein densities are the inter-neuronal cell adhesion molecules, the underpinning cytoskeletal networks, and the sub-cellular organization of the membrane trafficking and cell signaling architectures fundamental to synaptic function (Palay and Palade 1955; De Robertis and Bennet 1955). The classical images of synapses are currently being refined by modern high magnification analysis, strongly reinforcing that these densities contain many multimeric protein complexes (Harlow et al. 2001). These findings reveal the synapse as a dense meshwork of protein architectures, which on its own suggests that protein modulatory cascades, including chaperone- and protein homeostasis, will be at play in assembling and maintaining function (Yi and Ehlers 2005; Granata et al. 2008; Sebeo et al. 2009).

The beauty of the static morphology of synapses belies the dynamic nature of these structures. At their heart is a millisecond signaling involving presynaptic release of transmitter and postsynaptic signaling by activation of juxtaposed postsynaptic receptors. This speed strongly argues for pre-organization and a highly dynamic protein and membrane biology. Central to this signaling is the vesicle-mediated neurotransmitter release in which neurotransmitter pre-stored synaptic vesicles release their content extracellularly by stimulated exocytosis. Exocytosis is stimulated by an influx of Ca^{2+} through ion channels that are accumulated in the pre-nerve and localize close to synaptic vesicles. The rapidity of this coupling is ensured by a co-ordinated biochemical activity that ensures that the vesicles that execute exocytosis are pre-activated (primed). Several factors are pivotal in generating key intermediates in transmitter release and modulating protein conformations. Hence this highlights the potential for protein folding pathways in transmitter release.

Synapses are relatively isolated from the cell body of neurons. Historically, the cell body has been recognized for its importance as the basis for biosynthesis and degradation in neuronal homeostasis, supplying the key proteins and lipids that co-ordinate the pre- and postsynaptic function. Indeed, efficient transport from the soma to the synapse is well studied. These are based on motor proteins such as kinesins and dyneins that transport material to and from the synapses as part of their anabolic and catabolic activities. This appreciation has been extended by realizing that synapses and their distinct pre- and postsynaptic compartments are capable of a critical activity autonomous from the cell body. This general principle is exemplified by the local membrane recycling associated with presynaptic transmitter vesicles and postsynaptic membrane receptors. Autonomous synapse bioactivity now includes several processes such as posttranscriptional regulation of mRNA, local protein translation, protein maturation, and degradation. Such biochemical autonomy is likely to require local protein chaperone pathways to sustain its activity.

The book uses a broad definition for protein homeostasis and chaperones. A major focus is on activity that assists in the refolding of protein from one conformation to another. Classically, first, this involves the use of ATPase activities, but we have asked contributors to consider additionally activities of chaperones that bind or ameliorate misfolded states. Second, the context of our discussion has been the synapse. Our emphasis is that chaperone function is fundamental to synapse function and we have encouraged contributions that directly show the central role played by chaperone or co-chaperone activity for synaptic function. A major burden for neuronal chaperones is the misfolded proteins associated with neurodegeneration. Accordingly, we have encouraged the inclusion of chapters addressing major brain diseases.

The first section of this book with contributions from J. Ellis, J. Höhfeld, and L. Van den Bosch and colleagues is dedicated to introduce aspects of the regulation of protein folding and misfolding, and degradation of proteins my molecular chaperones. Hence the book opens with a synopsis of what protein chaperones are, and underlines the need to understand such molecular chaperone activity in the context of crowded protein environments that exist in cells relative to those artificially used in the test tube. In view of the dense proteinaceous nature of synaptic specializations, the general coverage provided by J. Ellis is worth thinking about in the specific context of the synapse (see above). Synaptic sites harbor the most concentrated membrane protein and/or cytomatrix sub-domains. The following chapter by J. Höhfeld and colleagues details how the chaperones introduced by J. Ellis couple to the degradative pathways involving ubiquitin. This work provides insight into a fundamental process of protein homeostasis that is reiterated in several subsequent chapters. The final chapter of this section focuses attention on a class of chaperones, the small heat shock proteins (sHSPs) that are collaborators in the pathways that use classic ATP-dependent refolding. Although mechanistic details of the cell biology and biochemistry of these proteins are currently not well defined, they play essential roles in neuronal function. As discussed by Van den Bosch and colleagues, mutations in sHSPs are associated with several diseases, highlighting the importance of proper functioning of chaperone networks to sustain human health.

Chapter 5 and 6 focus on the transport of organelles and vesicles to the synapse. The neuronal cell bodies' ability to supply most of the components for synaptic function and the contribution to its homeostasis require selective communication. In this part, the book's focus is on important protein complexes and describes the guiding principles of intraneuronal transport. Anne Stephenson and colleague discuss mitochondrial transport. Central to such an example of selective transport is the ability of a neuron to choose carefully a cargo for transport and utilize motor proteins with their underpinning cytoskeleton elements to move a designated cargo. Both processes make use of protein chaperoning activity to regulate the assembly and disassembly of the key protein complexes. This idea is then more distinctively discussed by C. Sinadinos and A. Mudher, also linking the role of chaperone pathways to dysfunctions causing protein misfolding disease.

The distribution and compartmentalization of molecular chaperones in the mammalian brain are not well defined, and with the exception of the trimeric chaperone complex (see below), their recruitment to the synapse and functioning in key synaptic processes are largely unknown. Hence, in four further contributions (Chaps. 7–10) we aimed at summarizing and discussing the distribution and compartmentalisation of molecular chaperones in the mammalian brain and the synapse and focus on key synaptic processes such as exo- and endocytosis, protein turnover by the ubiquitin–proteasome system (UPS), and the role of protein SUMOylation during neuronal signaling, an exciting emerging concept.

Many of the chaperones and their associated regulatory mechanism of protein homeostasis are variously recruited to key synaptic processes. Accordingly A. Wyttenbach, S. Quraishe, and V. O'Connor describe how some key classes of chaperones distribute to the synapse and highlight where they function in specific synaptic processes. We also included a discussion on what are now classical examples of proteins with a synaptic chaperone function (e.g., NSF and Hsc-70). How generic chaperone activity can selectively be used for synaptic function is directly addressed in the subsequent chapter by K. Zinzmaier and M. Imad. These authors describe how the synaptic vesicle protein cysteine string protein (CSP) recruits Hsc-70 activity. In many ways, the functional investigation of this co-chaperone/chaperone interaction has set important agendas revealing that deficiency in synaptic origami leads to neuronal degeneration. Zinzmaier's contribution highlights why CSP is a standout paradigm for the concept and potential significance of synaptic origami. Nevertheless, it also makes clear that there is still much to be learnt about this well-studied molecular cascade that is essential for synaptic function. Li and colleagues then go on to address specifically the synaptic role of the UPS. This work highlights the emergence of this arm of protein regulation in a synaptic context and also outlines its potential significance in a protein misfolding disease. The chapter that follows introduces a relatively novel player in the regulation of synaptic protein dynamics: K. Webster and J. Henley introduce the ways in which the covalent exchange of SUMO protein may contribute to synaptic function in health and disease.

The nine chapters highlight key facets of chaperone and protein homeostasis pathways, and where possible, authors have aimed at discussing how these impact on synaptic functions. However, it is in the major age-related diseases associated with protein misfolding for which we see an emerging significance for chaperone pathways. There are initial hints as to how such disease mechanism may impact on protein folding to disturb synaptic function (see above). However, it is the untapped potential significance of these pathways that provides a potential route to new therapeutic avenues. To promote thinking around this potential, we have invited contributions that focus on five different protein misfolding diseases (also called "proteinopathies"). These include three disorders that are primarily driven by misfolding of intracellular proteins and two disorders that are associated with the accumulation of extracellular misfolded protein aggregates. In all diseases, disruption of various cellular functions occurs and in all cases, the potential for synaptic dysfunction is addressed. The disruption of neuronal function in the various proteinopathies is not generic, but dysfunctions normally occur in sub-types of neurons and their distinct systems regulated by the particular neuronal cell types. Where possible, the authors have included some discussion on the current thinking of why selective neurodegeneration is observed during neuropathology, despite the main triggers of disease exhibiting a potential for targeting all neuronal and synaptic populations. A possible answer here is that a differential buffering of misfolding insults occurs in distinct types of neurons. Indeed, this possibility highlights the additional need to further understand the interactions between protein chaperones and disease.

The first chapter of this section (P. Skehel) considers a motor neuron disease associated with a mutation of the ubiquitously expressed ER protein VAP. This is a less studied model of motor neuron disease (also see Chap. 4 for consideration of motorneuropathies), but highlights the general principle that mutations leading to protein misfolding can selectively activate degeneration in the motor neurons. An interesting aspect of this model is that the disruptive signaling emanates from the ER involving consequences for transcription and perhaps downstream synaptic signaling. This is followed by a rigorous discussion on the role of intracellular protein misfolding and synaptic dysfunction that are involved in Huntington's Disease (HD). In this chapter, D. Glynn and A. J. Morton highlight the significance of synaptic dysfunction in HD. The chapter revisits many of the functionally important synaptic cascades discussed earlier in the book to address how a polyglutamine (polyQ) expansion mutation and its associated polyQ protein misfolding disrupt synaptic function. T. Outerio and F. Lopez then consider a human condition that emerges from several potential misfolding insults causing Parkinson's Disease (PD). The chapter discusses both the molecular basis of the misfolding focusing on the protein alpha-synuclein and how this disrupts the synaptic and neuronal signaling associated with PD. The final chapters describe diseases associated with the extracellular accumulation of a peptide (Aβ) and a protein (prion protein). Despite their different ontogenies, both Alzheimer's Disease (AD) and prion diseases share the common feature that disease progression is associated with increased loss of synapses. In prion disease, the misfolding insult arises from a widely expressed cell surface protein, and the reason why accumulated misfolded protein leads to selective synaptic dysfunction remains unclear. However, the present understanding and some likely possibilities that may underlie synaptic dysfunction are outlined by S. Zizkova, V. H. Perry, and A. Asuni. The final chapter addresses the key features of potentially the most burdensome disease of the modern age, AD. N Denham, J. Nicoll, and D. Boche outline the significance of dysfunction and loss of synapses in AD and discuss the current understanding of how misprocessing and misfolding of the amyloid precursor protein (APP) bring about dysfunction in disease. The authors also include their views on the recent interest in immunotherapy for AD.

Overall, the book lays down the core features of cellular chaperone functions and links these to the synaptic sub-compartment. It also outlines how disease processes might hijack or disrupt the efficient use of molecular chaperones in synaptic compartments. Individually, chapters should promote our understanding of the protein folding per se and reinforce the current understanding of some major diseases of the human nervous system. However, by assembling the distinct components into one tome, we hope to increase the reader's appreciation of how protein folding pathways and their associated activities converge to control synaptic function and dysfunction. It is likely that such an integrated understanding will help our research and ultimately accelerate the successful treatment of major brain diseases.

References

Behrends C, Langer CA, Boteva R, Böttcher UM, Stemp MJ, Schaffar G, Rao BV, Giese A, Kretzschmar H, Siegers K, Hartl FU. (2006) Chaperonin TRiC promotes the assembly of polyQ expansion proteins into nontoxic oligomers. Mol Cell. 23(6):887–97.

De Robertis, E.D.P., and Bennett H.S. (1955) Some features of the submicroscopic morphology of synapses in frog and earthworm. J. Biophys. Biochem. Cytol. 1:47–58.

Fonte V, Kipp DR, Yerg J 3rd, Merin D, Forrestal M, Wagner E, Roberts CM, Link CD. (2008) Suppression of in vivo beta-amyloid peptide toxicity by overexpression of the HSP-16.2 small chaperone protein. J Biol Chem. 283(2):784–91

Granata A, Watson R, Collinson LM, Schiavo G, Warner TT. (2008) The dystonia-associated protein torsinA modulates synaptic vesicle recycling. J Biol Chem. 283(12):7568–79.

Harlow ML, Ress D, Stoschek A, Marshall RM, McMahan UJ (2001) The architecture of active zone material at the frog's neuromuscular junction. Nature 409:479–484.

Palay SL., and Palade GE (1955) The fine structure of neurons. J Biophys Biochem Cytol. 1(1):69–88.

Sebeo J, Hsiao K, Bozdagi O, Dumitriu D, Ge Y, Zhou Q, Benson DL. (2009) Requirement for protein synthesis at developing synapses. J Neurosci. 29(31):9778–93

Wacker JL, Huang SY, Steele AD, Aron R, Lotz GP, Nguyen Q, Giorgini F, Roberson ED, Lindquist S, Masliah E, Muchowski PJ. (2009) Loss of Hsp70 exacerbates pathogenesis but not levels of fibrillar aggregates in a mouse model of Huntington's disease. J Neurosci. 29(28):9104–14

Wyttenbach A., and Arrigo AP. (2007) The Role of Heat Shock Proteins during Neurodegeneration in Alzheimer's, Parkinson's and Huntington's Disease. In: Heat shock proteins in neural cells. Edited by: C. Richter-Landesberg. Landes Bioscience. Springer.

Yi JJ., and Ehlers MD. (2005) Ubiquitin and protein turnover in synapse function. Neuron. 47(5):629–32.

Part I
The Regulation of Protein (mis)folding, Aggregation and Degradation of Proteins by Molecular Chaperones

Chapter 2
Protein Aggregation: Opposing Effects of Chaperones and Crowding

R. John Ellis

Abstract Each molecule of every protein runs the risk that at any time between its synthesis and its degradation, it will bind to one or more identical molecules to form a nonfunctional aggregate. Some protein aggregates are toxic to cells, including neurones, and are thus factors in the development of neurodegenerative and other human diseases. The incidence of such diseases is increasing, together with human longevity and obesity. The probability of protein aggregation is increased by the crowded state of most intracellular compartments, but is reduced by the activities of a diverse range of proteins acting as molecular chaperones. These chaperones use a variety of mechanisms to combat aggregation during the folding of newly synthesized protein chains, their transport into and across membranes, and their assembly into functional oligomers. This article discusses some of the key concepts and basic evidence underlying these conclusions.

2.1 History

While researching for my Ph. D. degree, I was fortunate enough to attend the famous lecture by Francis Crick entitled "*On protein synthesis*" that he gave in 1958 at a symposium of the Society for Experimental Biology (Crick 1958). In this talk, Crick explained that the folding of newly synthesized protein chains is "simply a function of the order of amino acids." This view was based on the pioneering studies of Christian Anfinsen, who discovered serendipitiously that pure denatured ribonuclease A will refold in the test tube into an active enzyme when the concentration of denaturing agent is lowered sufficiently (Anfinsen 1973). Moreover, the denatured chains refold *spontaneously*, i.e., in the absence of either other macromolecules or an added energy source. Many subsequent repetitions of this type of in vitro refolding experiment with a variety of denatured pure proteins have

---

R.J. Ellis (✉)
Department of Biological Sciences, University of Warwick, Coventry, CV4 7AL, UK
e-mail: R.J.Ellis@warwick.ac.uk

A. Wyttenbach and V. O'Connor (eds.), *Folding for the Synapse*,
DOI 10.1007/978-1-4419-7061-9_2,

established the principle of protein self-assembly; i.e., the idea that all the information required for a protein chain to fold into its functional conformation is contained within the primary structure of the initial translation product. This principle also applies to the association of folded protein monomers with one another or with other macromolecules, as demonstrated by the work of Caspar and Klug (1962) on virus assembly. The principle of parsimony ("Occam's razor") suggests that protein chains may also fold and assemble spontaneously inside the cell as they do in the test tube, but this conclusion ignores the fact that the conditions under which proteins fold and assemble inside cells are very different from those used to perform in vitro refolding experiments. What some of these conditions are, and their consequences for protein folding and assembly will be discussed later in this article. This will also be of importance in light of the other chapters in this book dealing with folding and misfolding of proteins within the synaptic compartment of neurons. We now know that many proteins require both the expenditure of energy in the form of ATP hydrolysis, and the assistance of pre-existing proteins, before they will fold and assemble correctly inside the cell. This change of view began with the discovery of molecular chaperones (reviewed in Ellis and Hemmingsen 1989; Ellis 2004).

A glance at the chaperone literature shows that it is a common misconception that the function of molecular chaperones is primarily, or even solely, to assist the folding of polypeptide chains into monomers in the cytosol. In fact, the term "molecular chaperone" was coined by Ron Laskey to describe the function of an acidic nuclear protein, called nucleoplasmin, that solves an aggregation problem during the assembly of nucleosomes from folded histones inside the nuclei of amphibian eggs (Laskey et al. 1978). Nucleosomes consist of histone proteins bound to DNA by electrostatic interactions. Disruption of these interactions requires high salt concentrations (>1 M NaCl), but exposing mixtures of isolated DNA and histones to the salt concentrations found inside the nucleus (~ 0.1 M NaCl) results in the formation of insoluble aggregates rather than the assembly of nucleosomes. In other words, nucleosomes fail to conform to the principle of protein self-assembly. Nucleoplasmin solves this aggregation problem by transiently binding its acidic groups to positively charged groups on the histones, thus lowering their overall surface charge and allowing the intrinsic self-assembly properties of the histones to predominate over the incorrect interactions favored by the high density of opposite charges (reviewed in Ellis 2006). Control experiments show that nucleoplasmin neither provides steric information essential for histones to bind correctly to DNA, nor is it a component of assembled nucleosomes. It is these two latter features that laid the basic foundations for our current general concept of the chaperone function.

The term "molecular chaperone" was later extended to include an abundant chloroplast protein called the rubisco large subunit-binding protein that functions to keep newly synthesized rubisco large subunits from aggregating with one another until they assemble into the rubisco holoenzyme inside plant chloroplasts (Musgrove and Ellis 1986). These rubisco subunits are notoriously prone to aggregation, not because of electrostatic interactions, but because they expose highly hydrophobic surfaces to the aqueous environment. For a while the term "molecular chaperone" was restricted to the two proteins that assist the assembly of amphibian nucleosomes and chloroplast rubisco, but its modern usage started when I suggested the term could be usefully extended to describe the function of a larger range of proteins postulated to assist folding and assembly/disassembly reactions in a wide range of cellular processes (Ellis 1987).

The identification of the chaperonin family of molecular chaperones in the following year (Hemmingsen et al. 1988) triggered a tidal wave of research in several laboratories aimed at unraveling how the GroEL/GroES chaperones, and later the DnaK/DnaJ chaperones, from *Escherichia coli* facilitate the folding of newly synthesized polypeptide chains and the refolding of mature proteins unfolded by stress. This wave of research continues to surge, with the result that much detailed information is now available about the structure and function of the families of chaperone that assist protein folding (reviewed in Chakraborty et al. in press). Other families assist the assembly of folded proteins into structures such as proteasomes and ribosomes, as well as nucleosomes, suggesting that we should distinguish *folding* chaperones from *assembly* chaperones (reviewed in Ellis 2006). Thus, chaperones are not defined by either their structure or by their mechanism of action but by their function in facilitating protein folding and protein assembly.

2.2 Current View of the Chaperone Function

The chaperone function is currently defined as the prevention and/or reversal of incorrect interactions that might occur when potentially interactive macromolecular surfaces are exposed to the intracellular environment (Ellis 2006). These surfaces occur on nascent and newly released protein chains, on proteins entering or crossing membranes, on mature proteins unfolded by environmental stress, and on folded proteins in native or near-native conformations. The same concept applies to other macromolecules that can undergo incorrect interactions, especially RNA. Incorrect interactions are defined as those that result in products that fail to carry out the biological functions which they were selected for in evolution.

The suggestions made in the first comprehensive postulate of the chaperone function have so far stood the test of time (Ellis and Hemmingsen 1989). Molecular chaperones are now defined as a large and diverse group of proteins that share the property of assisting the non-covalent assembly/disassembly of other macromolecular structures, but which are not permanent components of these structures when these are performing their normal biological functions (Ellis 2006). Assembly is used here in a broad sense and includes several universal intracellular processes, such as the folding of nascent polypeptide chains, both during their synthesis and after release from ribosomes; the unfolding and refolding of polypeptides during their transfer across membranes; and the association of polypeptides with one another and with other macromolecules to form oligomeric complexes. Molecular chaperones are also involved in macromolecular *dis*assembly processes such as the partial unfolding and dissociation of subunits when some proteins carry out their normal functions, and the resolubilization and/or degradation of proteins partially denatured and/or aggregated by mutation or by exposure to environmental stresses, such as high temperatures and oxidative conditions. Some, but not all, chaperones are also stress or heat shock proteins (Hsps) because the requirement for chaperone function increases under stress conditions that cause proteins to unfold and aggregate. Conversely, some, but not all, stress proteins are molecular chaperones. Chaperones and stress proteins are thus ovelapping sets, not identical sets. This distinction is often confused in the literature.

It is important to note that the definition of molecular chaperones given above is functional, not structural, and that it contains no constraints on the mechanisms by which different chaperones may act: this is the reason for the use of the imprecise term "assist." On my definition, only two criteria need be satisfied to designate a macromolecule a molecular chaperone. First, it must in some sense assist the non-covalent assembly/disassembly of other macromolecular structures, the mechanism being irrelevant, and second, it must not be a component of these structures when they are performing their normal biological functions. In all cases studied so far, chaperones bind non-covalently to regions of macromolecules that are inaccessible when these structures are correctly assembled and functioning, but that are accessible at other times. Such regions commonly include hydrophobic regions during protein folding and charged regions during protein assembly.

The term "non-covalent" is used in this definition to exclude proteins that catalyze co- or posttranslational covalent modifications. These modifications are often important for protein folding and assembly but are distinct from the proteins being considered here. Protein disulfide isomerase may appear to be an exception, but it is not. It is both a covalent modification enzyme *and* a molecular chaperone, but these activities lie in different parts of the molecule and can be functionally separated by mutation (Puig and Gilbert 1994; reviewed by Freedman 2008). Other examples include peptidyl-prolyl isomerase, which possesses both enzymatic and chaperone activities in different regions of the molecule, and the alpha-crystallins, which in the lens of the eye combine two essential functions in the same molecule – contributing to the transparency and refractive index required for vision, and the chaperone function that combats the loss of transparency as the protein chains aggregate with increasing age (Horwitz 2000). There is no reason in principle why molecular chaperones should not possess additional functions. For example, it has been proposed that some chaperones act to buffer phenotypic effects of mutations (Queitsch et al. 2002; Cowen and Lindquist 2005), while others may possess cell–cell signaling functions (Henderson and Pockley 2005). Table 2.1 lists the various types of specific functions suggested for molecular chaperones.

The number of distinct chaperone families continues to rise, and examples occur in all types of cell and in most intracellular compartments. The families are defined on the basis that members within each family have high sequence similarity but members in different families do not. Table 2.2 presents an incomplete list of proteins described as chaperones, but it must be emphasized that in many cases, this description rests on in vitro data only and needs confirmation by in vivo methods. There is evidence that some chaperones cooperate with one another in defined reaction

Table 2.1 Specific functions suggested for molecular chaperones

1. Prevention or modification of protein aggregation.
2. Disassembly of macromolecular assemblies and aggregates.
3. Regulation of intracellular signal transduction.
4. Helping decide when a protein should be degraded.
5. Prevention of premature folding during protein transport.
6. Buffering genetic variation.
7. Mediating cell–cell signaling.

Table 2.2 Incomplete list of proteins described as molecular chaperones

Family	Proposed roles
1. Nucleoplasmins/nucleophosmins	Nucleosome and ribosome assembly/disassembly
2. Chaperonins (Hsp60)	Folding of newly synthesized and denatured polypeptides
3. Hsp27/28	Prevention of stress-induced aggregation by adsorbing unfolded chains
4. Hsp40	Protein folding and transport, oligomer disassembly
5. Hsp47	Procollagen folding in the endoplasmic reticulum
6. Hsp70	Protein folding and transport, oligomer disassembly
7. Hsp90	Cell cycle, hormone activation, signal transduction
8. Hsp100	Dissolution of insoluble protein aggregates
9. Calnexin/calreticulin	Folding of glycoproteins in endoplasmic reticulum
10. SecB protein	Protein transport in bacteria
11. Lim protein	Folding of bacterial lipase
12. Syc proteins	Secretion of toxic YOP proteins by bacteria
13. Protein disulfide isomerase	Prevention of misfolding in endoplasmic reticulum
14. ExbB proteins (may be structural rather than chaperones)	Folding of TonB protein in bacteria
15. Ubiquitinated ribosomal proteins	Ribosome assembly in yeast
16. NAC	Folding of nascent polypeptides
17. Signal recognition particle	Arrest of translation and targeting to ER membrane
18. Trigger factor	Folding of nascent polypeptides in bacteria
19. Prefoldin	Cooperation with chaperonins in folding of newly synthesized polypeptides in *Archaea* and the eukaryotic cytosol

(continued)

Table 2.2 (continued)

Family	Proposed roles
20. Tim9/Tim10 complex	Prevention of aggregation of hydrophobic proteins during import across mitochondrial intermembrane space
21. 23S Ribosomal RNA	Folding of nascent polypeptides
22. PrsA protein	Secretion of proteins by *Bacillus subtilis*
23. Clusterin	Extracellular animal chaperone
24. Phosphatidylethanolamine	Folding of lactose permease
25. RNA binding proteins	Folding of RNA
26. P45	Protection against denaturation in halophilic *Archaea*
27. PapD proteins	Assembly of bacterial pili
28. Propeptides (Class I)	Folding of some proteases

sequences, but this, along with many other aspects of "chaperonology," is beyond the scope of this chapter. The vital roles of *co-chaperones*, defined as proteins that modulate the activity of chaperones, are discussed in the chapter by Jörg Höhfeld in this volume with respect to protein degradation.

It is now well established that a subset of proteins requires the chaperone function, not because chaperones provide steric information required for correct folding, but because chaperones inhibit deleterious side reactions. Our paradigm of protein folding and assembly in the cell has thus changed from the earlier view that it is a *spontaneous* self-assembly process, to the current view that it is an *assisted* self-assembly process. In this new view, the basic principle of self-assembly is retained, but is modified by the need for proteins to avoid deleterious side reactions inside the cell. The principal side reaction to avoid is protein aggregation (Ellis and Minton 2006).

2.3 Protein Aggregation

It has long been noted during protein refolding studies that a fraction of the protein fails to refold correctly but instead aggregates irreversibly into nonfunctional structures. The fraction that aggregates varies with the nature of the protein, but increases with temperature and protein concentration. These aggregates may be as small as dimers or large enough to be insoluble.

It used to be thought that protein aggregation is merely an annoying, uninteresting complication of in vitro experiments because the emphasis of refolding research was on determining the rules of folding. It is now clear, however, that protein aggregation is a serious and universal problem for all cells because it can potentially reduce the efficiency of folding as well as lead to the formation of toxic products. Protein aggregation not only produces inclusion bodies in bacteria but also contributes to distressing human diseases such as Alzheimer's disease, Parkinson's disease, Huntington's diseases, type 2 diabetes, sickle cell anemia, systemic amyloidoses, and the prion diseases (Dobson et al. 2001; Chiti and Dobson 2006).

2.3.1 *Definitions*

Aggregation is defined as the production of nonfunctional structures by the binding of two or more identical or very similar protein chains to one another. Aggregation should be distinguished from both precipitation and coagulation. Precipitation is the decrease in protein solubility due to changes in solvent, but in contrast to aggregation, it is not accompanied by changes in structure and is thus used to purify proteins. Coagulation is the increase in particle size caused by denaturation that is large enough to cause the particles to separate out. Coagulation is thus one extreme type of aggregation, the other extreme being the formation of dimers that remain soluble, but are nevertheless nonfunctional compared to the native protein.

One problem with defining protein aggregation in terms of the production of nonfunctional assemblies is that examples of functional assemblies with similar types of tertiary structure are being discovered in both prokaryotes and eukaryotes (Chiti and Dobson 2006). One solution to this semantic problem would be to refer to the nonfunctional assemblies as misassemblies rather than as aggregates. This usage has the advantage that is contrasts logically with the term "assembly," which refers to the formation of *functional* oligomers. However, the term "aggregate" is so embedded in the literature that it is probably more realistic to use the term "functional aggregates" to describe these recently discovered structures.

What about the term "misfolding"? Its exact definition is not agreed. The traditional view is that misfolded protein chains are those that are on the correct folding pathway but require some unfolding to reach the native conformation (Lazardis and Karplus 1997). This definition does not include a biological criterion, as does the definition of the term "aggregation." In my view, misfolding should be defined as meaning that a protein chain has reached a conformation that is stable enough not to be able to proceed to the functional conformation on a biologically relevant time scale.

Current models of protein folding include the view that misfolded chains form a subset of partially folded chains that are intermediates in the correct folding process, and, like them, vary in their tendency to aggregate with one another (Dinner et al. 2000). Thus all aggregates are by definition, misfolded but to what extent misfolding per se, as distinct from aggregation, causes cellular problems is not clear. There is one potential source of misfolding per se, and that is mutation. However, this is a very minor problem because mutation is minor – every time a human cell replicates its DNA, about three base pairs on average are changed out of a total of three billion basepairs. This is a very small effect compared to the risk of the thousands of copies of each polypeptide made during the lifetime of each cell aggregating together for the reasons described in Sect. 1.3.5. In addition, clear examples of misfolded, but monomeric, protein chains that are toxic to cells have not yet been found. In other words, the problem for cells appears to be protein aggregation, not protein misfolding as such.

2.3.2 Structure of Protein Aggregates

The diseases associated with protein aggregation are characterized by the accumulation of organized fibrillar structures called amyloid fibrils or plaques, visible in the light microscope. The term "amyloid" was suggested by the Russian cell biologist Virchow in 1853 because he thought from their staining properties with iodine that they are composed of carbohydrate. Ten years later, it was shown that they are made of protein, but the name has stuck. Some protein aggregates are amorphous however, such as the inclusion bodies formed when some eukaryote proteins are expressed in bacteria, and thus little is known about their structure.

Amyloid fibrils occur both inside and outside cells, and can also be formed in the laboratory by incubating pure proteins under partly denaturing conditions. These fibrils are insoluble in aqueous buffers; so until recently, there were no high resolution crystal structures, only low resolution electron microsco+pic or atomic force microscopic images. However, soluble and solid state NMR techniques have now been applied to the Aβ peptides found in Alzheimer's disease (Petkova et al. 2002; Luhrs et al. 2005), and high-resolution X-ray crystallography to crystals of peptides derived from the Sup35 amyloid found in yeast (Nelson et al. 2005).

Amyloid fibrils usually consist of 2–6 unbranched protofilaments, each about 2–5 nm in diameter. These protofilaments are either twisted around one another or associated laterally into fibrils of 4–13 nm diameter (Serpell et al. 2000). The fibrils give an X-ray reflection at 4.7 Å, indicating a cross-β structure, i.e., within each protofilament, the protein chains are arranged into β strands that are at right angles to the long axis of the fibril. This cross-β arrangement is thought to be a common molecular feature of amyloid fibrils, as is the ability of the fibrils to bind dyes such as thioflavin T and Congo red. How specific this dye binding actually is for amyloid fibrils, however, is under debate (Bousset et al. 2004). Not all the protein chains in an amyloid fibril are necessarily organized in a cross-β structure, but may retain some elements of the native structure, including enzymic activity (Nelson et al. 2005; reviewed by Rousseau et al. 2006).

The surprising fact that all amyloid fibrils, whether formed biologically or in the laboratory, have a similar cross-β structure, regardless of the sequence of the protein from which they are formed, has been interpreted to mean that the fibrils are held together by main chain interactions that are possible in principle for all proteins (Dobson 1999). Thus, the ability to form these structures could be a generic feature of all polypeptide chains, and the amyloid fibril state could be a stable default state for all proteins. This suggestion contrasts with the traditional view, derived from the study of functional globular proteins, that stable proteins result only from the formation of distinct conformations created by close stereospecific packing of side chains. The generic hypothesis is supported by the observation that many proteins, not known to be associated with disease, can be induced to form amyloid fibrils under partially denaturing conditions. However, sequence diversity can influence the rate and extent of fibril formation via stereospecific interactions between side chains, and there are computer algorithims that predict the tendency to aggregate from sequence data (reviewed by Idicula-Thomas and Balaji 2007; Rousseau et al. 2006). Data from a range of structural techniques are interpreted to indicate that in the case of the Alzheimer β 1–42 peptide, residues 17–42 adopt a strand-turn-strand motif that contains two intermolecular parallel in-register β sheets. Thus each peptide molecule contributes a pair of β strands to the core of the fibril. These strands participate in the formation of two distinct β sheets within each protofilament (Luhrs et al. 2005; Fig. 2.1). Thus β strands run perpendicular to the fibril axis, while β sheets run parallel to the fibril axis. The sheets are stabilized by hydrophobic side chain interactions between the sheets.

Knowledge of the structure of amyloid fibrils is being used in developing inhibitors of amyloid fibril growth that may have therapeutic potential. However, there is a growing body of evidence that the toxic factors in the development of Alzheimer's and Parkinson's disease are small, soluble aggregates rather than the insoluble amyloid fibrils (Chiti and Dobson 2006). According to this view, the fibrils are stable structures that function to sequester the intermediate aggregates in a relatively non-toxic state. The emphasis of research is thus shifting to studying the early stages in the formation of protein aggregates in the hope that therapeutic treatments targeted at these stages may be more effective. But such an approach requires the development of methods to diagnose the onset of protein aggregation disease well before symptoms are apparent, and this is probably the most challenging problem in this field.

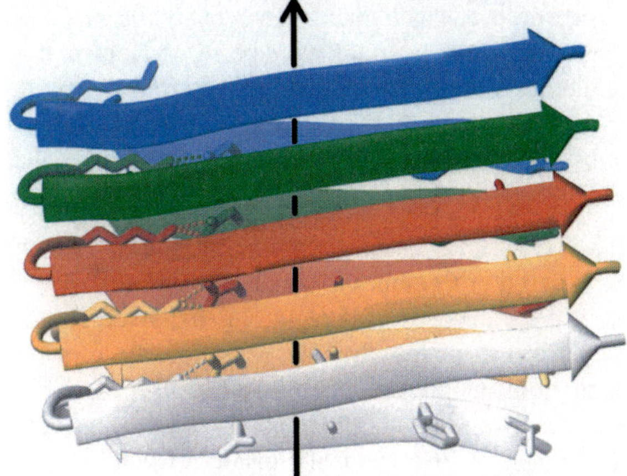

Fig. 2.1 Ribbon diagram of the core structure of residues 17–42 in the amyloid fibril formed from the Aβ (1–42) Alzheimer peptide. The arrows and colours indicate individual β strands. Reproduced from Luhrs et al. (2005) with the permission of the National Academy of Sciences, USA

2.3.3 Mechanisms of Protein Aggregation

A general outline of the process of protein aggregation, deduced from in vitro experiments with native globular proteins, is as follows.

1. The first step requires some partial unfolding of the protein structure; partial unfolding is induced by treatments such as low pH, high temperature, high pressure, or exposure to organic solvents or other chaotropes, especially urea and guanidinium chloride. These treatments share in common the ability to disrupt the various non-covalent interactions that stabilize native protein conformations.
2. In the second step, partially folded structures interact together in a highly specific fashion to produce the soluble assemblies that we call aggregates, because they have lost the biological properties characteristic of the native conformations. The high specificity of aggregate formation indicates that the interacting molecules have a degree of secondary/tertiary structure, i.e., that they are partly unfolded. This high specificity of protein aggregation was established first in 1974 by adding total crude extracts of *E. coli* cells to 8 M urea in a standard Anfinsen refolding experiment applied to tryptophanase. The recovery of enzymatically active tryptophanase was unaffected by the addition of 100 times as much total protein from *E. coli* to the urea-containing buffer (London et al. 1975). Thus the several thousand different unfolded proteins in the *E. coli* extract do not affect the competition between correct folding and aggregation. More recent work on the specificity of protein aggregation has revealed that sequence identities of adjacent homologous domains in multidomain proteins have evolved to be below the level that favors aggregation (Wright et al. 2005). Totally unfolded protein chains, such as those found in the presence of 8 M urea or 6 M guanidinium chloride, do not form aggregates until the concentration of denaturant is lowered. Because aggregation is a high-order process, its rate increases rapidly with protein concentration in a nonlinear fashion. This step is also characterized by a lag phase that can be abolished by adding small concentrations of fibrillar aggregates of the same protein. This feature is interpreted to indicate that this step proceeds by a nucleation event that limits the rate of the overall process.
3. The third step overlaps with the second step and involves the polymerization of the initial intermediates into either amorphous or fibrillar structures. This step proceeds much faster than the nucleation step, and is influenced by sequence-based properties such as hydrophobicity, net charge, and secondary structure propensities (Chiti et al. 2003; reviewed by Rousseau et al. 2006).

The above scenario is based on in vitro experiments in which purified native proteins are exposed in test tubes to conditions very different from those in which they have evolved to function inside cells. In such experiments, the protein concentration is almost always too low, the pH is often nonphysiological (some published experiments on amyloid stability were performed at pH 1.8), and the ionic strength is usually too low and always too simple in composition, e.g., the potassium ion concentration is often far below that occurring inside cells and magnesium ions are usually absent.

It is an unfortunate feature of this field that some protein chemists show a reluctance to include in their experiments attempts to mimic the intracellular environment in which the proteins they study have evolved (Ellis 2000). So it is not clear how relevant the above scenario is to understanding protein aggregation in human disease. We need to consider in what ways protein folding and assembly in the cell differ from studies of protein folding and assembly in the test tube.

2.3.4 *Protein Aggregation Inside the Cell*

The functioning of globular proteins often requires some degree of flexibility in the non-covalent interactions that stabilize these proteins – so-called conformational breathing. This breathing allows proteins to undergo dynamic and partial changes of conformation essential to their normal functioning. Thus, even globular proteins can exhibit a partly folded phenotype, and thereby run the risk of aggregating with identical or similar proteins, but to what extent this is a problem inside cells is not clear. What is clear, however, is that the problem of aggregation is very real when proteins are being synthesized.

There are two obvious respects in which the conditions under which proteins fold and assemble inside the cell differ from those in the in vitro systems used by protein chemists:

1. Protein synthesis in vivo takes place on polysomes. This fact inevitably brings identical, partly folded polypeptides within touching distance of one another (Fig. 2.2a). Because protein aggregation requires the interacting chains to be identical or highly similar, this propinquity optimizes the probability of aggregation. Moreover, these chains grow vectorially, so that they cannot fold correctly until a complete folding domain has been synthesized. The rate of polypeptide chain synthesis is typically between 5 and 20 amino acid residues added per second, but the rate of protein folding is often much faster. Thus, incomplete chains may fold into nonfunctional conformations before the entire chain is made. It appears that the advantage of making several polypeptide chains simultaneously from one molecule of mRNA has been bought at the risk of both misfolding and aggregation. Some chaperones function to reduce this risk by binding to incomplete and newly synthesized polypeptide chains before they complete their folding.
2. Protein folding and assembly invariably take place in environments that contain very high concentrations of macromolecule, especially protein, carbohydrate, lipid, and nucleic acid. Such environments are said to be "macromolecularly crowded." The term "crowded" is used rather than "concentrated" because, with some exceptions such as hemoglobin in erythrocytes, no single type of macromolecule occurs at very high concentration. Figure 2.2b shows a cartoon illustrating the crowded nature of the eukaryotic cytosol. Crowding is important because theory predicts that it affects in a dramatic and nonlinear fashion both the

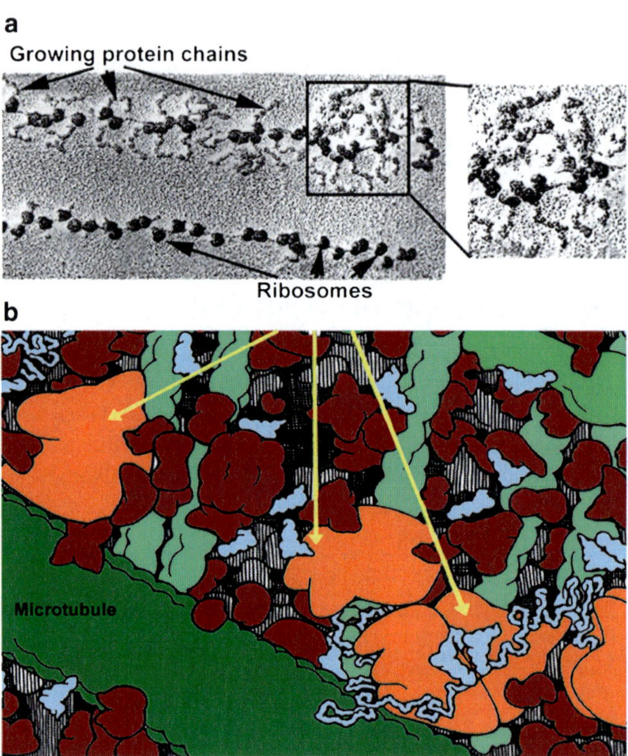

Fig. 2.2 Proteins fold in highly crowded environments. (**a**) Electron micrograph of a polysome isolated from the salivary gland of the insect *Chironomus* ×140,000. At the *bottom right* is the start of the polysome, i.e., where the ribosomes attach to the messenger RNA, while the end of the polysome is at the *top right*; the rest of the polysome is looped out of shot. Reprinted from Kisleva (1989). (**b**) Cartoon representation of part of the cytosol of an animal cell. The macromolecules are drawn to approximately the correct sizes and density of packing. Reprinted from Goodsell (1992) with permission

kinetics and thermodynamics of interactions between macromolecules, of which protein aggregation is just one example. This prediction of crowding theory has been known for over 40 years, but is largely ignored by biochemists and protein chemists (Ellis 2000). The next section discusses the implications of crowding theory for protein aggregation inside the cell.

2.4 Macromolecular Crowding

The degree of crowding inside eukaryotic cells ranges from 80 to 400 g/l of total macromolecules. These molecules occupy space, so this means that about 8–40% of the total volume is physically occupied by macromolecules, and therefore is unavailable to other molecules, just as in a football crowd most of the space is occupied by people and is unavailable to other people – other people are sterically excluded. This steric exclusion of part of the volume generates considerable energetic consequences, whose magnitude is not generally appreciated because it is so counter- intuitive – 8–40% does not sound very much and you might think it means that the effective concentration (i.e., the thermodynamic activity) is 8–40% greater than the actual concentration, but this is incorrect. Crowding theory predicts that the effective concentration is *one-to-three orders of magnitude* greater than the actual concentration.

There is a mathematically based physical theory of crowding developed largely by Allen Minton and coworkers (reviewed in Zhou et al. 2008). This theory predicts two major consequences:

1. The first consequence is self-evident – crowding reduces diffusion of both small and large molecules, by around three- to fourfold in eukaryotic cells. Actual measurements of diffusion inside cells support this conclusion, but it is the second consequence that is both much larger and more relevant to protein aggregation.
2. Crowding increases association constants of large molecules because its increases their effective concentration – their thermodynamic activity.

Consider two 40 kDa protein monomers associating into a dimer, and assume that the association constant in dilute buffer is 1.0. Crowding theory predicts that inside the crowded cytoplasm of *E. coli*, this constant will be 24. If we let this homodimer form a homotetramer, the association constant increases to 10,000; this effect of crowding on association is both large and very dependent on molecular size. This dramatic effect is relevant to protein aggregation because aggregation is a type of association (Ellis and Minton 2006).

This is theory, what about experiment? Figure 2.3a plots the activity coefficient, as calculated from the osmotic pressure, against actual concentration for

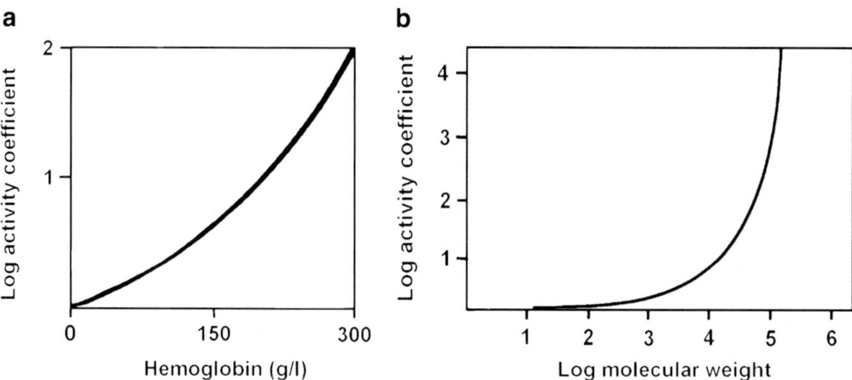

Fig. 2.3 Activity coefficients increase nonlinearly with cellular concentrations of macromolecule. (**a**) The log of the activity coefficient of hemoglobin is plotted against the actual concentration of hemoglobin. Reprinted from Minton (1983) with permission. (**b**) This log–log graph plots the change in activity of a test molecule introduced into a solution crowded with hemoglobin at 300 g/l against the molecular weight of the test molecule. Reprinted from Minton et al. (1992) with permission

increasing concentrations of hemoglobin. Note that this is a log/linear plot. The actual concentration of hemoglobin inside red blood cells is around 340 g per liter, so the activity of hemoglobin inside the cell is over two orders of magnitude greater than it is in the dilute buffer in which its properties are commonly studied.

The graph in Fig 2.3b plots the activity coefficient against molecular weight for a test molecule placed in a background of hemoglobin at 300 g/l. So in this experiment, hemoglobin is the macromolecule causing crowding, and we are asking how the activity of another molecule placed in this hemoglobin solution depends on the molecular weight of that molecule. Note that this is a log/log plot. You can see that the effect of crowding on activity becomes significant only after about 10,000 molecular weight. This is the reason that the term "macromolecular crowding" is used, because the effect on the activity of small molecules, such as metabolites and inorganic ions, is small by comparison.

Despite the fact that these predictions of crowding theory are both longstanding and undisputed, the vast majority of studies on isolated macromolecules, including those on protein folding, protein aggregation, and the functions of molecular chaperones, continue to be done in uncrowded buffers. In my view, this is an omission that needs to be rectified. Rectification is experimentally feasible because it is possible to mimic crowding in the cell by adding high concentration of suitable polymers – so-called crowding agents – to isolated proteins in the test tube. It would be of especial interest to test the effect of adding crowding agents to the increasing numbers of proteins that are found to lack ordered structure when studied in dilute uncrowded buffers. Crowding theory predicts that crowding favors compact conformations over extended ones, so it is possible that some of these "disordered" proteins are in fact ordered when inside the crowded cellular interior.

Figure 2.4 shows the dramatic effect of crowding agents on the in vitro formation of amyloid fibrils from two isolated proteins, one of them associated with

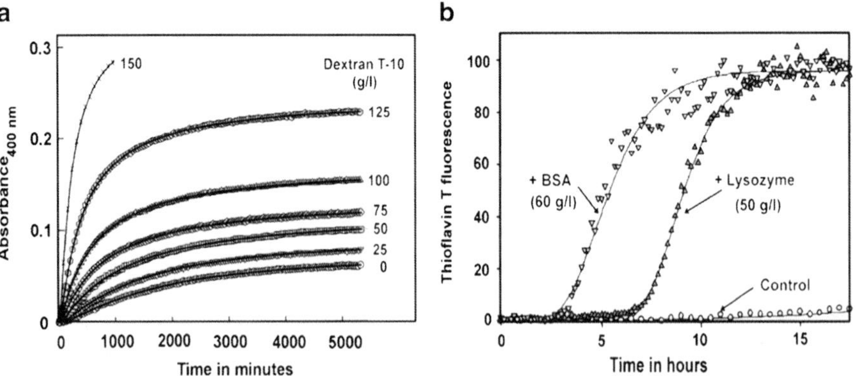

Fig. 2.4 Crowding agents promote the formation of amyloid fibrils. (**a**) Effect of increasing concentrations of Dextran T-10 on the rate of formation of amyloid fibrils from apoliprotein C-II. Reprinted from Hatters et al. (2002) with permission. (**b**) Effect of high concentrations of lysozyme and bovine serum albumin (BSA) on the formation of amyloid fibrils from α-synuclein, the protein associated with Parkinson's disease. Reprinted from Uversky et al. (2002) with permission

Parkinson's disease. So it is clear from both theory and experiment that the problem of protein aggregation inside the cell is likely to be much greater than it appears to be from studies done outside the cell in uncrowded buffers (Ellis and Minton 2006). Protein chemists often ignore this problem, but cells do not have this luxury. So what do cells do about this problem?

2.5 Prevention of Protein Aggregation by Molecular Chaperones

The generic ability of all polypeptides to form aggregates, combined with the stimulation of the rate of aggregation by macromolecular crowding, raises the question as to why this process is confined to a small number of proteins inside the cell. There are several answers to this question. One answer is that sequences evolve in directions that minimize the probability of aggregation. For example, there is evidence that some residues that flank aggregation-prone regions act as so-called gatekeepers that reduce the probability of aggregation (reviewed in Idicula-Thomas and Balaji 2007). Another answer to the problem of aggregation is the action of molecular chaperones. Some chaperones act at the level of protein folding, while others act at the level of protein assembly. Folding chaperones vary widely in their specificity for their protein substrates, but assembly chaperones are specific.

2.5.1 Folding Chaperones

Some chaperones assist the initial folding of both nascent chains bound to ribosomes and newly synthesized chains just released from ribosomes (i.e., in both co-translational and posttranslational modes), as well as the refolding of mature proteins unfolded by environmental stresses. A recent interesting discovery is that of four membrane-located chaperones, required for the folding of some intrinsic membrane proteins (Kota and Ljungdahl 2005). The chaperones working in these co-translational and posttranslational modes are distinct and can be usefully termed small and large chaperones, respectively, because this is a case where size is important for function.

Small chaperones are less than 200 kDa in size and include trigger factor, nascent chain-asssociated complex, prefoldin, the hsp70 and hsp40 families, and their associated co-chaperones. Co-chaperones are defined as proteins that bind to chaperones to modulate their activity; they may or may not also be chaperones in their own right. Large chaperones are more than 800 kDa in size and include the thermosome in Archaea, GroE proteins in Bacteria and the eukaryotic organelles evolutionarily derived from Bacteria, and the tailless complex polypeptide-1 (TCP-1) or TRiC complexes and associated co-chaperones in the cytosol of Eukarya. The large chaperones are evolutionarily related to one another, and are collectively referred to as the chaperonins. In the endoplasmic reticulum lumen of eukaryotic cells, there are no large chaperones, but there are small chaperones, such as BiP (an Hsp70 homolog), calnexin, calreticulin, and protein disulfide isomerase, that assist the folding of chains transported into the lumen after synthesis in the cytosol. Table 2.3 lists some of the chaperones that assist protein folding in various intracellular compartments. Many of these chaperones have been conserved throughout evolution; Fig. 2.5 illustrates chaperones that assist protein folding in the currently recognized three domains of life.

Table 2.3 Chaperones that assist protein folding

	Other names		
Family	Eukaryotes	Prokaryotes	Functions
Hsp100	Hsp104, 78	ClpA/B/X	Disassembly of oligomers and aggregates
Hsp90	Hsp82, Hsp83, Grp94	HtpG	Regulation of assembly of steroid receptors and signal transduction proteins
Hsp70	Hsc70, Ssa1–4, Ssb1–2, BiP, Grp75	DnaK, Hsc66 Absent from many Archaea	Prevention of aggregation of unfolded protein chains
Chaperonins	Hsp60, TRiC, CCT, TCP-1, Rubisco subunit-binding protein	GroEL, GroES	Sequestering partly folded chains inside central cage to allow completion of folding in absence of other folding chains
Hsp40	Ydj1, Sis1, Sec63p, auxilin, zuotin, Hdj2	DnaJ	Stimulation of ATPase, activity of Hsp70
Prefoldin	GimC	Absent from Bacteria Present in Archaea	Prevention of aggregation of unfolded protein chains
Trigger factor	Absent from Eukarya	Present	Binding to nascent chains as they emerge from ribosome
Calnexin, calreticulin	Present	Absent from prokaryotes	Binding to partly folded glycoproteins; located in ER membrane and lumen, respectively
Nascent chain-associated complex (NAC)	Present	Absent from prokaryotes	Binding to nascent chains as they emerge from ribosome
PapD	Absent from Eukarya	Present in some	Prevention of aggregation of subunits of pili
Membrane chaperones	Shr3p Gsf2p Pho68p Chs7p		Prevent aggregation of some integral polytopic membrane proteins

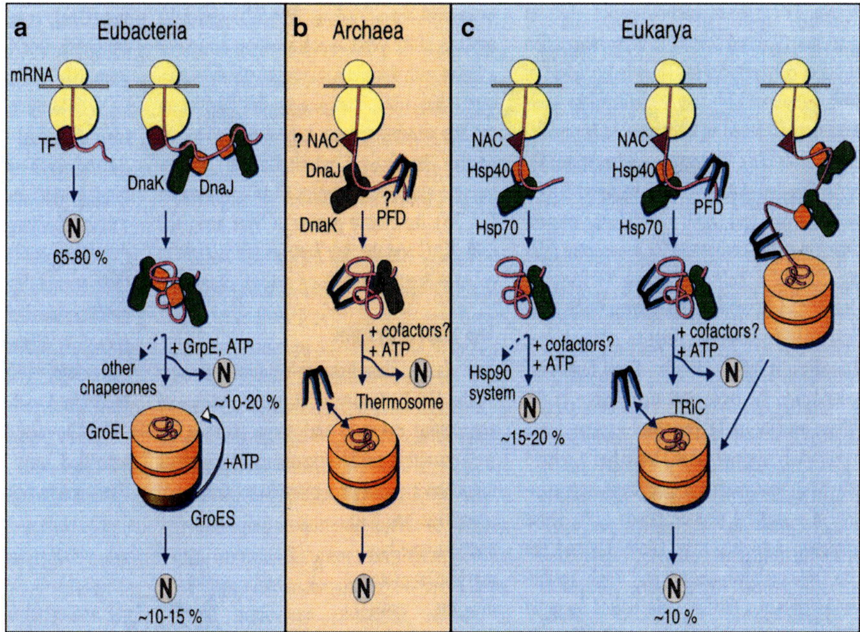

Fig. 2.5 Models for the chaperone-assisted folding of newly synthesized polypeptides in the cytosol. (**a**) Bacteria. TF, trigger factor; N, native protein. Most nascent chains probably interact with TF, and most small proteins (about 65–80% of total chain types) may fold rapidly upon synthesis without further chaperone assistance. Longer chains (10–20% of total chain types) interact subsequently with DnaK and DnaJ, and fold after one or several cycles of ATP-dependent binding and release. About 10–15% of total chains fold within the chaperonin GroEL/GroES system. GroEL does not bind to nascent chains and thus is likely to receive its substrates after their release from DnaK. (**b**) Archaea. PFD, prefoldin; NAC, nascent chain-associated complex. Only some archaeal species contain DnaK/DnaJ. The existence of a ribosome-bound NAC homolog and the binding of prefoldin to nascent chains are not shown. (**c**) Eukarya. Like TF, NAC probably interacts with many nascent chains. The majority of smaller chains may fold without further chaperone assistance. About 15–20% of chains reach their native states after assistance by hsp70 and hsp40, and a specific fraction of these are then transferred to hsp90. About 10% of chains are passed to the TRiC system in a reaction involving PFD. Reprinted from Young et al. (2004) with permission from the American Association for the Advancement of Science. Abbreviations defined in text

The small chaperones combat aggregation by buying time. They achieve this by binding transiently, but repeatedly, to small hydrophobic regions exposed on partly folded polypeptide chains, thus reducing the probability that these regions will interact with similar regions on nearby chains. Such a simple mechanism can be thought of as analogous to tossing a hot potato from hand-to-hand until it has cooled, an analogy suggested by Ulrich Hartl. Figure 2.6 illustrates the basic idea. Chaperone release is required to allow folding to continue, and so bury the hydrophobic regions in the interior of the protein. In some cases, but not all, this release is controlled by ATP-dependent conformational switching and is regulated by a variety of co-chaperones. A fraction of each population of identical chains folds

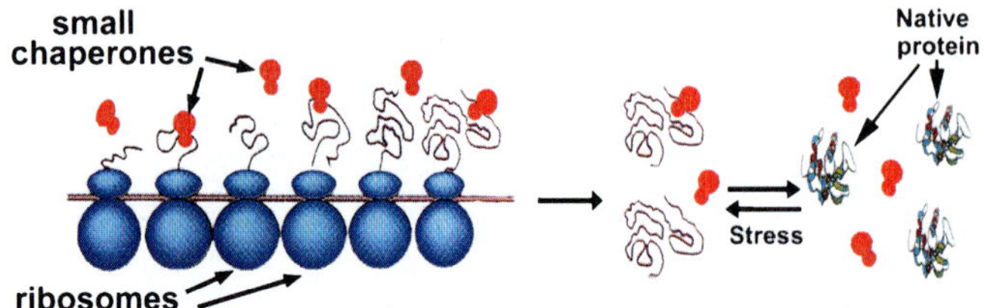

Fig. 2.6 The "hot potato" model for the action of small chaperones. These proteins bind and release from hydrophobic regions transiently exposed on nascent and newly released polypeptide chains, as well as on mature proteins partly unfolded by environmental stresses

more rapidly than the remainder, so it is thought that these repeated cycles of binding and release allow some chains to fold quickly and so avoid aggregation, while the slower folding fraction rebinds to the chaperones. Chains that are unable to fold rapidly enough on a biologically relevant time scale are then transferred to the large chaperones. Thus, the small chaperones act upstream of the large chaperones.

The large chaperones function by a much more sophisticated mechanism enabled by their large size, whereby entire partly folded chains are enclosed one at a time inside a closed chamber with a removable lid and a hydrophilic interior. Inside this "Anfinsen cage," each chain can complete its folding in the total absence of similar chains, and then be released into the cytosol. This mechanism also requires ATP binding and hydrolysis to regulate several distinct allosteric changes in conformation.

A summary of some of the key features of the small and large folding chaperones is presented in the next two sections. (More detailed information can be found in the reviews by Frydman 2001; Hartl et al. 2002; Young et al. 2004; Horwich et al. 2007 and Chakraborty et al. 2009).

2.5.1.1 Small Folding Chaperones

Trigger factor (48 kDa) is the first chaperone to bind to nascent chains in prokaryotes because it is associated with the ribosomal large subunit at the tunnel from which the chains emerge. A cell of *E. coli* contains about 20,000 copies of this chaperone, enough in principle to bind to all nascent chains; the majority (~80%) of cytosolic proteins in this organism are thought to interact with the trigger factor (Fig. 2.5). Trigger factor shows peptidyl-prolyl isomerase activity and contains a hydrophobic groove that binds transiently to regions of the nascent chain enriched in aromatic residues. It binds to nascent chains as short as 57 residues, and dissociates in an ATP-independent manner after the chain is released from the ribosome. This binding does not require prolyl residues in the nascent chain, so the isomerase activity may provide a means of keeping nascent chains containing prolyl residues in a flexible state. Structural studies are interpreted to mean that the trigger factor provides a partly shielded environment inside which a region of an elongating polypeptide chain may fold unhindered by aggregation with chains on neighboring ribosomes

(Merz et al. 2008), but other models are not excluded. The eukaryotic cytosol lacks trigger factor but its function may be replaced by that of a heterodimeric complex of 33 and 22 kDa subunits, termed the nascent chain-associated complex or NAC (Wegrzyn et al. 2006). Like trigger factor, this complex binds transiently to short nascent chains and acts independently of ATP, but unlike trigger factor, it does not possess peptidyl-prolyl isomerase activity.

Cells lacking trigger factor show no phenotype but this is because its function can be replaced by that of the other major small chaperone, Hsp70. The Hsp70 family has many ~70 kDa proteins distributed between the cytoplasm of bacteria and some, but not all, Archaea, the cytosol of Eukarya, and eukaryotic organelles such as the endoplasmic reticulum, mitochondria, and chloroplasts. Some, but not all, of these members are also stress proteins, because the probability of aggregation increases as proteins are unfolded as a result of stress. Unlike trigger factor, most of the Hsp70 members do not bind to ribosomes but do bind to short regions of hydrophobic residues exposed on nascent and newly synthesized chains. Such regions occur statistically about every 50 residues and are recognized by a peptide-binding cleft in hsp70. These hydrophobic regions are typically seven residues long and flanked by positively charged residues (Rudiger et al. 1997).

Most information is available about the Hsp70 member in *E. coli*, termed DnaK (Bukau and Horwich 1998). DnaK preferentially binds to polypeptides larger than 20–30 kDa; coimmunoprecipitation studies suggest that 15–20% by mass of *E. coli* proteins bind to DnaK at 30°C (Teter et al. 1999). Like all Hsp70 chaperones, DnaK contains an ATPase site, and occupation of this site by ATP promotes rapid but reversible peptide binding. ATP hydrolysis then tightens the binding through conformational changes in DnaK. The cycling of ATP between these states is regulated by a 41-kD co-chaperone of the Hsp40 family (termed DnaJ in *E. coli*), and GrpE, a nucleotide exchange factor that is a co-chaperone but not a chaperone. DnaJ binds to DnaK through its J domain and increases the rate of ATP hydrolysis, thus facilitating peptide binding. DnaJ, like all the Hsp40 proteins, acts as a chaperone in its own right since it also binds to hydrophobic peptides. Thus DnaK and DnaJ co-operate in binding each other to nascent chains; all Hsp70 chaperones are thought to co-operate with hsp40 chaperones. The role of GrpE is to stimulate the release of ADP from DnaK, allowing the latter to bind another molecule of ATP and so release the peptide. In the eukaryotic cytosol, the role of GrpE is fulfilled by three unrelated co-chaperones called Bag-1, Hsp BP1, and members of the Hsp110 family (Raviol et al. 2006). Some Archaea lack Hsp70 proteins, but it is speculated that their role in protein folding may be replaced by that of an unrelated chaperone called prefoldin.

There is enough DnaK in a cell of *E. coli* for one molecule to bind to each nascent chain. DnaK binds to longer chains than trigger factor and so probably binds after trigger factor. When the gene for trigger factor is deleted, the fraction of nascent and newly synthesized chains binding to DnaK increases from about 15% to about 40%. However, removal of the genes for both trigger factor and DnaK in the same cell causes the aggregation of many newly synthesized chains and is lethal to the cell (Deuerling et al. 1999). This observation suggests that redundancy of important control systems is as good a design principle for cells as it is for passenger planes. An exception to this principle is the GroEL/ES chaperone system (see Sect. 2.5.1.2), whose genes are essential for the survival of *E. coli.*

2.5.1.2 Large Folding Chaperones

Large chaperones, such as the chaperonin GroEL/GroES, function by a much more sophisticated mechanism that uses what I call the Anfinsen cage principle, worked out in the laboratories of Ulrich Hartl and Art Horwich. Their studies established that the chaperonins assist the folding of about 15% of proteins newly synthesized inside *E. coli* by encapsulating each chain one at a time inside a molecular nanocage provided by cavities inside the GroEL/ES complex.

GroEL (800 kDa) consists of two heptameric rings of identical 57-kDa ATPase subunits stacked back to back, containing a cage in each ring (Xu et al. 1997). The term "cage" is used since the walls surrounding each central cavity contain gaps, perhaps to allow entry and exit of water. Each subunit contains three domains. The equatorial domain contains the nucleotide-binding site and is connected by a flexible intermediate domain with the apical domain. The latter presents several hydrophobic side chains at the top of the ring orientated toward the cavity of the cage, an arrangement that permits either a partly folded polypeptide chain or a molecule of GroES to bind, but prevents binding to another GroEL oligomer.

GroES is a single heptameric ring of 10-kDa subunits that cycles on and off either end of the GroEL in a manner regulated by the ATPase activity of GroEL. At any one time, GroES is bound to only one end of GroEL, leaving the other end free to bind a partly folded polypeptide chain after its release from the ribosome. GroEL does not bind to nascent chains, but its homolog in the eukaryotic cytosol, TCP-1, may do so. The two rings of GroEL are coupled by negative allostery so that only one ring at a time binds nucleotide, but within each ring the binding of nucleotide is co-operative. When either ADP or ATP is bound to one GroEL ring, the GroES sits on top of this ring – now called the *cis* ring. The binding of GroES triggers a large rotation and upward movement of the apical domains, resulting in an enlarged cage and a change in its internal surface properties from hydrophobic to hydrophilic. This enlarged cage can accommodate a single partly folded compact polypeptide chain up to about 60 kDa in size, perhaps depending on shape.

The reaction cycle starts with a GroEL–GroES complex containing ADP bound to the *cis* ring (Fig. 2.7, step 1). The hydrophobic residues on the apical domains on the other ring, now called *trans*, bind to hydrophobic residues exposed on a partly folded polypeptide chain, presumably after the release of small chaperones from this chain. GroES and ATP then bind to this ring, thereby converting it into a new *cis* ring, and causing the release of GroES and ADP from the old *cis* ring (Fig. 2.7, step 2). This binding of GroES to the *trans* ring displaces the bound polypeptide into the cavity of the cage because some of the hydrophobic residues of the apical domains that bind the polypeptide are the same residues that bind GroES. Details of the conformational changes in the trapped chain as it is released into the cage are being revealed by fluorescent techniques applied to single molecules of GroEL (Sharma et al. 2008).

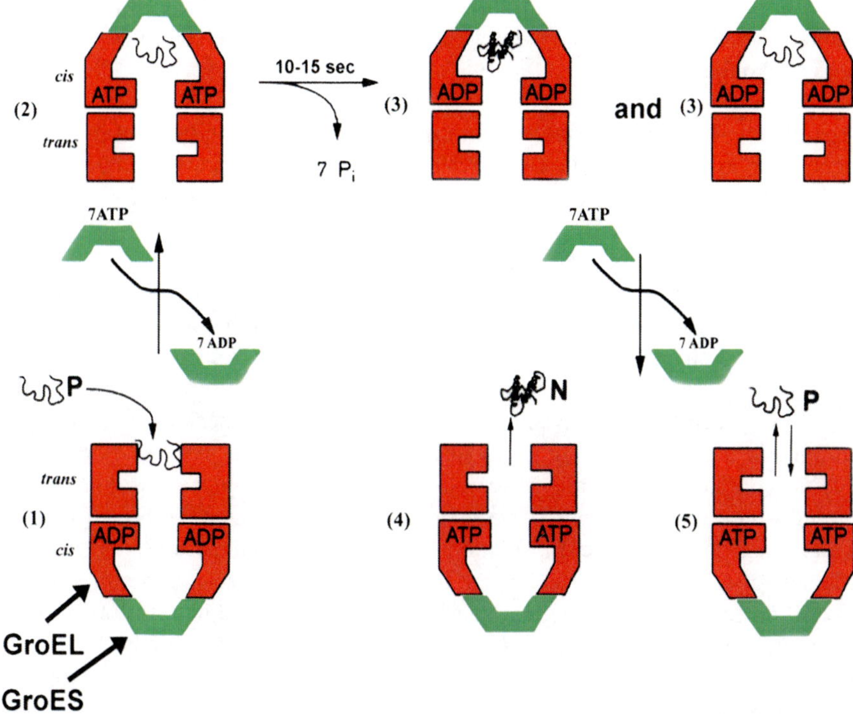

Fig. 2.7 Mechanism of action of the GroEL/GroES system in *E.coli*. P, partly unfolded polypeptide chain. N, native folded chain. For details, see text. Reprinted from Ellis (2003) with permission from Elsevier

The displaced chain lying free in the cavity of the cage now has 10–15 s to continue folding, a time set by the slow but co-ooperative ATPase activity of the seven subunits in each ring (Fig. 2.7, step 3). The chain thus continues its folding sheltered in a hydrophilic environment containing no other folding chain. Many denatured polypeptide chains will fold completely within 15 s in the classic Anfinsen renaturing experiment carried out in a test tube instead of inside GroEL. It is for this reason that I call this mechanism the "Anfinsen cage model."

The binding of ATP and GroES to the new *trans* ring then triggers the release of GroES and ADP from the *cis* ring containing the polypeptide chain, allowing the latter to diffuse out of the cage into the cytoplasm. If this chain has internalized its hydrophobic residues, it remains free in the cytoplasm (Fig. 2.7, step 4). But any chain that still exposes hydrophobic residues rebinds back to the same ring for another round of encapsulation (Fig. 2.7, step 5). Rebinding to the same ring rather than the ring of another GroEL oligomer is favored by the crowding effect created by the high concentration of macromolecules in the cytoplasm, and reduces the risk that partly folded chains will meet one another in the cytoplasm in a potentially disastrous encounter (Martin and Hartl 1997). Thus crowding is not all bad – while it promotes aggregation, it also promotes rebinding of partly folded polypeptides to GroEL, as well as favors the compaction of the partly folded chains that have managed to avoid aggregation.

This model was proposed to explain the results of many ingenious in vitro experiments, but genetic studies confirm the importance of the Anfinsen cage mechanism in intact cells. Mutants in which the mechanism is prevented by blockage of the entrance to one of the rings of each GroEL oligomer are viable, but the cells form colonies only 10% the size of the wild-type colonies (Farr et al. 2001). That these mutants are viable at all suggests that the ring whose entrance is not blocked is acting rather like the small chaperones, i.e., reducing aggregation by binding and releasing from the hydrophobic regions on newly synthesized chains. It is possible that the large chaperones evolved by the formation of oligomers from originally monomeric small chaperones.

An unexpected added advantage of the Anfinsen cage mechanism is that, for proteins in a certain size range, encapsulation in the cage increases the rate of folding compared to the rate observed in free solution under conditions where aggregation is not a problem. Thus the rate of folding of bacterial rubisco (50 kDa) is increased fourfold by encapsulation, while that of rhodanese (33 kDa) is not affected; alterations of the cavity size have differential effects, depending on the size of the encapsulated protein. This effect can be explained in terms of a type of macromolecular crowding called confinement, in which the promixity of the walls of the confining cage stabilizes compact conformations more than extended ones, and so enhances the rate of interactions, leading to compaction of the folding chain (Tang et al. 2006).

2.5.2 Assembly Chaperones

The nuclear compartment contains high concentrations of negatively charged nucleic acids bound to positively charged proteins that together form chromatin and various ribonucleoprotein particles. Besides nucleoplasmin, several other chaperones assist the assembly and disassembly of these nuclear structures by preventing incorrect ionic interactions between them, while cytosolic assembly chaperones assist the assembly of ribosomes, proteasomes, and proteins such as hemoglobin. Unlike folding chaperones, which vary widely in the specificity for their protein substrates and often require ATP to function, assembly chaperones are specific and do not require ATP.

For further information, reviews by Philpott et al. (2000), Akey and Luger (2003), Ellis (2006), and Kusmierczyk and Hochstrasser (2008) should be consulted.

The continuing emphasis on the role of chaperones in protein folding may explain the relative dearth of information about assembly chaperones in the literature. This omission should be rectified by examining with nondenaturing techniques the assembly of oligomeric structures in media designed to mimic crowded intracellular environments.

2.6 Chaperones and Disease

Protein aggregation is becoming associated with an increasing number of human diseases, including neurodegenerative disorders that are linked to the production of toxic protein aggregates. Accordingly, Alzheimer's, Parkinson's, Huntington's, and the prion diseases are characterized by the accumulation of amyloid fibrils both outside and inside neurones. In terms of numbers affected, Alzheimer's disease is the most important, with around 700,000 cases of this and related dementias in the UK in 2007 and around 25 million worldwide. The number of such patients is predicted to rise as more people survive for longer in the developed nations, and require significant increases in the cost of health care.

A growing view is that the primary cytotoxic agents in the neurodegenerative diseases are small soluble protein aggregates rather than amyloid fibrils (Chiti and Dobson 2006). How these small aggregates cause damage is not clear, but a popular idea is that they change the properties of other essential proteins, especially those in membranes (Barral et al. 2004). There have been several reports that increasing the levels of certain chaperones, especially members of the hsp70 and TriC families, in cultured cell model systems, inhibits the formation of the toxic species by changing the pathway of aggregation so that amorphous aggregates form rather than amyloid fibrils (Schaffar et al. 2004; Tam et al. 2006). These observations support the view that an important factor in the development of neurodegenerative diseases is the declining capacity with age of the chaperone systems associated with protein degradation (Balch et al. 2008; also see chapter by Höhfeld this volume).

Clinical trials are underway of small molecules, sometimes called pharmacological chaperones, that bind to and stabilize the folded functional state of proteins and thus reduce the probability of partial unfolding, leading to aggregation. An alternative approach is to use small interfering RNA technology to alter systems controlling protein folding and degradation via signaling pathways. For example, RNAi-induced reduction in insulin growth factor signaling extends the life span of a nematode worm-based model of Alzheimer's disease by increasing the levels of transcription factors regulating the expression of chaperones and other proteins (Cohen et al. 2006).

It is to be hoped that future studies on the mechanisms of protein aggregation will remedy the current neglect of the exact intracellular conditions under which this cause of so much human distress takes place.

References

Akey CW, Luger K (2003) Histone chaperones and nucleosome assembly. Curr Opin Struct Biol 13: 6–14

Anfinsen CB (1973) Principles that govern the folding of protein chains. Science 181: 223–230

Balch WE, Morimoto RI, Dillin A, Kelly JW (2008) Adapting proteostasis for disease intervention. Science 319: 916–919

Barral JM, Broadley SA, Schaffar G, Hartl FU (2004) Roles of molecular chaperones in protein misfolding diseases. Semin Cell Dev Biol 15: 17–29

Bousset L, Redeker V, Decottignies P, Dubois S, Marechal PL, Melki R (2004) Structural characterisation of th fibrillar form of the yeast *Saccharomyces cerevisiae* prion Ure2p. Biochemistry 43: 5022–5032

Bukau B, Horwich AL (1998) The Hsp70 and Hsp60 chaperone machines. Cell 92: 351–366

Caspar DLD, Klug A (1962) Physical principles in the construction of regular virsuses. Cold Spring Harbor Symp Quant Biol 27:1–24

Chakraborty K, Georgeacauld F, Hayer-Hartl M, Hartl FU (2010) Roles of molecular chaperones in protein folding. In: Ramirez-Alvarado M, Kelly JW, Dobson CM (eds) Protein Misfolding Diseases, Wiley pp. 42–72

Chiti F, Stefani M, Taddei N, Ramponi G, Dobson CM (2003) Rationalization of the effects of mutations on peptide and protein aggregation rates. Nature 424: 805–808

Chiti F, Dobson CM (2006) Protein misfolding, functional amyloid and disease. Annu Rev Biochem 75: 333–366

Cohen E, Bieschke J, Perciavalle RM, Kelly JW, Dillin A (2006) Opposing activities protect against age-onset proteotoxicity. Science 313: 1604–1610

Cowen LE, Lindquist S (2005) Hsp90 potentiates the rapid evolution of new traits: drug resistance in diverse fungi. Science 309: 2185–2189

Crick FHC (1958) On protein synthesis. Symp Soc Exp Biol 13: 138–163

Deuerling E, Schulze-Specking A, Tomoyasu A, Mogk A, Bukau B. (1999) Trigger factors and DnaK cooperate in the folding of newly synthesized proteins. Nature 400: 693–696

Dinner AR et al (2000) Understanding protein Folding via Free-energy Surfaces from theory and experiment. Trends Biochem Sc. 25 331–339

Dobson CM (1999) Protein misfolding, evolution and disease. Trends Biochem Sci 24: 329–332

Dobson CM, Ellis RJ, Fersht AR (eds) (2001) Protein Misfolding and Disease. Phil Trans R Soc B: 356: 129–131

Ellis RJ (1987) Proteins as molecular chaperones. Nature 328: 378–379

Ellis RJ (2000) Macromolecular crowding: obvious but underappreciated. Trends Biochem Sci 26: 597–604

Ellis RJ (2003) The importance of the Anfinsen cage. Current Biol 13: R881–R883

Ellis RJ (2004) From chloroplasts to chaperones: how one thing led to another. Photosyn Res 80: 333–343

Ellis RJ (2006) Molecular chaperones: assisting assembly in addition to folding. Trends Biochem Sci 31: 395–401

Ellis RJ, Hemmingsen SM (1989) Molecular chaperones: proteins essential for the biogenesis of some macromolecular structures. Trends Biochem Sci 14: 339–342

Ellis RJ, Minton AP (2006) Protein aggregation in crowded environments. Biol Chem 387: 485–497

Farr GW, Fenton WA, Rospert S, Horwich AR. Folding with and without encapsulation by cis and trans-only GroEL-GroES complexes (2001) EMBO J 22: 3220–3230

Freedman RB (2008) Eukaryotic protein disulfide-isomerases and their potential in the production of disulfide-bonded protein products: what we need to know but do not! In: Buchner J, Moroder L (eds) Oxidative folding of peptides and proteins, Royal Society of Chemistry pp. 121–157

Frydman J (2001) Folding of newly translated proteins in vivo: the role of molecular chaperones. Annu Rev Biochem 70: 603–647

Goodsell DS (1992) The Machinery of Life. Springer Varlaa New York Inc.

Hartl FU, Hayer-Hartl M (2002) Molecular chaperones in the cytosol: from nascent chain to folded protein. Science 295: 1852–1858

Hatters DM, Minton AP, Howlett GJ (2002) Macromolecular crowding accelerates amyloid formation by human apolipoprotein C-II. J Biol Chem 277: 7824–7830

Hemmingsen SM, Woolford C, van der Vies SM, Tilly K, Dennis DT, Georgopoulos GC, Hendrix RW, Ellis RJ (1988) Homologous plant and bacterial proteins chaperone oligomeric protein assembly. Nature 333: 330–334

Henderson B, Pockley AG (eds) (2005) Molecular chaperones and cell signalling. Cambridge University Press, New York

Horwich AL, Fenton WA, Chapman E, Farr GW (2007) Two families of chaperonin: physiology and mechanism. Annu Rev Cell Dev Biol 23: 115–145

Horwitz J (2000) The function of alpha-crystallin in vision. Semin Cell Develop Biol 11: 53–60

Idicula-Thomas S, Balaji PV (2007) Protein aggregation: a perspective from amyloid and inclusion-body formation. Current Science 92: 758–767

Kota J, Ljungdahl PO (2005) Specialized membrane-localized chaperones prevent aggregation of polytopic proteins in the ER. J Cell Biol 168: 79–88

Kusmierczyk AR, Hochstrasser M (2008) Some assembly required: dedicated chaperones in eukaryotic proteasome biogenesis, Biol Chem 389: 1143–1151

Laskey RA, Honda BM, Mills AD, Finch JT (1978) Nucleosomes are assembled by an acidic protein which binds histones and transfers them to DNA. Nature 275: 416–420

Lazardis T, Karplus M (1997) 'New view' of protein folding reconciled with the old through multiple unfolding simulations. Science 278: 1928–1931

London J, Skrzynia C, Goldberg M (1975) Renaturation of *Escherichia coli* tryptophanase in aqueous urea solutions. Eur J Biochem 47: 409–415

Luhrs T, Ritter C, Adrian M., Riek-Loher D, Bohrmann B, Dobeli H, Schubert D, Riel R (2005) 3D structure of Alzheimer's amyloid-ß(1–42) fibrils. Proc Nat Acad Sci 102: 17342–17347

Martin J, Hartl FU (1997) The effect of macromolecular crowding on chaperonin-mediated protein folding. Proc Nat Acad Sci 94: 1107–1112

Merz F, Hoffmann A, Rutkowska A, Zachmann-Brand B, Bukau B, Deuerling E (2006) The C-terminal domain of Escherichia coli trigger factor represents the central module of its chaperone activity. J Biol Chem 272: 21865–21871

Minton AP (1983) The effect of volume occupancy upon the thermodynamic activity of proteins: some biochemical consequences. Mol Cell Biochem 55: 119–140

Minton AP, (1992) Confinement as a determinant of macromolecular structure. Biophys J 63: 1090–1100

Musgrove JE, Ellis RJ (1986) The rubisco large subunit binding protein. Phil Trans. Roy Soc Lond B 313: 419–428

Nelson R, Sawaya MR, Balbirnie M, Madsen AO, Riekel C, Grothe R, Eisenberg D (2005) Structure of the cross-β spine of amyloid-like fibrils. Nature 435: 773–778

Petkova AT, Ishii Y, Balbach JJ, Antzutkin ON, Leapman RD, Delaglio F, Tycko R (2002) A structural model for Alzheimer's β-amyloid fibrils based on experimental constraints from solid state NMR. Proc Natl Acad Sci 99: 16742–16753

Philpott A, Krude T, Lasker RA (2000) Nuclear chaperones. Semin Cell Dev Biol 11: 7–14

Puig A, Gilbert HF (1994) Protein disulfide isomerase exhibits chaperone and antichaperone activities in the oxidative folding of lysozyme. J Biol Chem 269: 7764–7771

Queitsch C, Sangster TA, Lindquist S (2002) Hsp90 as a capacitor of phenotypic variation. Nature 417: 618–624

Raviol H, Sadlish H, Rodrigvez F, Mayer MP, Bukau B (2006) Chaperone network in the yeast cytosol=Hsp 110 is revealed as an Hsp 70 nucleotide exchange Factor. EMBO J. 2510–2518

Rousseau F, Schymkowitz J, Serrano L (2006) Protein aggregation and amyloidosis: confusion of the kinds? Curr Opin Struct Biol 16: 118–126

Rudiger S, Germeroth L, Schneider-Mergener J, Bukau B, (1997) Substrate specificity of the DnaK chaperone determined by screening cellulose-bound peptide libraries. EMBO J 16: 1501–1507

Schaffar G, Breuer P, Boteva R, Behrends C, Tzvetkov N, Strippel N, Sakahira H, Siegers K, Hayer-hartl M, Hartl FU (2004) Cellular toxicity of polyglutamine expansion proteins: mechanism of transcription factor inactivation. Molecular Cell 15: 95–105

Serpell LC, Sunde M, Benson MD, Tennent GA, Pepys MB, Fraser PE (2000) The protofilament substructure of amyloid fibrils. J Mol Biol 300: 1033–1039

Sharma S, Chakraborty K, Muller BK, Astola N, Tang YC, Lamb DC, Hayer-Hartl M, Hartl FU (2008) Monitoring protein conformation along the pathway of chaperonin-assisted folding. Cell 133: 142–153

Tam S, Geller R, Spiess C, Frydman J (2006) The chaperonin TriC controls polyglutamine aggregation and toxicity through subunit-specific interactions. Nature Cell Biology 8: 1155–1162

Tang Y-C, Chang H-C, Roeben A, Wischnewski D, Wischnewski N, Kerner MJ, Hartl FU, Hayer-Hartl M (2006) Structural features of the GroEL-GroES nanocage required for rapid folding of encapsulated protein. Cell 125: 903–914

Teter SA, Houry WA, Ang D, Tradler T, Rockabrand D, Fischer G, Blum P, Georgopoulos C, Hartl FU (1999) Polypeptide flux through bacterial hsp70: DnaK cooperates with trigger factor in chaperoning nascent chains. Cell 97: 755–765

Uversky VN, Cooper EM, Bower JI, Fink AL (2002) Accelerated α-synuclein fibrillation in crowded mileu. FEBS Lett 515: 99–103

Wegrzyn RD, Hofmann D, Merz F, Nikolay R, Rauch T, Graf C, Deuerling E (2006) A conserved motif is prerequisite for the interaction of NAC with ribosomal protein L23 and nascent chains. J Biol Chem 281: 2847–2857

Wright CF, Teichmann SA, Clarke J, Dodson CM (2005) Specificity in aggregation and the evolution of modular proteins. Nature 438: 878–881

Xu Z, Horwich AL, Sigler, P.(1997) The crystal structure of the asymmetric GroEL-GroES-$(ADP)_7$ chaperonin complex. Nature 388:741–750

Young JC, Agashe VR, Siegers K, Hartl FU (2004) Pathways of chaperone-mediated protein folding in the cytosol. Nature Revs Cell Biol 5:781–790

Zhou HX, Rivas G, Minton AP (2008) Macromolecular crowding and confinement: biochemical, biophysical and potential physiological consequences. Annu Rev Biophys 37: 375–397

Chapter 3
Molecular Chaperones as Facilitators of Protein Degradation

Jörg Höhfeld, Nikolaus Dick, and Verena Arndt

Abstract Molecular chaperones are widely recognized as intracellular folding factors. Yet, over the last couple of years more and more examples have emerged where chaperones actively participated in protein degradation. From the chaperone-bound state substrate proteins can be directed onto proteasome as well as autophagy-mediated degradation pathways. Chaperone-assisted degradation indeed seems to be a vital aspect of cellular protein quality control. Here we describe molecular mechanisms underlying chaperone-assisted degradation and discuss its physiological relevance in the context of neurodegenerative diseases and aging.

3.1 Introduction

Protein aggregation is a hallmark of many neurodegenerative diseases, including prion diseases, Alzheimer's and Parkinson's disease (Taylor et al. 2002). It seems that post-mitotic cells such as neurons are particularly prone to the toxic consequences of the aggregation process. As protective measures cells employ molecular chaperones that recognize aberrant protein conformers and facilitate their folding to the native state whenever possible (see the previous chapter by J. Ellis). Another line of defence involves the proteolytic removal of hazardous protein species by the ubiquitin/proteasome system or by autophagy (Rubinsztein 2006). For a long time chaperones and the constituents of the degradation machineries were viewed as opposing forces whose competition seemed to determine the balance between protein (re-) folding and degradation (Wickner et al. 1999). However, more recent data reveal an active involvement of certain chaperones, i.e. members of the Hsp70 chaperone family, in protein degradation (Arndt et al. 2007). These findings emphasize the central role of molecular chaperones in protein homeostasis. Hsp70-type chaperones apparently screen the

J. Höhfeld (✉)
Institut für Zellbiologie, Rheinische Friedrich-Wilhelms Universität Bonn, Ulrich-Haberland-Str. 61a D-53121 Bonn, Germany
e-mail: hoehfeld@uni-bonn.de

A. Wyttenbach and V. O'Connor (eds.), *Folding for the Synapse*,
DOI 10.1007/978-1-4419-7061-9_3,

cellular environment for non-native proteins, they stabilize aberrant conformers through direct binding, and subsequently folding as well as degradation is initiated from the chaperone-bound state. This functional concept provides a simple explanation of how the degradation machineries recognize non-native proteins: they recruit chaperone proteins to do the job. At the same time new questions arise. What determines the mode of action of the chaperone proteins? How is a decision reached between resurrection and destruction? In this chapter we will try to give some answers to these questions. In line with the focus of this book we will concentrate on findings that are relevant for our understanding of the role of 'degrading' chaperones in neuronal cells.

3.2 The Hsp70 Chaperone Machinery

Hsp70 proteins are highly conserved from bacteria to men and present in all compartments of eukaryotic cells (except for peroxisomes) (see also the chapter by J. Ellis). They recognize short hydrophobic patches, about 7 amino acids in length, which are exposed in non-native proteins and statistically occur every 40–50 residues. Accordingly, Hsp70s are engaged in the folding of newly synthesized and damaged proteins, in protein translocation and also in protein degradation. During all these processes the chaperones associate with the protein substrates in a dynamic manner regulated by cycles of ATP-binding and -hydrolysis (Mayer and Bukau 2005). The ATP-bound state of the chaperone represents an open conformation characterized by the rapid binding and release of substrates, resulting in a low overall binding affinity (Fig. 3.1). In the ADP conformation the carboxy terminus covers the peptide binding

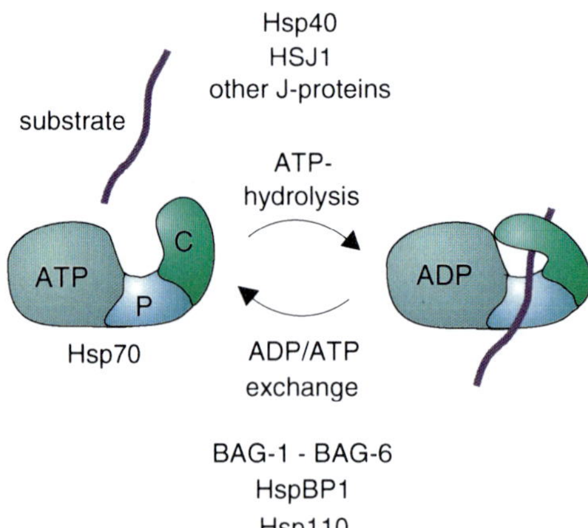

Fig. 3.1 The ATPase- and peptide-binding cycle of Hsp70 is regulated by co-chaperones. The ATP-bound state of the chaperone represents an open conformation with a low overall affinity for a substrate protein. ATP-hydrolysis leads to a closure of the carboxy-terminal lid domain, resulting in high affinity substrate binding. ATP-hydrolysis is stimulated by J-domain co-chaperones such as Hsp40 and HSJ1. Nucleotide exchange is facilitated by BAG-domain co-chaperones, HspBP1 and Hsp110. (*P* peptide binding domain of Hsp70; *C* carboxy-terminal lid)

pocket leading to a closed state and high affinity substrate binding. The ATPase cycle provides a basis for the regulation of Hsp70's chaperone activity through a stimulation of ATP hydrolysis and nucleotide exchange, respectively, mediated by co-chaperones. Members of the Hsp40/DnaJ co-chaperone family trigger the hydrolysis step and are therefore essential for efficient substrate loading onto Hsp70 (Laufen et al. 1999). Nucleotide exchange, which results in substrate release, is facilitated by diverse proteins in mammalian cells including BAG-domain co-chaperones, HspBP1 and Hsp110 (Brehmer et al. 2001; Dragovic et al. 2006; Höhfeld and Jentsch 1997; Polier et al. 2008; Shomura et al. 2005). Besides the regulation of the ATPase cycle there is a second important function of co-chaperones: they recruit Hsp70 family members to diverse subcellular locations and protein complexes. Hsp40 family members are, for example associated with the protein translocases operating in the endoplasmic reticulum and the inner mitochondrial membrane, and facilitate binding of Hsp70 proteins to the incoming polypeptide chain (Dudek et al. 2005; Mokranjac et al. 2006). In a different functional context the Hsp70- and Hsp90-organizing co-chaperone HOP stimulates the interaction between the two chaperones and thus coordinates their activity on certain folding pathways (Morishima et al. 2000; Young et al. 2001). These examples illustrate the essential role of co-chaperones in determining the specific functions of Hsp70 family members. Probably in most, if not all, cases where an involvement of Hsp70 chaperones has been observed we are looking in fact at the activity of a multi-component chaperone machinery that includes a defined set of regulating co-chaperones. It is therefore not surprising that the function of Hsp70 in protein degradation also relies on dedicated co-chaperones.

3.3 Cooperation of Chaperones with the Ubiquitin/Proteasome System

The ubiquitin/proteasome system represents the main degradation machinery for the removal of non-native proteins and for the turnover of short-lived regulatory proteins in eukaryotic cells (Ciechanover 2005; Hershko and Ciechanover 1998). Proteins destined for degradation are labelled with a chain of ubiquitin moieties through the concerted action of an E2 ubiquitin conjugating enzyme and an E3 ubiquitin ligase, after ubiquitin is activated by the E1 enzyme (Fig. 3.2). The attached ubiquitin chain directs the substrate to the 26S proteasome, which mediates cleavage into small peptides. During the sorting process additional factors can be employed to elongate or edit the attached ubiquitin chain or to facilitate interactions with the proteolytic complex (Hartmann-Petersen et al. 2003; Hershko and Ciechanover 1998; Newton et al. 2008).

A critical step on the degradation pathway is the initial recognition of the protein substrate by the ubiquitin conjugation machinery that is formed through the specific pairing of one of the 10s of E2 enzymes and one of the 100s of E3 enzymes, which exist in mammalian cells. In the case of chaperone-assisted degradation the conjugation machinery closely cooperates with molecular chaperones during initial substrate recognition (Arndt et al. 2007). The ability of certain E3 ubiquitin ligases

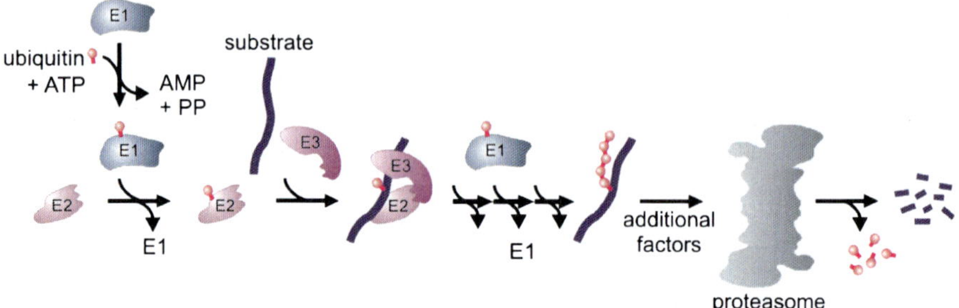

Fig. 3.2 Schematic presentation of the ubiquitin- and proteasome-mediated degradation pathway. Following ubiquitin activation mediated by the E1 enzyme, E2 ubiquitin conjugating enzymes and E3 ubiquitin ligases cooperate in the recognition of the substrate protein and mediate ubiquitin chain attachment. The ubiquitin chain directs the substrate to the 26S proteasome, where cleavage into small peptides occurs. During proteasomal sorting additional factors might be employed to elongate or edit the ubiquitin chain or to facilitate docking at the proteolytic complex

to interact directly with chaperones provides a molecular basis for this cooperation. The formed complex might actually be considered as a multi-subunit ubiquitin ligase complex with the bound chaperone acting as the main substrate recognition factor. A prominent player among the chaperone-associated ubiquitin ligases is the carboxy terminus of Hsp70-interacting protein CHIP. It was initially identified in the lab of Cam Patterson as a novel co-chaperone of Hsp70 and later shown to bind also Hsp90 (Ballinger et al. 1999; Connell et al. 2001). The interaction with the chaperones is mediated by an amino-terminal domain that comprises multiple tetratricopeptide repeats (TPR domain). Ubiquitin ligase activity is conferred by a carboxy terminal U-box, which recruits diverse E2 enzymes into the chaperone/co-chaperone complex (Fig. 3.3) (Connell et al. 2001; Demand et al. 2001; Grelle et al. 2006; Jiang et al. 2001; Zhang et al. 2005). The functional characterization of CHIP established that the co-chaperone switches chaperone activity from protein folding to protein degradation. It triggers the proteasomal degradation of diverse signalling proteins and apoptosis regulators from a chaperone-bound state and facilitates the disposal of aggregation-prone proteins during chaperone-mediated protein quality control (Arndt et al. 2007). The latter function appears to be of particular importance in neuronal cells. Indeed, CHIP was shown to trigger the degradation of major culprits observed to aggregate in neurodegenerative diseases, including hyperphosphorylated tau (Alzheimer's disease), α-synuclein (Parkinson's disease), mutant SOD1 (amyotrophic lateral sclerosis), SCA1 (spinocerebellar ataxia type 1) and polyglutamine-containing Huntingtin (Huntington's disease) (Al-Ramahi et al. 2006; Dickey et al. 2007; Shimura et al. 2004; Shin et al. 2005; Urushitani et al. 2004; Westhoff et al. 2005). CHIP deficiency in mice leads to a markedly reduced life span, along with an increased accumulation of toxic protein oligomers (Min et al. 2008). Chaperone-assisted degradation mediated by CHIP thus emerges as a major cytoprotective strategy to counteract protein aggregation.

While CHIP-mediated ubiquitylation of the chaperone presented substrate protein initiates sorting to the proteasome, additional co-chaperones seem to regulate

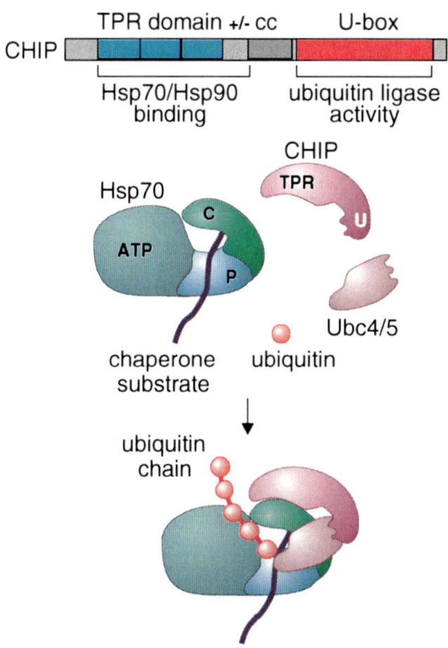

Fig. 3.3 CHIP is a chaperone-associated ubiquitin ligase. CHIP possesses multiple TPR domains for binding to Hsp70 and Hsp90. The U-box confers ubiquitin ligase activity. Interaction of CHIP with the carboxy terminus of Hsp70 triggers recruitment of Ubc4/5 ubiquitin conjugating enzymes into the chaperone complex, followed by the ubiquitylation of the chaperone-bound substrate protein. (+/- charged α-helical domain of CHIP; *cc* coiled coil domain of the co-chaperone; *P* peptide binding pocket of Hsp70; *C* carboxy terminus of the chaperone; *TPR* tetratricopeptide repeat domain of CHIP; *U* U-box of the ubiquitin ligase)

chaperone function during the targeting process. The co-chaperone HSJ1, for example, is able to bind ubiquitylated substrates after their initial modification by CHIP through ubiquitin interaction motifs (UIMs) (Westhoff et al. 2005). In addition, HSJ1 possesses a J-domain that enables the co-chaperone to facilitate substrate loading onto Hsp70 (Fig. 3.4). Because the chaperone is likely to go through multiple cycles of substrate binding and release during the sorting process, the combined activities of HSJ1 could ensure efficient re-loading of ubiquitylated substrates onto Hsp70. In this way the co-chaperone would ensure an engagement of Hsp70 until docking at the proteasome is achieved (Fig. 3.4). Consistent with such a function HSJ1 stimulates the CHIP-mediated degradation of pathogenic forms of Huntingtin in neuronal cells (Westhoff et al. 2005).

At the proteasome the co-chaperone BAG-1 seems to exert a chaperone-regulating activity (Alberti et al. 2002; Lüders et al. 2000). BAG-1 belongs to a family of BAG domain co-chaperones (BAG-1 to BAG-6) that all share the ability to stimulate nucleotide exchange on Hsp70 (Brehmer et al. 2001; Höhfeld and Jentsch 1997; Sondermann et al. 2001). In this way the co-chaperones facilitate substrate release from the chaperone. Remarkably, BAG-1 possesses, in addition to the BAG domain, an integrated ubiquitin-like domain that mediates an interaction with the proteasome (Fig. 3.5) (Alberti et al. 2002; Lüders et al. 2000). It is therefore conceivable that BAG-1 triggers

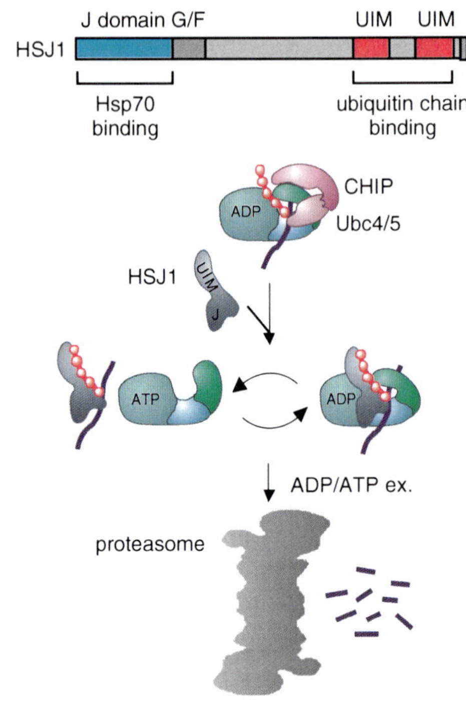

Fig. 3.4 HSJ1 accompanies ubiquitylated chaperone substrates to the proteasome. HSJ1 possesses two ubiquitin interaction motifs for binding to ubiquitylated chaperone substrates after their encounter with CHIP. During sorting to the proteasome HSJ1 stimulates reloading of the substrate onto Hsp70 when the chaperone goes through peptide binding and release cycles. (*G/F* glycine and phenylalanine rich region; *UIM* ubi-quitin interaction motif; *J* J-domain; *ex* nucleotide exchange)

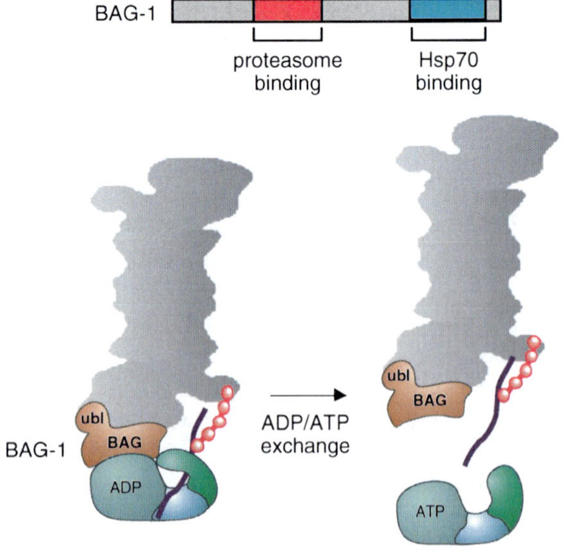

Fig. 3.5 BAG-1 could stimulate substrate release at the proteasome. The co-chaperone possesses a ubiquitin-like domain for binding to the proteasome. Based on its nucleotide exchange activity, BAG-1 stimulates the dissociation of the substrate protein from Hsp70. (*ubl* ubiquitin like domain of BAG-1; *BAG* BAG-domain)

the release of substrates from Hsp70 at the proteasome as a prerequisite for efficient transfer into the proteolytic cylinder. The degradation of a subset of CHIP substrates was indeed stimulated by BAG-1 (Demand et al. 2001).

The characterization of CHIP, HSJ1 and BAG-1 provides a conceptual framework of how co-chaperones could regulate chaperone activity at different stages of an ubiquitin-mediated degradation pathway. However, these co-chaperones are certainly not the only components that are involved in chaperone-assisted degradation. Regarding the initial ubiquitylation step, for example, the ubiquitin ligase Parkin, mutations of which cause familial forms of Parkinson's disease, may fulfill functions similar to CHIP. Parkin is able to associate with Hsp70 and displays a substrate specificity that partially overlaps with the one of CHIP (Imai et al. 2002; Morishima et al. 2008). Moreover, Hsp70 does not seem to be the only chaperone that presents substrates for proteasomal degradation. Targeting might also be initiated from an Hsp90-bound state, because CHIP binds directly to Hsp90 (Connell et al. 2001). In addition, the interaction of Hsp90 with SGT1, an interactor of the hetero oligomeric SCF ubiquitin ligase complex, points to yet another ubiquitin conjugation machinery that could potentially be involved in chaperone-assisted degradation (Kitagawa et al. 1999; Mayor et al. 2007; Zhang et al. 2008). Finally, the small heat shock protein α-B crystallin that is overexpressed in many neurological diseases was recently shown to associate with the SCF complex adaptor Fbx4 and to stimulate the Fbx4-mediated ubiquitylation of yet to be identified substrate proteins (den Engelsman et al. 2003). Taken together it becomes apparent that chaperone-assisted degradation is not mediated by only one highly defined pathway but rather relies on different chaperone machineries and on a range of diverse chaperone-associated ubiquitin ligases and ancillary factors, which divert chaperone substrates from a folding onto a degradation pathway.

3.4 Controlling Chaperone-Assisted Degradation

Chaperone-assisted degradation needs to be carefully controlled to avoid the disposal of protein substrates that associate with chaperones during their folding or conformational regulation. In the regulation of the chaperone-associated ubiquitin ligase CHIP other co-chaperones are very prominently involved and exert important control functions. Some co-chaperones that promote the folding activity of Hsp70 are able to block the binding of CHIP to Hsp70. Among these co-chaperones is the Hsp70- and Hsp90-organizing protein HOP that binds to the carboxy-terminus of Hsp70 in a manner competitive with CHIP (Fig. 3.6) (Connell et al. 2001; Demand et al. 1998). The co-chaperones HspBP1 and BAG-2, on the other hand, are able to interact with Hsp70 simultaneously with CHIP and inhibit the ubiquitin ligase activity of CHIP in the formed chaperone/co-chaperone complex (Alberti et al. 2004; Arndt et al. 2005; Dai et al. 2005). In the case of HspBP1 inhibition seems to involve a shielding of the

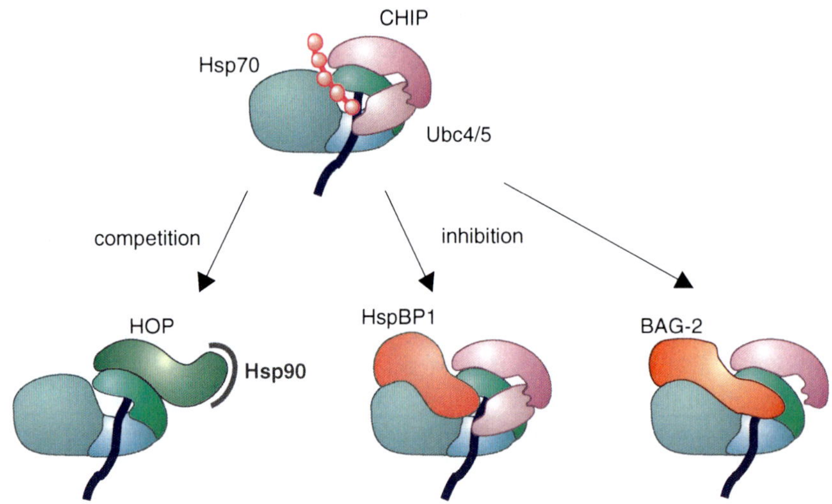

Fig. 3.6 The degradation-inducing activity of CHIP is controlled by other co-chaperones. The Hsp70- and Hsp90-organizing protein HOP binds to the carboxy terminus of Hsp70 in a manner competitive with CHIP. HspBP1 and BAG-2 hook up to the Hsp70/CHIP complex and interfere with CHIP-mediated ubiquitylation. HspBP1 seems to shield the substrate against the ubiquitin ligase. BAG-2 inhibits the interaction between CHIP and Ubc4/5

chaperone-bound substrate protein against the ubiquitin ligase, whereas the presence of BAG-2 in the complex interferes with the interaction of CHIP with its partner ubiquitin conjugating enzyme UbcH5b (Fig. 3.6). How the interplay between different co-chaperones influences the biogenesis of a chaperone-bound substrate was recently revealed in a study investigating the cytoplasmic fate of the toxin Ricin (Spooner et al. 2008). After uptake into mammalian cells through endocytosis, the toxin initially enters the endoplasmic reticulum from where it is retrotranslocated in the cytoplasm. Following retrotranslocation the toxin is recognized by the Hsp70 and Hsp90 chaperone systems. Its eventual fate is then determined by the balanced activity of diverse co-chaperones, with CHIP and BAG-1 triggering degradation and BAG-2 exerting a stabilizing activity (Spooner et al. 2008). The findings reveal an intricate network of competing and collaborating co-chaperones in mammalian cells, which determines folding and degradation activities.

3.5 Chaperone-Mediated Autophagy

Besides the ubiquitin/proteasome system the lysosomal compartment represents a major site for protein degradation (Rubinsztein 2006). Cellular components are targeted for lysosomal degradation in a process termed autophagy (Mizushima et al. 2008; Yorimitsu and Klionsky 2005). Dependent on the type of cellular waste different targeting mechanisms are employed. During macroautophagy, damaged mitochondria or other compartments and protein complexes become engulfed by an autophagosomal membrane, prior to fusion with the lysosomal compartment. On the other hand, cytoplasmic Hsp70 can target single protein molecules for lysosomal degradation without initial membrane engulfment in a process termed chaperone-mediated autophagy (CMA) (Cuervo et al. 1995; Kaushik et al. 2007). CMA substrate proteins contain a lysosomal targeting motif related to the pentapeptide KFERQ. This motif is present in ~30% of cytoplasmic proteins and is recognized by Hsp70. The chaperone/substrate complex docks at the lysosomal membrane through binding to LAMP2A, which initiates the dynamic formation of a translocase within the membrane for substrate transfer (Bandyopadhyay et al. 2008; Cuervo and Dice 1996; Kaushik et al. 2006).

It remains to be analysed how CMA is regulated at the level of the chaperone-bound state. In analogy to the findings for the regulation of chaperone-assisted degradation one might speculate that dedicated co-chaperones could be involved (see above). However, folding- as well as degradation-inducing co-chaperones were detected at the lysosomal membrane, including the Hsp70- and Hsp90-organizing protein HOP and the proteasome docking factor BAG-1 (Agarraberes and Dice 2001). It, therefore, seems possible that binding of Hsp70 to LAMP2A is itself the decisive step for directing a chaperone substrate onto the CMA pathway.

CMA represents an important degradation pathway in neuronal cells, which is illustrated by the fact that it mediates the degradation of wild-type α-synuclein (Cuervo et al. 2004). The protein contains a CMA recognition motif and is therefore directed towards lysosomal degradation in a chaperone-dependent manner. Remarkably, mutant forms of α-synuclein that cause Parkinson's disease are found in association with LAMP2A but are not internalized and degraded in the lysosomal compartment. The accumulation of these proteins at the receptor actually impairs the CMA-mediated degradation of other protein substrates, which may significantly contribute to the observed neuropathology (Cuervo et al. 2004). In this context, it is also noteworthy that CMA activity declines in aged organisms (Zhang and Cuervo 2008). A decreased ability to remove proteins via chaperone-mediated autophagy may thus aggravate the pathology of age-dependent diseases.

3.6 Involvement of Chaperones in Macroautophagy

Several findings link molecular chaperones to macroautophagy. The BAG-domain co-chaperone BAG-3 was recently shown to stimulate the proteolytic removal of pathologic forms of Huntingtin through macroautophagy (Carra et al. 2008). Remarkably, this activity depends on an interaction of BAG-3 with the small heat-shock protein Hsp22/HspB8. The co-chaperone regulates the oligomeric state of Hsp22 and in this way seems to control the chaperone activity of the small heat shock protein (Carra et al. 2008).

The BAG-3/Hsp22 chaperone complex apparently recognizes aggregation prone forms of Huntingtin and stimulates their lysosomal uptake. The molecular details of this chaperone-assisted degradation pathway remain to be established, however.

A stimulating activity in macroautophagy was also recently assigned to BAG-1, previously shown to facilitate the CHIP-induced degradation of chaperone substrates due to its ability to bind to the proteasome (see above). Gurusamy et al. (2008) observed an upregulation of BAG-1 together with LC3 and Beclin-1, main components of the autophagy pathway, in the heart of mice subjected to ischemia and reperfusion injury. Importantly, depletion of BAG-1 attenuated LC3 induction under these conditions (Gurusamy et al. 2008). Whether these findings reflect a cooperation of BAG-1 with the machinery that mediates macroautophagy needs to be clarified in future experiments. In any case, the data indicate an extensive crosstalk between diverse degradation pathways. Evidence for such a crosstalk also stems from recent findings regarding the function of Parkin in the degradation of damaged mitochondria through macroautophagy (Narendra et al. 2008). Like BAG-1, Parkin is a binding partner of Hsp70 and stimulates the proteasome-mediated degradation of chaperone substrates (Imai et al. 2002; Morishima et al. 2008). Yet, Narendra et al. (2008) demonstrate that Parkin is selectively recruited to dysfunctional mitochondria with low membrane potential and triggers the engulfment of these mitochondria by autophagosomes for subsequent disposal. Again, it is unclear how chaperone- and autophagy-regulating functions correlate and how a decision is reached between proteasomal and lysosomal degradation. An important aspect in this case might be that Parkin is able to cooperate with different E2 ubiquitin conjugating enzymes and can therefore assemble ubiquitin chains with different architectures, i.e. involving different lysine linkages between ubiquitin moieties (Lim et al. 2005). In this way, Parkin could generate different ubiquitin-based sorting signals on the modified substrate protein. Indeed, Parkin-generated lysine-63-linked ubiquitin chains trigger binding to the histone deacetylase HDAC6, which acts as an adaptor between ubiquitylated proteins and the dynein motor during the microtubule-dependent sequestration of protein aggregates in mammalian cells (Kawaguchi et al. 2003; Olzmann et al. 2007). Conceivably, Parkin cooperates with Hsp70 during the recognition of misfolded proteins and damaged mitochondria (that may expose an increased amount of misfolded molecules), and the type of ubiquitin chain that becomes attached to the recognized substrate protein may influence sorting onto a proteasomal or lysosomal degradation pathway.

A molecular mechanism that links protein ubiquitylation with lysosomal uptake was recently revealed with the characterization of the p62 protein (Bjorkoy et al. 2006; Komatsu et al. 2007). The protein combines a ubiquitin-binding UBA domain and a binding site for LC3, a posttranslational modifier essential for autophagosome formation. The domain arrangement enables p62 to co-ordinate the capture and sequestration of ubiquitylated substrate proteins with the formation of the autophagosomal membrane.

3.7 Concluding Remarks

Work over the last couple of years has clearly established the physiological importance of chaperone-assisted degradation for protein homeostasis. From a chaperone-bound state diverse degradation pathways can be initiated, including macroautophagy, chaperone-mediated autophagy and proteasomal degradation. At the same time it has become apparent that degradation pathways do not operate independently but are highly interconnected (Cuervo et al. 2004; Pandey et al. 2007). In part this might be attributed to the intricate co-chaperone network that determines chaperone activity in mammalian cells and directs substrate proteins onto diverse folding and degradation pathways. The identification and characterization of co-chaperones certainly revealed an unexpected organization of these pathways. Through binding to multiple partner proteins co-chaperones are able to induce the assembly of multi-component protein machines, in which the activity of chaperones and degradation factors is coordinated. This may contribute to the formation of subcellular compartments dedicated to the processing of non-native proteins such as aggresomes (Kaganovich et al. 2008). The further analysis of the co-chaperone network will advance our understanding of protein homeostasis and its impairment during disease and aging.

References

Agarraberes FA, Dice JF. 2001. A molecular chaperone complex at the lysosomal membrane is required for protein translocation. J Cell Sci 114: 2491–2499.

Al-Ramahi I, et al. 2006. CHIP protects from the neurotoxicity of expanded and wild-type ataxin-1 and promotes their ubiquitination and degradation. J Biol Chem 281: 26714–26724.

Alberti S, Bohse K, Arndt V, Schmitz A, Höhfeld J. 2004. The cochaperone HspBP1 inhibits the CHIP ubiquitin ligase and stimulates the maturation of the cystic fibrosis transmembrane conductance regulator. Mol Biol Cell 15: 4003–4010.

Alberti S, Demand J, Esser C, Emmerich N, Schild H, Höhfeld J. 2002. Ubiquitylation of BAG-1 suggests a novel regulatory mechanism during the sorting of chaperone substrates to the proteasome. J Biol Chem 277: 45920–45927.

Arndt V, Rogon C, Höhfeld J. 2007. To be, or not to be–molecular chaperones in protein degradation. Cell Mol Life Sci 64: 2525–2541.

Arndt V, Daniel C, Nastainczyk W, Alberti S, Höhfeld J. 2005. BAG-2 acts as an inhibitor of the chaperone-associated ubiquitin ligase CHIP. Mol Biol Cell 16: 5891–5900.

Ballinger CA, Connell P, Wu Y, Hu Z, Thompson LJ, Yin LY, Patterson C. 1999. Identification of CHIP, a novel tetratricopeptide repeat-containing protein that interacts with heat shock proteins and negatively regulates chaperone functions. Mol Cell Biol 19: 4535–4545.

Bandyopadhyay U, Kaushik S, Varticovski L, Cuervo AM. 2008. The chaperone-mediated autophagy receptor organizes in dynamic protein complexes at the lysosomal membrane. Mol Cell Biol 28: 5747–5763.

Bjorkoy G, Lamark T, Johansen T. 2006. p62/SQSTM1: a missing link between protein aggregates and the autophagy machinery. Autophagy 2: 138–139.

Brehmer D, Rudiger S, Gassler CS, Klostermeier D, Packschies L, Reinstein J, Mayer MP, Bukau B. 2001. Tuning of chaperone activity of Hsp70 proteins by modulation of nucleotide exchange. Nat Struct Biol 8: 427–432.

Carra S, Seguin SJ, Lambert H, Landry J. 2008. HspB8 chaperone activity toward poly(Q)-containing proteins depends on its association with Bag3, a stimulator of macroautophagy. J Biol Chem 283: 1437–1444.

Ciechanover A. 2005. Intracellular protein degradation: from a vague idea thru the lysosome and the ubiquitin-proteasome system and onto human diseases and drug targeting. Cell Death Differ 12: 1178–1190.

Connell P, Ballinger CA, Jiang J, Wu Y, Thompson LJ, Höhfeld J, Patterson C. 2001. The co-chaperone CHIP regulates protein triage decisions mediated by heat-shock proteins. Nat Cell Biol 3: 93–96.

Cuervo AM, Dice JF. 1996. A receptor for the selective uptake and degradation of proteins by lysosomes. Science 273: 501–503.

Cuervo AM, Knecht E, Terlecky SR, Dice JF. 1995. Activation of a selective pathway of lysosomal proteolysis in rat liver by prolonged starvation. Am J Physiol 269: C1200–1208.
Cuervo AM, Stefanis L, Fredenburg R, Lansbury PT, Sulzer D. 2004. Impaired degradation of mutant alpha-synuclein by chaperone-mediated autophagy. Science 305: 1292–1295.
Dai Q, et al. 2005. Regulation of the cytoplasmic quality control protein degradation pathway by BAG2. J Biol Chem 280: 38673–38681.
Demand J, Lüders J, Höhfeld J. 1998. The carboxy-terminal domain of Hsc70 provides binding sites for a distinct set of chaperone cofactors. Mol Cell Biol 18: 2023–2028.
Demand J, Alberti S, Patterson C, Höhfeld J. 2001. Cooperation of a ubiquitin domain protein and an E3 ubiquitin ligase during chaperone/proteasome coupling. Curr Biol 11: 1569–1577.
den Engelsman J, Keijsers V, de Jong WW, Boelens WC. 2003. The small heat-shock protein alpha B-crystallin promotes FBX4-dependent ubiquitination. J Biol Chem 278: 4699–4704.
Dickey CA, et al. 2007. The high-affinity HSP90-CHIP complex recognizes and selectively degrades phosphorylated tau client proteins. J Clin Invest 117: 648–658.
Dragovic Z, Broadley SA, Shomura Y, Bracher A, Hartl FU. 2006. Molecular chaperones of the Hsp110 family act as nucleotide exchange factors of Hsp70s. Embo J 25: 2519–2528.
Dudek J, Greiner M, Muller A, Hendershot LM, Kopsch K, Nastainczyk W, Zimmermann R. 2005. ERj1p has a basic role in protein biogenesis at the endoplasmic reticulum. Nat Struct Mol Biol 12: 1008–1014.
Grelle G, et al. 2006. Identification of VCP/p97, carboxyl terminus of Hsp70-interacting protein (CHIP), and amphiphysin II interaction partners using membrane-based human proteome arrays. Mol Cell Proteomics 5: 234–244.
Gurusamy N, Lekli I, Gorbunov N, Gherghiceanu M, Popescu LM, Das DK. 2008. Cardioprotection by adaptation to ischemia augments autophagy in association with BAG-1 protein. J Cell Mol Med. 12: 1–15.
Hartmann-Petersen R, Seeger M, Gordon C. 2003. Transferring substrates to the 26S proteasome. Trends Biochem Sci 28: 26–31.
Hershko A, Ciechanover A. 1998. The ubiquitin system. Annu Rev Biochem 67: 425–479.
Höhfeld J, Jentsch S. 1997. GrpE-like regulation of the hsc70 chaperone by the anti-apoptotic protein BAG-1. Embo J 16: 6209–6216.
Imai Y, Soda M, Hatakeyama S, Akagi T, Hashikawa T, Nakayama KI, Takahashi R. 2002. CHIP is associated with Parkin, a gene responsible for familial Parkinson's disease, and enhances its ubiquitin ligase activity. Mol Cell 10: 55–67.
Jiang J, Ballinger CA, Wu Y, Dai Q, Cyr DM, Höhfeld J, Patterson C. 2001. CHIP is a U-box-dependent E3 ubiquitin ligase: identification of Hsc70 as a target for ubiquitylation. J Biol Chem 276: 42938–42944.
Kaganovich D, Kopito R, Frydman J. 2008. Misfolded proteins partition between two distinct quality control compartments. Nature 454: 1088–1095.
Kaushik S, Massey AC, Cuervo AM. 2006. Lysosome membrane lipid microdomains: novel regulators of chaperone-mediated autophagy. Embo J 25: 3921–3933.
Kaushik S, Kiffin R, Cuervo AM. 2007. Chaperone-mediated autophagy and aging: a novel regulatory role of lipids revealed. Autophagy 3: 387–389.
Kawaguchi Y, Kovacs JJ, McLaurin A, Vance JM, Ito A, Yao TP. 2003. The deacetylase HDAC6 regulates aggresome formation and cell viability in response to misfolded protein stress. Cell 115: 727–738.
Kitagawa K, Skowyra D, Elledge SJ, Harper JW, Hieter P. 1999. SGT1 encodes an essential component of the yeast kinetochore assembly pathway and a novel subunit of the SCF ubiquitin ligase complex. Mol Cell 4: 21–33.
Komatsu M, et al. 2007. Homeostatic levels of p62 control cytoplasmic inclusion body formation in autophagy-deficient mice. Cell 131: 1149–1163.
Laufen T, Mayer MP, Beisel C, Klostermeier D, Mogk A, Reinstein J, Bukau B. 1999. Mechanism of regulation of hsp70 chaperones by DnaJ cochaperones. Proc Natl Acad Sci U S A 96: 5452–5457.
Lim KL, et al. 2005. Parkin mediates nonclassical, proteasomal-independent ubiquitination of synphilin-1: implications for Lewy body formation. J Neurosci 25: 2002–2009.

Lüders J, Demand J, Höhfeld J. 2000. The ubiquitin-related BAG-1 provides a link between the molecular chaperones Hsc70/Hsp70 and the proteasome. J Biol Chem 275: 4613–4617.

Mayer MP, Bukau B. 2005. Hsp70 chaperones: cellular functions and molecular mechanism. Cell Mol Life Sci 62: 670–684.

Mayor A, Martinon F, De Smedt T, Petrilli V, Tschopp J. 2007. A crucial function of SGT1 and HSP90 in inflammasome activity links mammalian and plant innate immune responses. Nat Immunol 8: 497–503.

Min JN, Whaley RA, Sharpless NE, Lockyer P, Portbury AL, Patterson C. 2008. CHIP deficiency decreases longevity, with accelerated aging phenotypes accompanied by altered protein quality control. Mol Cell Biol 28: 4018–4025.

Mizushima N, Levine B, Cuervo AM, Klionsky DJ. 2008. Autophagy fights disease through cellular self-digestion. Nature 451: 1069–1075.

Mokranjac D, Bourenkov G, Hell K, Neupert W, Groll M. 2006. Structure and function of Tim14 and Tim16, the J and J-like components of the mitochondrial protein import motor. Embo J 25: 4675–4685.

Morishima Y, Kanelakis KC, Silverstein AM, Dittmar KD, Estrada L, Pratt WB. 2000. The Hsp organizer protein HOP enhances the rate of but is not essential for glucocorticoid receptor folding by the multiprotein Hsp90-based chaperone system. J Biol Chem 275: 6894–6900.

Morishima Y, Wang AM, Yu Z, Pratt WB, Osawa Y, Lieberman AP. 2008. CHIP deletion reveals functional redundancy of E3 ligases in promoting degradation of both signaling proteins and expanded glutamine proteins. Hum Mol Genet 17: 3942–3952.

Narendra D, Tanaka A, Suen DF, Youle RJ. 2008. Parkin is recruited selectively to impaired mitochondria and promotes their autophagy. J Cell Biol 183: 795–803.

Newton K, et al. 2008. Ubiquitin chain editing revealed by polyubiquitin linkage-specific antibodies. Cell 134: 668–678.

Olzmann JA, Li L, Chudaev MV, Chen J, Perez FA, Palmiter RD, Chin LS. 2007. Parkin-mediated K63-linked polyubiquitination targets misfolded DJ-1 to aggresomes via binding to HDAC6. J Cell Biol 178: 1025–1038.

Pandey UB, et al. 2007. HDAC6 rescues neurodegeneration and provides an essential link between autophagy and the UPS. Nature 447: 859–863.

Polier S, Dragovic Z, Hartl FU, Bracher A. 2008. Structural basis for the cooperation of Hsp70 and Hsp110 chaperones in protein folding. Cell 133: 1068–1079.

Rubinsztein DC. 2006. The roles of intracellular protein-degradation pathways in neurodegeneration. Nature 443: 780–786.

Shimura H, Schwartz D, Gygi SP, Kosik KS. 2004. CHIP-Hsc70 complex ubiquitinates phosphorylated tau and enhances cell survival. J Biol Chem 279: 4869–4876.

Shin Y, Klucken J, Patterson C, Hyman BT, McLean PJ. 2005. The co-chaperone carboxyl terminus of Hsp70-interacting protein (CHIP) mediates alpha-synuclein degradation decisions between proteasomal and lysosomal pathways. J Biol Chem 280: 23727–23734.

Shomura Y, Dragovic Z, Chang HC, Tzvetkov N, Young JC, Brodsky JL, Guerriero V, Hartl FU, Bracher A. 2005. Regulation of Hsp70 function by HspBP1: structural analysis reveals an alternate mechanism for Hsp70 nucleotide exchange. Mol Cell 17: 367–379.

Sondermann H, Scheufler C, Schneider C, Höhfeld J, Hartl FU, Moarefi I. 2001. Structure of a Bag/Hsc70 complex: convergent functional evolution of Hsp70 nucleotide exchange factors. Science 291: 1553–1557.

Spooner RA, Hart PJ, Cook JP, Pietroni P, Rogon C, Höhfeld J, Roberts LM, Lord JM. 2008. Cytosolic chaperones influence the fate of a toxin dislocated from the endoplasmic reticulum. Proc Natl Acad Sci U S A 105: 17408–17413.

Taylor JP, Hardy J, Fischbeck KH. 2002. Toxic proteins in neurodegenerative disease. Science 296: 1991–1995.

Urushitani M, Kurisu J, Tateno M, Hatakeyama S, Nakayama K, Kato S, Takahashi R. 2004. CHIP promotes proteasomal degradation of familial ALS-linked mutant SOD1 by ubiquitinating Hsp/Hsc70. J Neurochem 90: 231–244.

Westhoff B, Chapple JP, van der Spuy J, Höhfeld J, Cheetham ME. 2005. HSJ1 is a neuronal shuttling factor for the sorting of chaperone clients to the proteasome. Curr Biol 15: 1058–1064.

Wickner S, Maurizi MR, Gottesman S. 1999. Posttranslational quality control: folding, refolding, and degrading proteins. Science 286: 1888–1893.

Yorimitsu T, Klionsky DJ. 2005. Autophagy: molecular machinery for self-eating. Cell Death Differ 12 Suppl 2: 1542–1552.

Young JC, Moarefi I, Hartl FU. 2001. Hsp90: a specialized but essential protein-folding tool. J Cell Biol 154: 267–273.

Zhang C, Cuervo AM. 2008. Restoration of chaperone-mediated autophagy in aging liver improves cellular maintenance and hepatic function. Nat Med 14: 959–965.

Zhang M, Windheim M, Roe SM, Peggie M, Cohen P, Prodromou C, Pearl LH. 2005. Chaperoned ubiquitylation–crystal structures of the CHIP U box E3 ubiquitin ligase and a CHIP-Ubc13-Uev1a complex. Mol Cell 20: 525–538.

Zhang M, Boter M, Li K, Kadota Y, Panaretou B, Prodromou C, Shirasu K, Pearl LH. 2008. Structural and functional coupling of Hsp90- and Sgt1-centred multi-protein complexes. Embo J 27: 2789–2798.

Chapter 4
The Small Heat-Shock Proteins: Cellular Functions and Mutations Causing Neurodegeneration

C. d'Ydewalle, J. Krishnan, V. Timmerman, and L. Van Den Bosch

Abstract Small heat-shock proteins (small Hsps) are a family of highly conserved proteins involved in multiple cellular mechanisms. Apart from their central role as chaperones in protecting cells during stressful conditions (as outlined in the previous two chapters), small Hsps also function to maintain cellular homeostasis in physiological conditions. Correct protein refolding to avoid aggregation, targeting misfolded proteins for degradation, proper cytoskeletal organization, and anti-apoptotic functions are some of the extensively studied attributes of small Hsps. One or more of these cellular mechanisms may malfunction in specific sets of neurons leading to neurodegenerative diseases such as Alzheimer's disease, Parkinson's disease, polyglutamine disorders, and amyotrophic lateral sclerosis. Many in vitro models of these diseases have demonstrated the beneficial roles of small Hsps pointing out their protective role in attenuating the neurodegenerative phenotype. Interestingly, mutations in small Hsps themselves were linked to other degenerative disorders like inherited peripheral neuropathies and familial myopathies. Although not much is known regarding the exact patho-mechanism ("loss of function" or "gain of function") of mutations in causing disease, these discoveries reiterate the importance of small Hsps in maintaining neuronal health and indicate that the small Hsp family of proteins might have more functions than meets the eye. This chapter reviews the current knowledge regarding these enigmatic proteins, including their structure and function and how mutations in these once "forgotten proteins" might alter their functions and cause neurodegeneration.

L. Van Den Bosch (✉)
Laboratory of Neurobiology, Department of Experimental Neurology, VIB and University of Leuven, Campus Gasthuisberg O&N2 PB 1022, Herestraat 49, Leuven B-3000, Belgium
e-mail: ludo.vandenbosch@vib-kuleuven.be

A. Wyttenbach and V. O'Connor (eds.), *Folding for the Synapse*,
DOI 10.1007/978-1-4419-7061-9_4,

4.1 Introduction

Originally described by Ritossa in 1962, the heat-shock response was characterized by a puffing pattern in the salivary glands of *Drosophila busckii* (Ritossa 1962), which were active sites of increased transcription and translation of a number of proteins (Tissières et al. 1974, Ashburner and Bonner 1979). These proteins were called "heat-shock proteins" (Hsps) produced during a transient, sublethal heat shock, which elicited a heat-shock response characterized by the induction and increased synthesis of constitutively present proteins. This event was followed by a transient tolerance to heat, which was later found as a phenomenon of resistance toward other stressful conditions, including hypoxia, ischemia, inflammation, and exposure to cellular toxins such as heavy metals, endotoxins, and reactive oxygen species (Jäättelä 1999b; Lindquist and Craig 1988; Snoeckx et al. 2001). Given the variety of stimuli that can give rise to the "heat-shock response," it is nowadays referred to as the "stress response" (Jäättelä 1999a; Lindquist and Craig 1988).

The human Hsp superfamily comprises eight main families based on molecular mass (summarized in Table 4.1). These proteins are encoded by genes that, according to the HUGO gene nomenclature, are designated HSPH, HSPC, HSPA, HSPD, HSPF, HSPB, and HSPE. This chapter focuses on the small Hsps during neurodegeneration particularly because of the diversity in actions attributed to the various members of this family. Furthermore, we review the consequences of mutations within small (and other) Hsps to illustrate how malfunctioning of several (chaperone) mechanisms, also described in Chaps. 1 and 2, lead to human disease.

4.1.1 Structure of Small Hsps

To date, the human small Hsp family consists of ten proteins and only a few of them are well characterized (for an overview of small Hsps, see Table 4.2). Small Hsps are structurally related to lens crystallins. Although crystallins were discovered in 1894 as exclusive lens proteins, it was not until 1982 that it was recognized that specific crystallins shared sequence homology with, and were themselves, small Hsps (Ingolia and Craig 1982).

Table 4.1 Overview of the heat-shock protein superfamily

Protein family (alias)	HUGO gene name	Number of known members
Hsp100	*HSPH*	1
Hsp90	*HSPC*	2
Hsp70	*HSPA*	9
Hsp60	*HSPD*	1
Hsp40	*HSPF*	4
Small Hsps	*HSPB*	10
Hsp10	*HSPE*	1

Table 4.2 Overview of the small Hsp family

HUGO gene name	Protein alias(es)	Gene location
HSPB1	CMT2F; HMN2B; HSP27; HSP28; Hsp25; HS.76067; DKFZp586P1322	7q11.23
HSPB2	MKBP; Hs.78846; LOH11CR1K; MGC133245	11q22-q23
HSPB3	HSPL27	5q11.2
HSPB4	αA-crystallin (CRYAA), CRYA1	21q22.3
HSPB5	αB-crystallin (CRYAB); CRYA2; CTPP2	11q22.3–q23.1
HSPB6		19q13.12
HSPB7	cvHSP; FLJ32733; DKFZp779D0968	1p36.23–p34.3
HSPB8	H11; HMN2; CMT2L; DHMN2; E2IG1; HMN2A; HSP22	12q24.23
HSPB9	FLJ27437	17q21.2
HSPB10	ODF; RT7; ODF2; ODFP; SODF; ODF27; ODFPG; ODF1; ODFPGA; ODFPGB; MGC129928; MGC129929	8q22.3

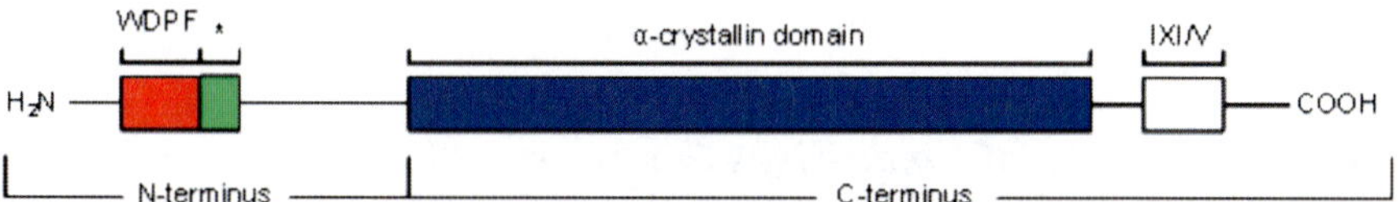

Fig. 4.1 General structure of small Hsps. A schematic overview of the general structure of small Hsps is shown. The variable N-terminal part of the protein comprises a hydrophobic WDPF motif (*red box*) which assists in oligomerization. Only in HSPB1, a MAP kinase recognition sequence (*green box*) is located immediately adjacent to the WDPF motif. The conserved α-crystallin domain (*blue box*), the signature motif of all small Hsps, lies within the C-terminal part. The more variable C-terminal tail contains an IXI/V motif (*white box*) which is important for interactions necessary for the tertiary structure of the protein

The monomeric, highly conserved small Hsps have a molecular weight of 12–43 kDa (de Jong et al. 1993). They contain a variable N-terminal sequence, a conserved core α-crystallin domain, and a short variable C-terminal tail (Fig. 4.1). The α-crystallin domain represents the signature motif of small Hsps, where the amino acid sequence (80–100 residues) is constant. This motif forms a β-sheet which is composed of two layers of three and five antiparallel strands, respectively, connected by a short interdomain loop. The α-crystallin domains can dimerize through the formation of an intersubunit composite β-sheet, also a conserved feature of small Hsps (Kim et al. 1998). Within the variable N-terminal tail lies a WDPF motif that is hydrophobic and helps in the oligomerization process (Lambert et al. 1999). Specifically for HSPB1, there is a MAP kinase recognition sequence present adjacent to the α-crystallin domain (James et al. 2008). The C-terminal tail contains an IXI/V motif that is important for interactions necessary for its structure; it interacts either with itself or with the IXI/V motif from another monomer (Pasta et al. 2004).

Small Hsps exist in active and inactive states (Haslbeck et al. 2005) in relation to their chaperone function. While larger Hsps like Hsp70 shift between the two states by ATP binding and hydrolysis (for more detail see Chap. 3), in general small Hsps are ATP-independent except for few reports on HSPB5 (Biswas and Das 2004; Haslbeck and Buchner 2002; Muchowski and Clark 1998). Some studies suggest that phosphorylation is one of the modulators of small Hsps. An experimental phosphorylation-mimicking mutant HSPB5 appears to have an enhanced chaperone activity in vitro, but is less stable to urea-induced denaturation (Ahmad et al. 2008). Likewise, Muchowski and Clark (1998) reported that refolding of denatured substrates by HSPB5 was enhanced in the presence of ATP. Phosphorylated HSPB1 interacts with thermally denatured F-actin, preventing the formation of F-actin aggregates (Pivovarova et al. 2007), the ability of which depends also on macromolecular (high oligomeric complexes) organization of HSPB1 (Benndorf et al. 1994). However, other studies have suggested that phosphorylation does not influence the activity of the small Hsps like anti-apoptotic function and oligomerization, or its ability to reduce intracellular reactive oxygen species (Bruey et al. 2000b; Preville et al. 1998a, b).

4.1.2 Functions of Small Hsps

The induction of small Hsps is regulated by heat-shock factors (HSFs) binding to a heat-shock element (HSE) in the promotor region of heat-shock genes (Morimoto et al. 1997). There are three different HSFs that are functionally distinct: HSF1 is essential for the heat-shock response and is also required for developmental processes, whereas HSF2 and HSF4 are important for development and differentiation (Akerfelt et al. 2007).

More recently, hypoxia has been proposed to induce an increased expression of (small) Hsps at specific sites in the brain and in the heart in porcine models (Chiral et al. 2004; Louapre et al. 2005; David et al. 2006). The increased expression level of stress proteins, including small Hsps such as Hsp20 and HSPB1, resulted from an increase in hypoxia-inducible transcription factor (HIF1α) activity and initiated a hypoxia-induced stress response (Chiral et al. 2004; Louapre et al. 2005; David et al. 2006). These observations indicate that elevated levels of small Hsps can be induced in a HSF-independent manner.

Interestingly, a genetic variant (c.-217T>C) targeting a conserved nucleotide in the HSE region of the *HSPB1* gene has been identified in a patient affected with amyotrophic lateral sclerosis (ALS) (Dierick et al. 2007). This HSE variant affected both constitutive and heat-induced transcriptional activity in transiently transfected neuronal and nonneuronal cells (Dierick et al. 2007). The decreased transcriptional activity was caused by a reduced HSF binding to the HSE variant, as confirmed by an electrophoretic mobility shift assay (Dierick et al. 2007).

The combination of the highly conserved α-crystallin domain and the universal presence of small Hsps indicate a crucial, evolutionary role in cellular homeostasis.

During the past few decades, extensive research has been performed regarding the function(s) of small Hsps. In vitro studies in the early 1990s showed that small Hsps bind (partially) denatured proteins and prevent their misfolding and/or irreversible aggregation, i.e., that they act as molecular chaperones (Horwitz 1992; Jakob et al. 1993) (also see Fig. 1.6 in 117 Chap. 2). But small Hsps are also involved in several other (unrelated) cellular processes, such as modulation of the actin cytoskeleton and the intermediate filaments, cell growth and differentiation, and apoptosis (Arrigo 2005; Gusev et al. 2002; Mehlen et al. 1997; Mounier and Arrigo 2002; Parcellier et al. 2003). As early as 1986, small Hsps were discovered to be associated with cytoskeletal structures (Leicht et al. 1986). HSPB1, for instance, inhibits actin polymerization by capping the plus end of the actin filament, thereby participating in the regulation of actin assembly. It also protects the cytoskeletal filaments from disruption due to filament-severing proteins activated by the stress response by coating these filaments (Mounier and Arrigo 2002). Similarly, HSPB5 is also involved in cytoskeletal dynamics, as has been shown by both in vitro and in vivo models. In cardiomyocytes, HSPB5 is distributed similar to desmin and vimentin at the cytoskeleton, while phosphorylated HSPB5 binds to and inhibits depolymerization of actin fibers (Djabali et al. 1997; Leach et al. 1994; Singh et al. 2007).

Cell differentiation is a further process in which a role for small Hsps has been suggested. Using an in vitro model, HSPB1 expression level was shown to be much lower in undifferentiated compared with differentiated cells (Kindås-Mügge and Trautinger 1994). HSPB5 protein level also followed a differentiation-dependent evolution (Ito et al. 2001). An earlier report demonstrated that phosphorylated HSPB1 was completely absent in differentiating murine embryonic stem cells compared with undifferentiated stem cells. Moreover, the process of differentiation was paralleled by an increased level of oligomerization and accumulation of HSPB1, suggesting a crucial role for higher order structures of HSPB1 in this process (Mehlen et al. 1997).

Both HSPB1 and HSPB5 have been demonstrated to have anti-apoptotic activities. The group of Arrigo demonstrated that HSPB1 interferes with the cytochrome-c activation of the caspase cascade, thereby also inhibiting procaspase-9 and other downstream apoptotic enzymes (Bruey et al. 2000a). Using a leukemic cell culture model, HSPB1 was shown to inhibit phosphorylation of p38 and c-Jun, as well as cytochrome-c release (Schepers et al. 2005). It has also been shown that phosphorylated dimers of HSPB1 interact with Daxx. Expression of HSPB1 prevents the translocation of Daxx from the nucleus to the cytoplasm, a movement induced upon expression of Ask1 or stimulation of Fas. Subsequently, its interaction with Ask1 and Fas is prevented, thereby blocking the Daxx-mediated apoptotic signal (Charette and Landry 2000). HSPB5 also interacts with p53, preventing the translocation of this transcription factor from the cytoplasm to mitochondria, blocking an early stage of the mitochondrial apoptotic pathway (Liu et al. 2007). The p53–HSPB5 interaction also leads to a very strong decrease or complete suppression of the transcriptional function of p53 and subsequent induction of its pro-apoptotic target genes (Liu et al. 2007). Boellmann et al. reported that Daxx interacts with HSF1, thereby increasing its activation. Eventually, this interaction leads to a higher transcriptional activity of HSF1, raising the (small) Hsp protein levels (Boellmann et al. 2004). The above mentioned observations are all in line with the general view that small Hsps play an anti-apoptotic role.

Thus, the many seemingly unrelated functions that small Hsps are involved in do point to their role as a general housekeeping protein family. As mentioned above, such house-keeping functions are necessary for proper protein production and targeted translocation as part of the cellular homeostasis, cellular growth, development, and maintenance in both stressful and normal conditions.

4.2 Role of Small Heat-Shock Proteins in Neurological Disease

Some members of the small Hsp family, notably HSPB1, have long been known to be expressed in the CNS tissue of mammals and are known to be involved in normal growth and development as well as in aging brains (Armstrong et al. 2001; Krueger-Naug et al. 2002; Loones et al. 2000). Recently however, the expression of the ten small Hsps in mouse CNS has been mapped (Quraishe et al. 2008). This study confirmed the presence of HSPB1, HSPB5, HSPB6, HSPB7, and HSPB8 in mouse brains. A distinct white matter specific expression pattern was observed for HSPB5, and an overlapping pattern of expression for HSPB1 and HSPB8 was detected. Some members were also associated with the synaptic compartment, suggesting that synapse regulation or homeostasis in the CNS could be part of the repertoire of functions that these proteins perform.

Since small Hsps function as chaperones, they are deemed to play a role in several neurodegenerative disorders in which protein aggregation is a prime feature. The formation of protein aggregates is a prominent neuropathological feature common to many neurodegenerative disorders including Alzheimer's disease (AD), Parkinson's disease (PD), polyglutamine disorders such as Huntington's disease (HD), and amyotrophic lateral sclerosis (ALS). In most cases, if not all, the aggregation mechanism is not well understood; hence its role in disease pathogenesis remains controversial. There are currently three hypotheses that try to unravel this controversy (Wood et al. 2003). First, protein aggregation could be a primary event in disease pathogenesis and underlie neurotoxicity. Second, the formation of aggregates could occur secondary to the primary neurotoxic event; this would mean that aggregates simply provide a marker of neuronal dysfunction. Finally, aggregate formation could reflect a cellular defense mechanism and represent an attempt to actively reduce the concentration of (a) toxic-free protein species. Thus, it is unclear if aggregation represents a toxic function or a beneficial clearing system (Ross and Poirier 2005). Even as the primary insult in neurodegeneration remains elusive, many in vitro models for neurodegenerative disorders have shown that small Hsps (but also other members of the Hsp superfamily such as Hsp70 and Hsp40) offer protection against the toxic effects of some aggregation prone proteins.

The presence of aggregates consisting of amyloid β (Aβ) peptide in typical senile plaques is the histological hallmark of AD. Upregulation of small Hsps is seen in autopsy samples of AD brains (Wilhelmus et al. 2006a, b). Both in vitro and in vivo models show that small Hsps colocalize with Aβ aggregates (Fonte et al. 2002). HSPB5 acts as a molecular chaperone by preventing proper fibril formation; but by doing so, it induces the formation of nonfibrillar Aβ-HSPB5 aggregates that are highly toxic (Raman et al. 2005; Stege et al. 1999). Thus, in AD HSPB5 seems to aggravate the neurodegenerative phenotype.

The loss of dopaminergic neurons in the *substantia nigra* and the cytoplasmic inclusion of Lewy bodies and Lewy neurites, containing α-synuclein aggregates, is the neuropathological feature of PD. Although there has not been much research regarding the role of small Hsps in PD, autopsy samples show an increase of HSPB1 in PD brains, which seems to be colocalized with α-synuclein inclusions similar to HSPB5 (Outeiro et al. 2006). The increased levels of small Hsps in PD brains might reflect the upregulation of these chaperones in a futile attempt by neurons to counteract the neurotoxicity of α-synuclein. In a primary midbrain cell culture, HSPB1 has a protective effect against α-synuclein toxicity (Outeiro et al. 2006). Although in vitro models for PD suggest a protective role for small Hsps in PD pathology, there are no in vivo studies to resolve the exact role of small Hsps in this neurodegenerative disorder.

Selective loss of GABAergic neurons of the striatum and intracellular aggregates of huntingtin protein (htt) with a polyglutamine (poly(Q)) tract in various areas of the brain is typical of HD. In an in vitro model for HD, HSPB1 reduces cell death without suppressing the formation of poly(Q) htt aggregates, possibly by acting on cytochrome-c-independent cell survival/death pathways (Wyttenbach et al. 2002). In vivo, endogenous HSPB1 as well as Hsp70 were up-regulated in HD rat brains, which altered the morphology of poly(Q) htt aggregates (Perrin et al. 2007). However, others found no phenotypical improvement in a mouse model for HD overexpressing HSPB1 (Zourlidou et al. 2007). The functional role of small/other HSPs in HD may have to be demonstrated by HSP knockout studies in the future.

Similarly, in vitro models of ALS have shown that small Hsps like HSPB1 and HSPB5 and others like Hsp40 and Hsp70 are protective against toxic effects of mutant SOD1 (Bruening et al. 1999; Patel et al. 2005). Samples from ALS brains are positive for HSPB1, HSPB5, and Hsp70 (Vleminckx et al. 2002, Batulan et al. 2003; Maatkamp et al. 2004), suggesting that Hsps might be connected to the pathogenesis of ALS. However, recent studies showed that ALS mice overexpressing HSPB1 did not have increased life span even though mutant SOD1 aggregates were decreased (Krishnan et al. 2008; Sharp et al. 2008).

As stated above, the genetic variant c.-217T>C in the HSE region of HSPB1 has been identified in a sporadic ALS patient (Dierick et al. 2007). This HSE variant abolished HSF binding and consequently impaired the HSPB1 heat-shock response (Dierick et al. 2007). It is plausible that a reduced stress mediated protection mechanism could aggravate motorneuron malfunction in ALS. Therefore, functional HSE or HSPB1 variants might contribute to ALS pathogenesis.

Thus, despite promising results in vitro and in contrast to acute neuronal injury models where overexpression of HSPB1 was protective against kainic acid induced seizures and nerve crush injuries (Akbar et al. 2003; Kalwy et al. 2003), chronic neurodegenerative mouse models do not show abrogation of disease with a single chaperone (Liu et al. 2005; Zourlidou et al. 2007). It might be plausible that multiple chaperones or a combination of small and large chaperones is required to mitigate complex and multifactorial disorders like HD and ALS, as also shown by our observation both in vitro and in vivo (Krishnan et al. 2006; 2008).

4.3 Inherited Degenerative Diseases: Mutations in Small Heat-Shock Proteins"

4.3.1 Myofibrillar Myopathies

Although the exact role of small Hsps is still enigmatic in neurodegenerative disease, mutations in genes encoding these porteins are associated with a number of inherited diseases (Fig. 4.2). The first mutation identified in this family of stress proteins is a mutation in the conserved domain of HSPB5 associated with desmin-related myopathy (DRM: myofibrillar myopathy, also called αB-crystallinopathy (Vicart et al. 1998)). Desmin-related myopathy has been described as an autosomal dominant, adult-onset neuromuscular disorder characterized by weakness of proximal and distal limb muscles (including neck, velopharynx, and trunk muscles), signs of cardiomyopathy and cataract. Pathologically this group of disorders shows delayed accumulation of aggregates of desmin, a protein belonging to the type III intermediate filament family, in the sarcoplasma of skeletal and cardiac muscles (Goebel and Fardeau 1996; Fuchs and Cleveland 1998; Vicart et al. 1998). The overexpression of both human and mouse Arg120 (R120G) missense mutation in mice were shown to cause a DRM-like phenotype (Wang et al. 2001; Sanbe et al. 2004; Rajasekaran et al. 2007).

The amino acid R120 corresponds to R116 in the HSPB4 moiety, the missense mutation of which causes congenital cataract in humans (Litt et al. 1998; Richter et al. 2008). These findings underscore the importance of structural stability of small Hsps for maintenance of normal function. Recently, two novel mutations also leading to myofibrillar myopathies (Q151X and 464delCT) have been identified in the terminal part of the HSPB5 coding sequence (Selcen and Engel 2003). Although some patients showed DRM-like phenotype without mutations in the HSPB5 gene (Pou Serradell et al. 2001; Kostera-Pruszczyk et al. 2006), it is now accepted that αB-crystallinopathies result from the misfolding and progressive aggregation of mutated HSPB5 to which desmin filaments subsequently associate to form HSPB5/desmin/amyloid positive aggresomes (Sanbe et al. 2005).

Thus, αB-crystallinopathies are a special case of protein conformation diseases in which the destabilizing mutations at the origin of the disorder occurs in a molecular chaperone, which is itself potentially involved in the protein quality control of the cell (Arrigo et al. 2007). Although other disease-causing mutations in the αB-crystallin gene (P20S, 464delCT, and D140N for dominant cataract (Liu et al. 2006a, b); R157H and G154S for cardiomyopathy (Inagaki et al. 2006)) have also

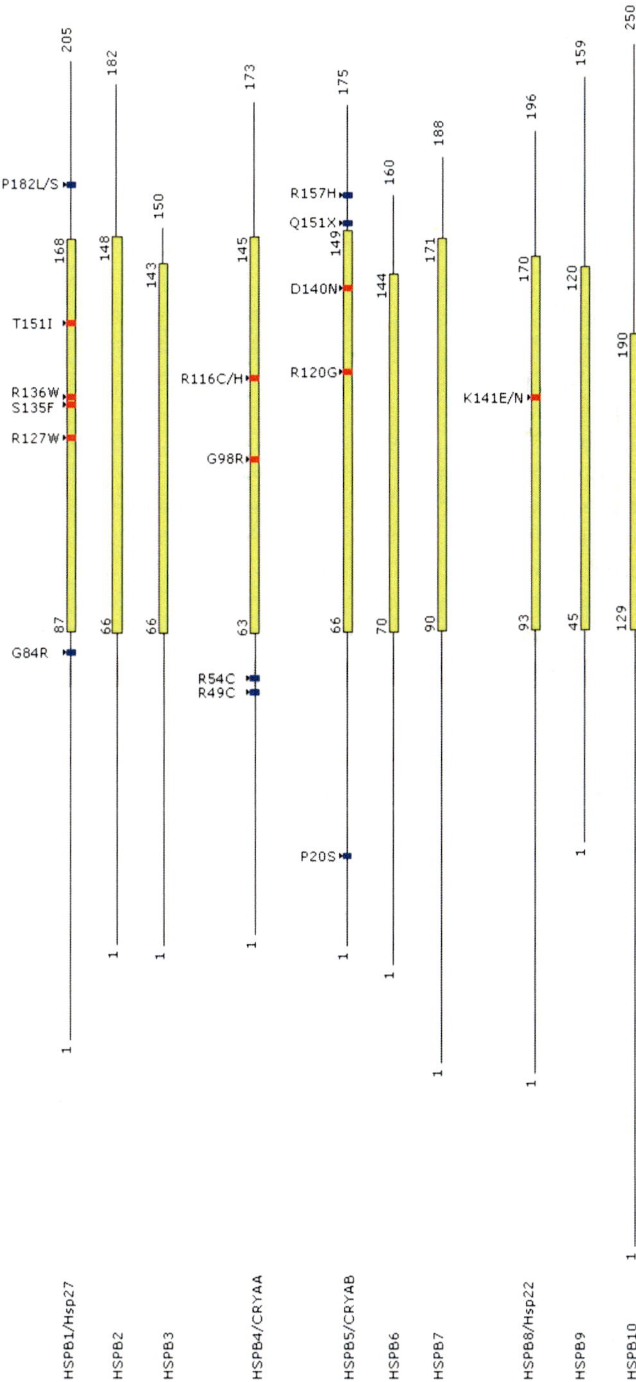

Fig. 4.2 Overview of small Hsps and localization of the different mutations. The ten members of the small Hsp protein family are shown to scale and allign to the α-crystallin domain. This domain (*yellow box*) is the signature motif of small Hsps. Missense mutations giving rise to amino acid substitutions either in- or outside the α-crystallin domain are shown in *red* and *blue*, respectively

been found, the role of the R120G mutation in the HSPB5 protein in aiding the phenotype of DRM remains the most studied to date.

4.3.2 Hereditary Motor Neuropathies

The R120 residue in HSPB5 corresponds to the K141 residue in HSPB8 and to R140 in HSPB1. Mutations in K141 in HSPB8 (K141E and K141N) and R127 (R127W), S135 (S135F) and R136 (R136W) – residues that are close to R140 in HSPB8 – are implicated in a group of clinically heterogeneous hereditary peripheral neuropathies (Evgrafov et al. 2004; Irobi et al. 2004). These neuropathies are often referred to as distal hereditary motor neuropathies (distal HMN) when predominantly motor neurons are affected. However, the neuropathies are classified as Charcot-Marie-Tooth (CMT) disease when both motor and sensory neurons are involved (Irobi et al. 2004, 2006). The considerable overlap between the different disorders and the exact nosology is still being redefined in the face of new molecular genetic studies (Irobi et al. 2006). Likewise, clinically, these patients are not thoroughly distinguishable and both distal HMN and CMT are known to coexist in some families (Van Den Bosch and Timmerman 2006).

The common clinical picture is distal weakness of the lower and upper limbs, sensory loss, decreased reflexes, and foot deformities. Less-frequent symptoms include cranial nerve involvement, scoliosis, vocal cord paresis, and glaucoma (Zuchner and Vance 2006). Harding classified the distal HMNs into seven subtypes, based upon age at onset, mode of inheritance, and presence of additional clinical features (Harding 1993). Mutations in genes other than HSPB1 and HSPB8 have also been described in both autosomal and recessive distal HMN (Irobi et al. 2006).

As information about these disorders has increased, it has become evident that the distinction between distal HMN and axonal CMT2 is less clear-cut than was originally believed. Mutations in HSPB8 occur in both distal HMN and CMT2 patients (Dierick et al. 2008). Among other genes, the identification of small Hsp gene mutations offer an intriguing connection into the role of the small Hsps in maintaining normal function in both the central and peripheral nervous system. Initially, two distinct missense mutations (K141N and K141E) involving the same amino acid (Lys141) in the core α-crystallin domein of HSPB8 were found in four distal HMN type II families (Irobi et al. 2004). Subsequently, the heterozygous amino acid change (K141N) in HSPB8 was also reported in a large Chinese CMT2 family (CMT2L) which was previously linked to the same chromosomal region as distal HMN II, as well as in a Korean family (Chung et al. 2008; Tang et al. 2004, 2005). Originally, five missense mutations in HSPB1, the molecular partner of HSPB8, were also associated with distal HMN and CMT2F (Evgrafov et al. 2004). Four of these mutations target the highly conserved α-crystallin domain (R127W, S135F, R136W, and T151I) and one is positioned in the C-terminal part (P182L). All five mutations occur in amino acids that are conserved in HSPB1 orthologs (Evgrafov et al. 2004). To date, six other missense mutations have been identified.

The P39L and G84R lie within the N-terminal part, while L99M, R140G, and K141Q are located in the crystallin domain (James et al. 2008; Houlden et al. 2008; Ikeda et al. 2008). The P182S missense mutation that has been identified in a Japanese patient lies within the C-terminal part (Kijima et al. 2005). The increasing number of mutations found in HSPB1, all associated to distal HMN and/or CMT2, emphasizes the importance of small Hsps in neuronal integrity.

4.3.3 How Do Mutations Cause Disease?

Mutations in ubiquitously expressed, house-keeping genes that generate proteins involved in the maintenance of cellular homeostasis are known to cause neurodegeneration (e.g., familial ALS caused by mutations in SOD1). Such mutations can be pathogenic by altering the normal functioning of the protein leading to breakdown of cellular homeostasis and hence degenerative changes: "loss of function" mutations. Alternatively, these mutations can endow the protein with a hitherto novel, unique, uncharacterized function that can be deleterious to the cell: "gain of function" mutations. The exact mechanism of small Hsp mutations in causing disease is unknown as yet. A few of the possible mechanisms are discussed below.

4.3.3.1 Inefficient Chaperoning Activity Due to Structural Changes

The identification of mutations in small Hsps associated with disease conditions demonstrates that retaining a positively charged amino acid in the highly conserved domain of the crystallin molecule is probably necessary for its cytoprotective functions (Bera et al. 2002). Site-directed mutagenesis of the core α-crystallin domain of HSPB5 altered its chaperone activity and reduced cell viability in vitro, while similar experimental mutations in the N-terminus did not have an effect (Muchowski et al. 1999). Interestingly, none of these experimental mutations affected the secondary, tertiary, or quaternary structure of the protein, indicating that the functional failure of chaperoning ability was due to the mutation in the conserved region, rather than a global perturbation in structure. These studies also proved that the core α-crystalline domain is responsible for the normal functioning of the protein.

The R120G mutation linked to DRM in the HSPB5 protein was found to have altered quaternary structure, oligomerization pattern, and reduced chaperoning activity in vitro (Bova et al. 1999; Perng et al. 1999; Simon et al. 2007). Similarly, R116C mutations in HSPB4 affected the structure and functional integrity of the protein and also reduced the protective ability in stress-induced lens epithelial cell apoptosis (Andley et al. 2002; Bera and Abraham 2002). Mutations in HSPB5 have been found to make the protein intrinsically unstable (Treweek et al. 2005). Since HSPB5 has a role in maintaining the intermediate filament proteins, mutation in the conserved domain of this protein leads to aggregation of desmin possibly due to its altered chaperone activity or due to altered interaction between these two proteins (Vicart et al. 1998; Wang et al. 2003). It has also been shown that HSPB5 modulates the aggregation of abnormal desmin in vivo (Wang et al. 2003). Indeed, a mutation in the core domain was shown to increase its affinity to desmin (Perng et al. 2004), thus explaining the occurrence of desmin aggregates in the muscle cells of DRM patients. These aggregates can by themselves be toxic, inhibiting the ubiquitin–proteasomal system of protein degradation (Bence et al. 2001) and causing deficits in mitochondrial function (Maloyan et al. 2005; Arrigo et al. 2007; Pinz et al. 2008). Release of pro-apoptotic molecules from damaged mitochondria might also lead to degeneration of the affected tissues (Maloyan et al. 2005), while the ability of HSPB5 to interact with members of the cytoskeletal polypeptides or with the other small Hsps may also be modified (Arrigo et al. 2007). Such alterations in normal functioning of the protein might constitute "loss of function" actions, the evidence of which is yet unclear.

Distal HMN-linked HSPB8 mutant K141E along with experimental K137E and K137+141E double mutants were also demonstrated to have lowered chaperoning activity in vitro in preventing the aggregation of insulin and rhodanase compared with wild-type (WT) HSPB8 (Kasakov et al. 2007). These mutations in the conserved β5-β7 strand of the molecule also affected its structure, suggesting that deleterious changes in the structure of the protein can give rise to congenital diseases (Kim et al. 2006). Although alteration in chaperone activity of HSPB1 mutants have not yet been demonstrated, it is likely that mutation in the conserved core α-crystalline domain might affect this function. The C-terminal region (especially the conserved IXI/V motif; see Fig. 4.1) seems to be critical for the chaperone activity of several bacterial small Hsps (Li et al. 2007; Saji et al. 2008). Forced experimental mutations in this motif was found to generate Hsps that had improved chaperone activity against amyloid forming target proteins (Treweek et al. 2007), while deletions in this region render the protein with no oligomerization and chaperone activity in vitro (Studer et al. 2002). Thus, the P182L mutation in the C-terminal of HSPB1 might reduce the interaction between the different domains of the protein or the interaction with other partners, affect its oligomerization status, and reduce chaperone activity, again suggesting a "loss of function" scenario.

4.3.3.2 Formation of Aggregates and Nuclear Transport

Whether the aggregates themselves are toxic is still open to debate, but in general it is agreed that the capacity of proteins to aggregate (and not per se the aggregated species) is correlated with toxicity (Rubinsztein 2006). Expression of HSPB1 and HSPB8 mutants in vitro lead to formation of aggregates (Evgrafov et al. 2004; Irobi et al. 2004; James, Rankin and Talbot 2008). In cortical neurons the WT HSPB1 co-segregated with aggregates formed by the P182L mutant HSPB1 and did not reduce the aggregates formed by the mutant (Ackerley et al. 2006). Co-immunoprecipitation studies demonstrated that mutant HSPB8 proteins pulled down more HSPB1 than WT HSPB8, suggesting that first HSPB1 and HSPB8 proteins interact, and that second the interaction seems to be more pronounced in the presence of mutations in HSPB8. Immunohistochemistry experiments also demonstrated that

WT HSPB1 co-localized with mutant HSPB8 (Irobi et al. 2004). These findings suggest that probably WT HSPB1 is sequestered into the toxic aggregates, thereby making the cell devoid of the WT protein to perform other cytoprotective functions, a suggestion also made for ALS (Pasinelli and Brown 2006).

Although toxic protein aggregation is a major contributor to a variety of neurodegenerative disorders (Wood et al. 2003), the exact role of mutant Hsp aggregation in inherited motor neuropathies is not yet clear. The aggregates of both HSPB1 and HSPB8 mutant proteins were observed in the cytoplasm, surrounding the perinuclear region of nonneuronal COS cells (Irobi et al. 2004), neuronal N2a cells, and primary motor neurons (unpublished observation), a phenomenon that might be important in two ways. First, mutation might affect the interaction between HSPB1 and other nuclear targeted molecules to enter the nucleus, as with HSPB5. The DRM-linked R120G mutation in HSPB5 is known to affect its interaction with survival motor neuron proteins that helps in the nuclear translocation of WT HSPB5 (den Engelsman et al. 2005). Second, non-entry of these mutant aggregates into the nucleus might mean that these proteins have lost the capacity to be phosphorylated, a condition necessary for the nuclear translocation of Hsp70/40 (Michels et al. 1997; Nollen et al. 2001). WT HSPB1 in the nucleus has been demonstrated to be colocalized with proteasomal components, possibly by directing nuclear proteins, that are beyond repair/refolding, for degradation (Bryantsev et al. 2007). Thus mutations can render the protein unable to enter the nucleus where it can repair denatured nuclear proteins, or impair nuclear import and export of other cytoskeletal molecules. Indeed, WT HSPB5 is co-localized with subnuclear structures like the SC35 speckles (den Engelsman et al. 2004) and its nuclear import is impaired in the R120G HSPB5 mutant (den Engelsman et al. 2005). In contrast, Simon et al. (2007) showed that the R120G mutant of HSPB5 was hyperphosphorylated and segregated additionally into nuclear and cytoskeletal subcellular fractions apart from the cytoplasm and organelle fractions. However, another dominant mutation R49C in HSPB4 (which occurs outside the conserved α-crystallin core and also causes cataract) was also found to segregate in the nucleus (Mackay et al. 2003). Thus, the exact role of phosphorylation in the context of subcellular localization and the implications of such altered segregation if any of the mutant small Hsps are still controversial.

4.3.3.3 Cytoskeletal Disorganization and Axonal Transport

Mutant HSPB1 affected the organization of neurofilament molecules (Evgrafov et al. 2004) suggesting that cytoskeletal disorganization of neurons, especially in the axons, might lead to premature axonal degeneration characteristic of distal HMN and CMT2 disorders. The disorganization of neurofilament M (NF-M) might affect transport of specific cargoes like the p150 molecule of the dynein/dynactin complex in the axons as reported for the P182L HSPB1 mutant (Ackerley et al. 2006). This effect on axonal transport demonstrated by immunohistochemical stainings seemed to be a specific cargo-related event, since localization of either mitochondria (which travel bidirectionally) or synaptotagmin 1 (anterograde movement) was not affected in the cortical neurons. However, whether this is true for all the mutants of HSPB1 and/or mutant HSPB8 is not yet known.

Point mutations in neurofilament L (NF-L) have been described to be associated with CMT2E (Mersiyanova et al. 2000). Studies have shown that these mutations affect the assembly of mutant NF-L with itself, with mid-sized NF-M and with NF-H (Brownlees et al. 2002; Perez-Olle et al. 2002, 2004, 2005). They also affect subcellular mitochondrial distribution and axonal transport of several markers including themselves (Perez-Olle et al. 2005). Hence disassembly of NF-L and subsequent axonal transport deficits due to mutations might represent a "loss of function" mechanism of mutations. However, the aggregates formed by mutant HSPB1 might also physically block axonal transport, the clear evidence for which is currently lacking. This might then constitute a novel "gain of function" mechanism, of which the exact nature is unclear. The afore-mentioned studies also suggest the disorganization of neurofilaments as a probable unifying mechanism of neurodegeneration in both CMT and distal HMN.

Interestingly, Zhai et al. (2007) showed that both CMT-linked NF-L mutants (P8R and Q333P) and HSPB1 mutant (S135F) caused neurofilament disorganization in primary motor neurons in culture. Both these mutant proteins decreased the viability of motor neurons. Wild-type HSPB1 was able to reverse the NF-L aggregation caused by the NF-L mutant. Both WT and mutant HSPB1 proteins associated with WT and NF-L mutants, suggesting that neurofilaments are one of the primary interacting partners of HSPB1. The injection of mutant HSPB1 also caused aggregation and disorganization of endogenous NF-L protein, thus demonstrating an inherent interaction between these proteins. Mutations in either of the interacting proteins seem to cause a common pathogenic mechanism of defective axonal transport in motor neurons, leading to their degeneration.

4.3.3.4 Altered Protein–Protein Interaction

The Hsp superfamily of porteins interact with each other as demonstrated for HSPB8. WT HSPB8 interacts with itself, HSPB1, HSPB5, and HSPB6 (Benndorf et al. 2001; Sun et al. 2004; Fontaine et al. 2005). Changes in these interactions may be a culprit mechanism in pathological conditions. The C-terminus of several small Hsps contains a conserved IXI/V motif (Fig. 4.1). Mutations in the conserved IXI/V motif (like P182L for HSPB1) might induce a conformational change and it might expose additional interaction sites or increase the interaction with other proteins. Such an altered interaction was demonstrated for the cataract-linked HSPB4 mutant R116C, which seemed to interact more with WT HSPB1 and HSPB5 compared with the nonmutated protein. Interestingly, the interaction of R116C HSPB4 with WT HSPB4 remained unchanged, suggesting that altered interactions within members of the superfamily might be the weakest link in the chain leading to neurodegeneration (Fu and Liang 2003a). Indeed, WT HSPB1 was found to stabilize the structure of HSPB5 similar to the role of HSPB4 in stabilizing HSPB5, thus making the propensity of mutations to alter these interactions a highly probable scenario (Fu and Liang 2003b). WT HSPB1 was also shown to rescue the toxic phenotype generated by the DRM-linked HSPB5 mutant R120G (Ito et al. 2003).

CMT-linked mutations in HSPB8 were demonstrated to interact more with HSPB1 and affect its activity (Irobi et al. 2004). Similar increased interactions of these mutants were also observed with WT HSPB8 and HSPB5, but not between WT HSPB8 and HSPB6, suggesting that specific interactions might be affected because of mutations rather than a general conformational change of the protein (Fontaine et al. 2006). Mutation-specific altered interaction was also found for DRM-linked HSPB5 mutants, where the R120G mutant had increased, Q151X mutant had decreased, and 464delCT mutant had unchanged interaction, respectively, with HSPB8 (Simon et al. 2007).

The interaction of small Hsps with non-Hsp proteins might also be altered because of mutations. There are now many reports available describing the association of HSPB4, HSPB5, HSPB1, and HSPB6 with cytoskeletal elements such as actin, desmin, vimentin, laminin, and glial fibrillary acidic protein (Lavoie et al. 1993; Benndorf et al. 1994; Nicholl and Quinlan 1994; Wang and Spector 1996; Perng et al. 1999; Der Perng et al. 2006; Brown et al. 2007). Frequently, this association was found to stabilize the cytoskeletal structures, e.g., HSPB1, HSPB4, and HSPB5 stabilize actin filaments (Lavoie et al. 1993, 1995; Wang and Spector 1996; Brown et al. 2007). Nonstabilization of these cytoskeletal proteins might have disastrous consequences for the neurons. Moreover, altered interactions with neurofilaments – which also belong to the large family of cytoskeletal proteins – might be deleterious in the axons leading to premature degeneration, as stated above (Perng et al. 1999). Such deleterious new functions might qualify for a novel "gain of function" mechanism.

4.3.3.5 Apoptosis and Autophagy

Finally, since both HSPB1 and HSPB8 have been shown to possess anti-apoptotic function in vitro, mutations might affect these functions (Bruey et al. 2000a, b; Garrido et al. 2001; Kamradt et al. 2001, 2002, 2005; Havasi et al. 2008; Lanneau et al. 2008). Mutations might not only decrease such anti-apoptotic role of these chaperones, but might actually induce apoptotic death when expressed in tissues. This has been demonstrated for both HSPB4 and HSPB5, where the R49C and R120G mutants lead to apoptotic cell death in cell lines and transgenic mice, respectively (Mackay et al. 2003; Maloyan et al. 2005).

As an alternative to the programmed pathway of cell death in apoptosis, autophagy involves direct dissolving of cellular contents by the resident lysosomal enzymes. This system is also known to degrade aggregation prone proteins as also done by the ubiquitin-proteasome system. Recently, it was shown that HSPB8 along with the co-chaperone Bag3 protein forms a complex and stimulates the degradation of proteins by macroautophagy (Carra et al. 2005, 2008a, b). This complex formation and subsequent macroautophagy was necessary to reduce aggregations formed by huntingtin protein (Htt43Q). As expected, both mutations K141E and K141N in HSPB8 reduced the aggregation clearing capacity of HSPB8 (Carra et al. 2005). Similarly, a recent report suggested that macroautophagy is an adaptive response in DRM caused by the R120G mutation of HSPB5 (Tannous et al. 2008a, b). The cardiomyocyte overexpression of R120G caused an increase in autophagic activity in the cardiac tissue and blunting of which increased the intracellular aggregates, accelerated ventricular dysfunction, and caused early mortality. Hence, mutations in small Hsps not only cause apoptosis, but also seem to avoid their own destruction by a self-propagating milieu decreasing the autophagic activity, and increase the formation of intracellular inclusions.

Thus, mutations in small Hsps might cause/potentiate degenerative changes in cells (especially neurons) via "gain of function" or "loss of function" mechanisms. The last word regarding the exact nature of the pathogenic mechanism of the point mutation in these small proteins is yet to be delivered.

4.4 Other Chaperonopathies

Mutations in chaperones other than small Hsps were also discovered to be involved in neurodegeneration. Here we shortly summarize such findings, but an extensive report on other chaperonopathies is beyond the scope of this review (Macario and Conway de Macario 2007).

The higher mass Hsp60 and its co-chaperonin Hsp10 are mitochondrial chaperones, protecting mitochondrial DNA and assisting in protein quality control. A mutation (V72I) in Hsp60 has been identified in a French family affected with autosomal dominant inherited spastic paraplegia (Hansen et al. 2002). This clinically and genetically heterogeneous group of neurodegenerative disorders is characterized by a progressive spasticity and weakness of the lower limbs. In vitro analysis revealed that the mutant Hsp60 decreased *E. coli* growth, compromised refolding of denatured proteins, and had decreased ATPase activity compared with WT Hsp60 (Bross et al. 2008; Hansen et al. 2002). A recent study identified a second mutation (D29G) in Hsp60, which causes an autosomal recessive neurodegenerative disease linked to brain hypomyelination and leukodystrophy (Magen et al. 2008). This newly identified disease was characterized by rotatory nystagmus, progressive spastic paraplegia, severe motor impairment, and mental retardation. In vitro, the D29G mutant Hsp60 showed less profound effects on cell viability compared with the V98I mutant. Taken together, these data indicate that mutations in Hsp60 can lead to a variety of neurodegenerative disorders, and that mitochondrial integrity plays a major role in the pathogenesis of neurodegeneration.

McKusick-Kaufman syndrome (MKKS) is a human developmental anomaly comprising hydrometrocolpos, postaxial polydactyly, and congenital heart disease (Stone et al. 2000). MKKS has been associated with two mutations (H84Y and A242S) in the MKKS gene, which shows similarity to the thermosome of *Thermoplasma acidophilum*, a member of the chaperonin family and has been shown to facilitate proper protein folding in an ATP-dependent manner (Stone et al. 2000). Other MKKS gene alterations have been associated with Bardet-Biedl syndrome, which is an autosomal recessive disorder characterized by a variety of symptoms including obesity, retinal dystrophy, polydactyly, and learning difficulties (Katsanis et al. 2000). The alterations include two frameshift (F94fsX103 and D143fsX157) and three missense mutations (Y37C, T57A and L227P). Despite the identification of mutations in MKKS and their association with diseases, the exact underlying pathological mechanism remains elusive.

Animal models for hereditary sensory neuropathy (HSN) are available. One of them is an autosomal, recessive, early-onset sensory neuropathy in rat. Clinical features of the disease include ataxia, insensitivity to pain and ulceration of the feet while pathology reveals severe reduction in the number of sensory ganglia and fibers (Lee et al. 2003). Genetic screening identified a C450Y mutation in the gene encoding the delta subunit of the cytosolic chaperonin-containing t-complex peptide-1 (Cct4). The chaperonin-containing tailless complex polypeptide (CCT) is a cytosolic molecular chaperone involved in folding tubulin, actin, and other cytosolic proteins (also see Chap. 2). Genetic profiling of a consanguineous Morrocan family affected with recessive mutilating sensory neuropathy with spastic paraplegia identified a homozygous mutation (H147R) in the epsilon subunit of CCT (Cct5) (Bouhouche et al. 2006). The identification of a mutated subunit of the CCT protein in patients confirms the involvement of this protein in the pathogenicity of HSN. Further research is needed to assess the consequences of the mutations on this protein's functions and whether these are "loss of function" or "gain of function" mutations.

4.5 Conclusions and Perspectives

Once termed the "forgotten chaperones," the family of small Hsps is far from forgotten in the intriguing field of neurodegeneration. That these are evolutionarily highly conserved proteins shows the antiquity of these proteins and the many chances they have had to interact with other newly emerging proteins. Small Hsps, by nature of their functions, appear to be significant integrators or coordinators of various cellular signals and responses (Arya et al. 2007). They provide "weak" but very important links between many "hubs," and thus, help transmission of information/signals across different paths in cellular networks (Csermely et al. 2007; Korcsmaros et al. 2007). Hsps are not just reversing stress-induced cellular damage or helping nascent polypeptides to fold appropriately, but have more global roles in cell metabolism like cytoskeletal maintenance and anti-apoptotic function.

Since neurons seem to be more vulnerable to toxic insults than most other cell types, owing to their nondividing nature and cellular homeostasis, maintenance issues assume paramount importance. Hence elucidating the role of small Hsps in modulating neurodegenerative phenotypes – where protein aggregation is a unifying theme in – via its chaperoning ability is a promising strategy toward understanding neuropathology. Knowledge regarding the functions of small Hsps and their interacting partners has almost always been obtained via biochemical assays using pure recombinant proteins, yeast-two-hybrid systems, or in vitro using cell lines. Extrapolation of data derived from the above-stated in vitro studies to in vivo situations in transgenic mice has not yielded similar results both in HD and ALS models. These studies nevertheless demonstrate a selective "context-based" action by small Hsps. The fact that despite protective actions by HSPB1 overexpression in mice against acute injuries, failure of HSPB1 to protect mice in chronic models points to the inherent differences in the mechanism of injury caused to the same neuron. It is possible that slow incremental accumulation of protein damage in misfolding chronic disease models, as opposed to a high flux of damaged proteins in acute stress models, is either not detected by cellular stress-response mechanisms or induced tolerance (Zourlidou et al. 2007). The failure of an individual chaperone to improve a neurodegenerative condition could reflect the requirement for a finely balanced concerted action of several chaperone systems, along with the ability to fine-tune their levels in response to specific cellular requirements.

Many classic neurodegenerative diseases like familial ALS are classified as multifactorial disorders, where more than one function of the neuron is compromised by an unexplained series of mutations in one or more genes. Thus, finding mutations in small Hsps suggests that more than one house-keeping function is either lost or one or more new toxic functions seem to be gained. Hence, although degenerative changes seem tissue specific (cardiac cells due to HSPB5 mutations, neuronal cells due to mutant HSPB1), multiple common mechanisms, like defects in cell differentiation and growth, deficits in anti-apoptotic machinery, disruption of cytoskeletal elements leading to defective cargo transport, and improper protein targeting, might all precipitate the observed phenotypes (Fig. 4.3).

The contribution of nonneuronal cells like glia is currently a highly debated topic in classic neurodegenerative diseases owing to their disease-modifying capacities (Lobsiger and Cleveland 2007). Since the discovery of mutations of Hsps in HMN and CMT is relatively recent, experimental data addressing such non-cell autonomous nature of disease propagation is lacking. Although restrictive, available in vitro studies of overexpression of HSPB1 in cortical neurons and HSPB5 in cardiac tissue of mice suggest that effects of mutations in small Hsps might be cell autonomous without the need for other cell types.

The exact mechanism of such deleterious effects accumulating to cause typically adult-onset degenerative phenotype is as yet unclear. As stated in the previous sections, although some experimental evidence argues for a "loss of function" mechanism (where the known function of the protein is lost or reduced) many recent developments also lay claim for a "gain of function" hypothesis. First, both distal HMN and CMT with mutations in small Hsps are dominantly inherited. Second, except for the 464delCT mutation in HSPB5 generating a truncated protein, none of the mutations discovered in small Hsps lead to a total loss of the proteins, but give rise to full-length proteins with missense mutations (see Table 4.3). Third, mice with targeted silencing of the HSPB1 gene (similar to a knock-out model) developed normally without any adverse phenotypical changes and remained viable

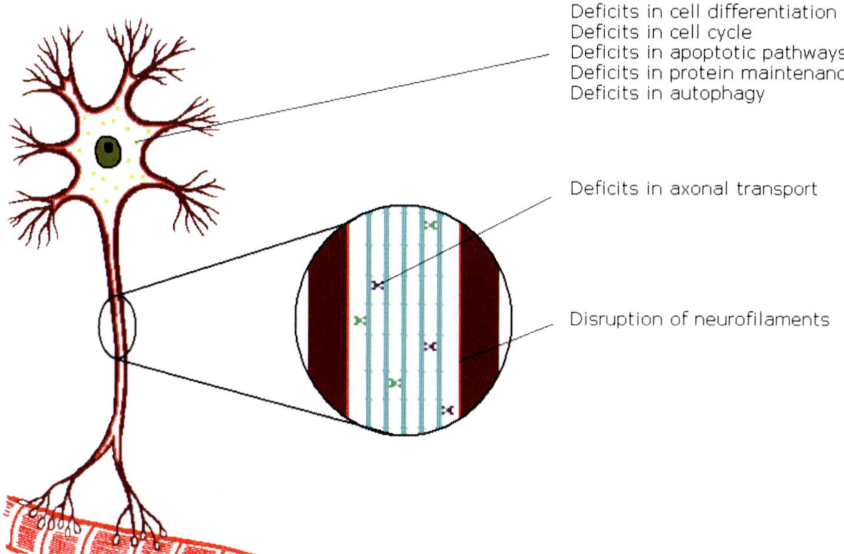

Fig. 4.3 Mutations in small Hsps can affect different cellular mechanisms. At the nuclear/cytoplasmic level, deficits in cell differentiation, cell cycle, apoptotic pathways, protein maintenance, and/or autophagy are plausible pathological mechanisms caused by mutant small Hsps. In the axon (*enlarged view*), defective antero- (*purple molecules*) or retrograde (*green molecules*) transport and/or disruption of neurofilaments are possible causes of mutant small Hsp-induced neurodegenerative diseases. At the neuromuscular junction, defective terminal branches could give rise to symptoms of the degenerative neuromuscular diseases

and fertile (Huang et al. 2007). This study proves that at least HSPB1 is dispensable during embryonic growth and development. More importantly all the above observations suggest that the presence of the mutant protein is necessary for the deleterious effects seen in these disorders tilting the balance toward a toxic "gain of function" scenario.

In the case of the inherited neurodegenerative disorders, identification of several genes whose proteins might share a similar domain of action will be helpful in developing cellular and animal models. These models can then be used to study the exact pathogenesis of this heterogeneous group of disorders and evaluate therapeutic options. However, there is currently a lack of transgenic animals harboring HSPB1 and HSPB8 mutations that recapitulate the neuronopathy phenotype of patients. Mice overexpressing the R120G mutation of HSPB5 (Wang et al. 2001) and R116C, R49C mutation of HSPB4 (Hsu et al. 2006; Bai et al. 2007; Xi et al. 2008) have been created and yield cardiac, lens phenotype as occurring in humans, respectively. Interestingly, overexpression of WT HSPB4 in nonlenticular tissues resulted in severe hind limb paralysis in mice (De Rijk et al. 2000; Van Rijk et al. 2003). Although ALS and HMSN phenotypes were excluded, the authors suggest that this model can be used to evaluate neuroaxonal dystrophy. This study, strengthening the "gain of function"

Table 4.3 Mutations in small Hsps and their diseases

Small Hsps	Mutations	Position	Associated disorders	Reference
HSPB1 (Hsp27)	c.-217T>C	5'UTR (HSE element)	ALS (single patient)	(Dierick et al. 2007)
	P39L	N-terminus	Distal hereditary motor neuropathy and Charcot-Marie-Tooth type 2F	(Houlden et al. 2008)
	G84R	N-terminus	Distal hereditary motor neuropathy	(James et al. 2008)
	L99M	N-terminus	Distal hereditary motor neuropathy and Charcot-Marie-Tooth type 2F	(Houlden et al. 2008)
	R127W	Core	Distal hereditary motor neuropathy and Charcot-Marie-Tooth type 2F	(Evgrafov et al. 2004)
	S135F	Core	Distal hereditary motor neuropathy and Charcot-Marie-Tooth type 2F	(Evgrafov et al. 2004)
	R136W	Core	Charcot-Marie-Tooth type2F	(Evgrafov et al. 2004)
	R140G	Core	Distal hereditary motor neuropathy	(Houlden et al. 2008)
	K141Q	Core	Distal hereditary motor neuropathy and Charcot-Marie-Tooth type 2F	(Houlden et al. 2008)
	T151I	Core	Distal hereditary motor neuropathy	(Evgrafov et al. 2004)
	P182L	C-terminus	Distal hereditary motor neuropathy	(Evgrafov et al. 2004)
	P182S	C-terminus	Distal hereditary motor neuropathy	(Kijima et al. 2005)
HSPB4 (CRYAA)	R116C	Core	Cataract	(Litt et al. 1998)
	R116H	Core	Cataract	(Richter et al. 2008)
HSPB5 (CRYAB)	R120G	Core	DRM-linked myopathy (also called myofibrillar myopathy), cardiomyopathy, cataract	(Vicart et al. 1998)
	Q151X	C-terminus	Myofibrillar myopathy	(Selcen and Engel 2003)
	464delCT	C-terminus	Myofibrillar myopathy	(Selcen and Engel 2003)
	R157H	C-terminus	Cardiomyopathy	(Inagaki et al. 2006)
	P20S	N-terminus	Cataract	(Liu et al. 2006a)
	D140N	Core	Cataract	(Liu et al. 2006b)
	450delA	C-terminus	Cataract	(Berry et al. 2001)
HSPB8 (Hsp22)	K141E	Core	Distal hereditary motor neuropathy	(Irobi et al. 2004)
	K141N	Core	Distal hereditary motor neuropathy	(Irobi et al. 2004)

action due to mutations emphasizes that small Hsps indeed have intriguing and enigmatic functions in the central and peripheral nervous system.

In conclusion, small Hsps are capable of modulating neuronal functions owing to their inherent house-keeping nature. Thus, it is conceivable that mutations in these set of proteins might be deleterious in many ways to the same cells including toxic self aggregation or nonclearance of other aggregation prone proteins, by inducing premature death due to loss of anti-apoptotic function or causing axonal anomalies due to cytoskeletal and transport deficits. Until the exact "Jackle-Hyde" nature of these proteins are deduced, there is bound to be an enigma in the puzzling field of neurodegeneration.

References

Ackerley S, James P A, Kalli A et al (2006) A mutation in the small heat-shock protein HSPB1 leading to distal hereditary motor neuronopathy disrupts neurofilament assembly and the axonal transport of specific cellular cargoes. Hum Mol Genet 15:347–354

Ahmad M F, Raman B, Ramakrishna T et al (2008) Effect of phosphorylation on alpha B-crystallin: differences in stability, subunit exchange and chaperone activity of homo and mixed oligomers of alpha B-crystallin and its phosphorylation-mimicking mutant. J Mol Biol 375:1040–1051

Akbar M T, Lundberg A M, Liu K et al (2003) The neuroprotective effects of heat shock protein 27 overexpression in transgenic animals against kainate-induced seizures and hippocampal cell death. J Biol Chem 278:19956–19965

Akerfelt M, Trouillet D, Mezger V et al (2007) Heat shock factors at a crossroad between stress and development. Ann N Y Acad Sci 1113:15–27

Andley U P, Patel H C, Xi J H (2002) The R116C mutation in alpha A-crystallin diminishes its protective ability against stress-induced lens epithelial cell apoptosis. J Biol Chem 277: 10178–86

Armstrong C L, Krueger-Naug A M, Currie R W et al (2001) Constitutive expression of heat shock protein HSP25 in the central nervous system of the developing and adult mouse. J Comp Neurol 434:262–274

Arrigo A P (2005) In search of the molecular mechanism by which small stress proteins counteract apoptosis during cellular differentiation. J Cell Biochem 94:241–246

Arrigo A P, Simon S, Gibert B et al (2007) Hsp27 (HspB1) and alphaB-crystallin (HspB5) as therapeutic targets. FEBS Lett 581:3665–3674

Arya R, Mallik M, and Lakhotia S C (2007) Heat shock genes – integrating cell survival and death. J Biosci 32:595–610

Ashburner M and Bonner J J (1979) The induction of gene activity in drosophilia by heat shock. Cell 17:241–254

Bai F, Xi J H, Andley U P (2007) Up-regulation of tau, a brain microtubule-associated protein, in lens cortical fractions of aged alphaA-, alphaB-, and alphaA/B-crystallin knockout mice. Mol Vis 13:1589–600

Batulan Z, Shinder G A, Minotti S et al (2003) High threshold for induction of the stress response in motor neurons is associated with failure to activate HSF1. J Neurosci 23:5789–5798

Bence N F, Sampat R M, Kopito R R (2001) Impairment of the ubiquitin-proteasome system by protein aggregation. Science 292:1552–5

Benndorf R, Hayess K, Ryazantsev S et al (1994) Phosphorylation and supramolecular organization of murine small heat shock protein HSP25 abolish its actin polymerization-inhibiting activity. J Biol Chem 269:20780–20784

Benndorf R, Sun X, Gilmont R R et al (2001) HSP22, a new member of the small heat shock protein superfamily, interacts with mimic of phosphorylated HSP27 ((3D)HSP27). J Biol Chem 276:26753–61

Bera S, Thampi P, Cho W J et al (2002) A positive charge preservation at position 116 of alpha A-crystallin is critical for its structural and functional integrity. Biochemistry 41:12421–12426

Bera S, Abraham E C (2002) The alphaA-crystallin R116C mutant has a higher affinity for forming heteroaggregates with alphaB-crystallin. Biochemistry 41:297–305

Berry V, Francis P, Reddy M A et al (2001) Alpha-B crystallin gene (CRYAB) mutation causes dominant congenital posterior polar cataract in humans. Am J Hum Genet 69:1141–1145

Biswas A and Das K P (2004) Role of ATP on the interaction of alpha-crystallin with its substrates and its implications for the molecular chaperone function. J Biol Chem 279:42648–42657

Boellmann F, Guettouche T, Guo Y et al (2004) DAXX interacts with heat shock factor 1 during stress activation and enhances its transcriptional activity. Proc Natl Acad Sci U S A 101:4100–4105

Bouhouche A, Benomar A, Bouslam N et al (2006) Mutation in the epsilon subunit of the cytosolic chaperonin-containing t-complex peptide-1 (Cct5) gene causes autosomal recessive mutilating sensory neuropathy with spastic paraplegia. J Med Genet 43:441–443

Bross P, Naundrup S, Hansen J et al (2008) The Hsp60-(p.V98I) mutation associated with hereditary spastic paraplegia SPG13 compromises chaperonin function both in vitro and in vivo. J Biol Chem 283:15694–15700

Brown D D, Christine K S, Showell C et al (2007) Small heat shock protein Hsp27 is required for proper heart tube formation. Genesis 45:667–78

Brownlees J, Ackerley S, Grierson A J et al (2002) Charcot-Marie-Tooth disease neurofilament mutations disrupt neurofilament assembly and axonal transport. Hum Mol Genet 11:2837–44

Bruening W, Roy J, Giasson B et al (1999) Up-regulation of protein chaperones preserves viability of cells expressing toxic Cu/Zn-superoxide dismutase mutants associated with amyotrophic lateral sclerosis. J Neurochem 72:693–699

Bruey J M, Ducasse C, Bonniaud P et al (2000a) Hsp27 negatively regulates cell death by interacting with cytochrome c. Nat Cell Biol 2:645–652

Bruey J M, Paul C, Fromentin A et al (2000b) Differential regulation of HSP27 oligomerization in tumor cells grown in vitro and in vivo. Oncogene 19:4855–4863

Bryantsev A L, Chechenova M B and Shelden E A (2007) Recruitment of phosphorylated small heat shock protein Hsp27 to nuclear speckles without stress. Exp Cell Res 313:195–209

Bova M P, Yaron O, Huang Q et al (1999) Mutation R120G in alphaB-crystallin, which is linked to a desmin-related myopathy, results in an irregular structure and defective chaperone-like function. Proc Natl Acad Sci U S A 96:6137–42

Carra S, Seguin S J, Lambert H et al (2008a) HspB8 chaperone activity toward poly(Q)-containing proteins depends on its association with Bag3, a stimulator of macroautophagy. J Biol Chem 283:1437–1444

Carra S, Seguin S J and Landry J (2008b) HspB8 and Bag3: a new chaperone complex targeting misfolded proteins to macroautophagy. Autophagy 4:237–239

Carra S, Sivilotti M, Chavez Zobel A T et al (2005) HspB8, a small heat shock protein mutated in human neuromuscular disorders, has in vivo chaperone activity in cultured cells. Hum Mol Genet 14:1659–1669

Charette S J and Landry J (2000) The interaction of HSP27 with Daxx identifies a potential regulatory role of HSP27 in Fas-induced apoptosis. Ann N Y Acad Sci 926:126–131

Chiral M, Grongnet J F, Plumier J C, David JC (2004) Effects of hypoxia on stress proteins in the piglet brain at birth. Pediatr Res 56:775–782

Chung K W, Kim S B, Cho S Y et al (2008) Distal hereditary motor neuropathy in Korean patients with a small heat shock protein 27 mutation. Exp Mol Med 40:304–312

Csermely P, Söti C, Blatch G L (2007) Chaperones as parts of cellular networks. Adv Exp Med Biol 594:55–63

David J C, Boelens W C, Grongnet J F (2006) Up-regulation of heat shock protein HSP 20 in the hippocampus as an early response to hypoxia of the newborn. J Neurochem 99:570–581

de Jong W W, Leunissen J A and Voorter C E (1993) Evolution of the alpha-crystallin/small heat-shock protein family. Mol Biol Evol 10:103–126

den Engelsman J, Bennink E J, Doerwald L et al (2004) Mimicking phosphorylation of the small heat-shock protein alphaB-crystallin recruits the F-box protein FBX4 to nuclear SC35 speckles. Eur J Biochem 271:4195–203

den Engelsman J, Gerrits D, de Jong W W et al (2005) Nuclear import of {alpha}B-crystallin is phosphorylation-dependent and hampered by hyperphosphorylation of the myopathy-related mutant R120G. J Biol Chem 280:37139–37148

Dierick I, Irobi J, Janssens S et al (2007) Genetic variant in the HSPB1 promotor region impairs the HSP27 stress response. Hum Mutat 28:830

Dierick I, Baets J, Irobi J et al (2008) Relative contribution of mutations in genes for autosomal dominant distal hereditary motor neuropathies: a genotype-phenotype correlation study. Brain 131:1217–27

Djabali K, de Nechaud B, Landon F et al (1997) AlphaB-crystallin interacts with intermediate filaments in response to stress. J Cell Sci 110 (Pt 21):2759–2769

Der Perng M, Su M, Wen S F et al (2006) The Alexander disease-causing glial fibrillary acidic protein mutant, R416W, accumulates into Rosenthal fibers by a pathway that involves filament aggregation and the association of alpha B-crystallin and HSP27. Am J Hum Genet 79:197–213

De Rijk E P, Van Rijk A F, Van Esch E et al (2000) Demyelination and axonal dystrophy in alpha A-crystallin transgenic mice. Int J Exp Pathol 81:271–82

Evgrafov O V, Mersiyanova I, Irobi J et al (2004) Mutant small heat-shock protein 27 causes axonal Charcot-Marie-Tooth disease and distal hereditary motor neuropathy. Nat Genet 36:602–606

Fontaine J M, Sun X, Benndorf R et al (2005) Interactions of HSP22 (HSPB8) with HSP20, alphaB-crystallin, and HSPB3. Biochem Biophys Res Commun 337:1006–11

Fontaine J M, Sun X, Hoppe A D et al (2006) Abnormal small heat shock protein interactions involving neuropathy-associated HSP22 (HSPB8) mutants. FASEB J 20:2168–2170

Fonte V, Kapulkin V, Taft A et al (2002) Interaction of intracellular beta amyloid peptide with chaperone proteins. Proc Natl Acad Sci U S A 99:9439–9444

Fu L and Liang J J (2003a) Alteration of protein-protein interactions of congenital cataract crystallin mutants. Invest Ophthalmol Vis Sci 44:1155–1159

Fu L and Liang J J (2003b) Enhanced stability of alpha B-crystallin in the presence of small heat shock protein Hsp27. Biochem Biophys Res Commun 302:710–714

Fuchs E, Cleveland D W (1998) A structural scaffolding of intermediate filaments in health and disease. Science 279:514–9

Garrido C, Gurbuxani S, Ravagna n L (2001) Heat shock proteins: endogenous modulators of apoptotic cell death. Biochem Biophys Res Commun 286:433–42

Goebel H H, Fardeau M. Familial desmin-related myopathies and cardiomyopathies--from myopathology to molecular and clinical genetics. 36th European Neuromuscular Center (ENMC)-Sponsored International Workshop 20-22 October, 1995, Naarden, The Netherlands. Neuromuscul Disord 6:383–8

Gusev N B, Bogatcheva N V and Marston S B (2002) Structure and properties of small heat shock proteins (sHsp) and their interaction with cytoskeleton proteins. Biochemistry (Mosc) 67:511–519

Hansen J J, Durr A, Cournu-Rebeix I et al (2002) Hereditary spastic paraplegia SPG13 is associated with a mutation in the gene encoding the mitochondrial chaperonin Hsp60. Am J Hum Genet 70:1328–1332

Haslbeck M and Buchner J (2002) Chaperone function of sHsps. Prog Mol Subcell Biol 28: 37–59

Haslbeck M, Franzmann T, Weinfurtner D et al (2005) Some like it hot: the structure and function of small heat-shock proteins. Nat Struct Mol Biol 12:842–846

Horwitz J (1992) Alpha-crystallin can function as a molecular chaperone. Proc Natl Acad Sci U S A 89:10449–10453

Harding A E (1993) Inherited neuronal atrophy and degeneration predominantly of lower motor neurons. In: Peripheral Neuropathy, 3rd Edition (Dyck PJ, Thomas PK, Griffin JW, Low PA, Poduslo JF, eds), pp 1051–1064

Havasi A, Li Z, Wang Z et al (2008) Hsp27 inhibits Bax activation and apoptosis via a phosphatidylinositol 3-kinase-dependent mechanism. J Biol Chem 283:12305–13

Houlden H, Laura M, Wavrant-De Vrièze F et al (2008) Mutations in the HSP27 (HSPB1) gene cause dominant, recessive, and sporadic distal HMN/CMT type 2. Neurology 71:1660–8

Hsu C D, Kymes S, Petrash J M (2006) A transgenic mouse model for human autosomal dominant cataract. Invest Ophthalmol Vis Sci 47:2036–44

Huang L, Min J N, Masters S et al (2007) Insights into function and regulation of small heat shock protein 25 (HSPB1) in a mouse model with targeted gene disruption. Genesis 45:487–501

Inagaki N, Hayashi T, Arimura T et al (2006) Alpha B-crystallin mutation in dilated cardiomyopathy. Biochem Biophys Res Commun 342:379–386

Ingolia T D and Craig E A (1982) Four small Drosophila heat shock proteins are related to each other and to mammalian alpha-crystallin. Proc Natl Acad Sci U S A 79:2360–2364

Irobi J, Dierick I, Jordanova A et al (2006) Unraveling the genetics of distal hereditary motor neuronopathies. Neuromolecular Med 8:131–146

Irobi J, Van Impe K, Seeman P et al (2004) Hot-spot residue in small heat-shock protein 22 causes distal motor neuropathy. Nat Genet 36:597–601

Ito H, Kamei K, Iwamoto I et al (2001) Regulation of the levels of small heat-shock proteins during differentiation of C2C12 cells. Exp Cell Res 266:213–221

Ito H, Kamei K, Iwamoto I et al (2003) Hsp27 suppresses the formation of inclusion bodies induced by expression of R120G alpha B-crystallin, a cause of desmin-related myopathy. Cell Mol Life Sci 60:1217–1223

Ikeda Y, Abe A, Ishida C et al (2008) A clinical phenotype of distal hereditary motor neuronopathy type II with a novel HSPB1 mutation. J Neurol Sci 277:9–12

Jäättelä M (1999a) Heat shock proteins as cellular lifeguards. Ann Med 31:261–271

Jäättelä M (1999b) Escaping cell death: survival proteins in cancer. Exp Cell Res 248:30–43

Jakob U, Gaestel M, Engel K et al (1993) Small heat shock proteins are molecular chaperones. J Biol Chem 268:1517–1520

James P A, Rankin J and Talbot K (2008) Asymmetrical late onset motor neuropathy associated with a novel mutation in the small heat shock protein HSPB1 (HSP27). J Neurol Neurosurg Psychiatry 79:461–463

Kalwy S A, Akbar M T, Coffin R S et al (2003) Heat shock protein 27 delivered via a herpes simplex virus vector can protect neurons of the hippocampus against kainic-acid-induced cell loss. Brain Res Mol Brain Res 111:91–103

Kasakov A S, Bukach O V, Seit-Nebi A S et al (2007) Effect of mutations in the beta5-beta7 loop on the structure and properties of human small heat shock protein HSP22 (HspB8, H11). FEBS J 274:5628–5642

Katsanis N, Beales P L, Woods M O et al (2000) Mutations in MKKS cause obesity, retinal dystrophy and renal malformations associated with Bardet-Biedl syndrome. Nat Genet 26:67–70

Kamradt M C, Chen F, Cryns V L (2001) The small heat shock protein alpha B-crystallin negatively regulates cytochrome c- and caspase-8-dependent activation of caspase-3 by inhibiting its autoproteolytic maturation. J Biol Chem 276:16059–63

Kamradt M C, Chen F, Sam S et al (2002) The small heat shock protein alpha B-crystallin negatively regulates apoptosis during myogenic differentiation by inhibiting caspase-3 activation. J Biol Chem 277:38731–6

Kamradt M C, Lu M, Werner M E et al (2005) The small heat shock protein alpha B-crystallin is a novel inhibitor of TRAIL-induced apoptosis that suppresses the activation of caspase-3. J Biol Chem 280:11059–66

Kijima K, Numakura C, Goto T et al (2005) Small heat shock protein 27 mutation in a Japanese patient with distal hereditary motor neuropathy. J Hum Genet 50:473–476

Kim K K, Kim R and Kim S H (1998) Crystal structure of a small heat-shock protein. Nature 394:595–599

Kim M V, Kasakov A S, Seit-Nebi A S et al (2006) Structure and properties of K141E mutant of small heat shock protein HSP22 (HspB8, H11) that is expressed in human neuromuscular disorders. Arch Biochem Biophys 454:32–41

Kindås-Mügge I and Trautinger F (1994) Increased expression of the M(r) 27,000 heat shock protein (hsp27) in in vitro differentiated normal human keratinocytes. Cell Growth Differ 5:777–781

Kostera-Pruszczyk A, Goudeau B, Ferreiro A et al (2006) Myofibrillar myopathy with congenital cataract and skeletal anomalies without mutations in the desmin, alphaB-crystallin, myotilin, LMNA or SEPN1 genes. Neuromuscul Disord 16:759–62

Korcsmáros T, Kovács I A, Szalay M S et al (2007) Molecular chaperones: the modular evolution of cellular networks. J Biosci 32:441–6

Krishnan J, Lemmens R and Robberecht W (2006) Role of heat shock response and Hsp27 in mutant SOD1-dependent cell death. Exp Neurol 200:301–310

Krishnan J, Vannuvel K, Andries M et al (2008) Overexpression of Hsp27 does not influence disease in the mutant SOD1(G93A) mouse model of amyotrophic lateral sclerosis. J Neurochem 106:2170–2183

Krueger-Naug A M, Plumier J C, Hopkins D A et al (2002) Hsp27 in the nervous system: expression in pathophysiology and in the aging brain. Prog Mol Subcell Biol 28:235–251

Lanneau D, Brunet M, Frisan E et al (2008) Heat shock proteins: essential proteins for apoptosis regulation. J Cell Mol Med 12:743–61

Lambert H, Charette S J, Bernier A F et al (1999) HSP27 multimerization mediated by phosphorylation-sensitive intermolecular interactions at the amino terminus. J Biol Chem 274:9378–9385

Lavoie J N, Hickey E, Weber L A et al (1993) Modulation of actin microfilament dynamics and fluid phase pinocytosis by phosphorylation of heat shock protein 27. J Biol Chem 268:24210–4

Lavoie J N, Lambert H, Hickey E et al (1995) Modulation of cellular thermoresistance and actin filament stability accompanies phosphorylation-induced changes in the oligomeric structure of heat shock protein 27. Mol Cell Biol 15:505-16

Leach I H, Tsang M L, Church R J et al (1994) Alpha-B crystallin in the normal human myocardium and cardiac conducting system. J Pathol 173:255–260

Lee M J, Stephenson D A, Groves M J et al (2003) Hereditary sensory neuropathy is caused by a mutation in the delta subunit of the cytosolic chaperonin-containing t-complex peptide-1 (Cct4) gene. Hum Mol Genet 12:1917–1925

Leicht B G, Biessmann H, Palter K B et al (1986) Small heat shock proteins of Drosophila associate with the cytoskeleton. Proc Natl Acad Sci U S A 83:90–94

Li Y, Schmitz K R, Salerno J C et al (2007) The role of the conserved COOH-terminal triad in alphaA-crystallin aggregation and functionality. Mol Vis 13:1758–68

Lindquist S and Craig E A (1988) The heat-shock proteins. Annu Rev Genet 22:631–677

Litt M, Kramer P, LaMorticella D M et al (1998) Autosomal dominant congenital cataract associated with a missense mutation in the human alpha crystallin gene CRYAA. Hum Mol Genet 7:471–474

Liu J, Shinobu L A, Ward C M et al (2005) Elevation of the Hsp70 chaperone does not effect toxicity in mouse models of familial amyotrophic lateral sclerosis. J Neurochem 93:875–882

Liu M, Ke T, Wang Z et al (2006a) Identification of a CRYAB mutation associated with autosomal dominant posterior polar cataract in a Chinese family. Invest Ophthalmol Vis Sci 47: 3461–3466

Liu S, Li J, Tao Y et al (2007) Small heat shock protein alphaB-crystallin binds to p53 to sequester its translocation to mitochondria during hydrogen peroxide-induced apoptosis. Biochem Biophys Res Commun 354:109–114

Liu Y, Zhang X, Luo L et al (2006b) A novel alphaB-crystallin mutation associated with autosomal dominant congenital lamellar cataract. Invest Ophthalmol Vis Sci 47:1069–1075

Lobsiger C S and Cleveland D W (2007) Glial cells as intrinsic components of non-cell-autonomous neurodegenerative disease. Nat Neurosci 10:1355–1360

Loones M T, Chang Y and Morange M (2000) The distribution of heat shock proteins in the nervous system of the unstressed mouse embryo suggests a role in neuronal and non-neuronal differentiation. Cell Stress Chaperones 5:291–305

Louapre P, Grongnet J F, Tanguay R M et al (2005) Effects of hypoxia on stress proteins in the piglet heart at birth. Cell Stress Chaperones 10:17–23

Maatkamp A, Vlug A, Haasdijk E et al (2004) Decrease of Hsp25 protein expression precedes degeneration of motoneurons in ALS-SOD1 mice. Eur J Neurosci 20:14–28

Macario A J and Conway de Macario E (2007) Molecular chaperones: multiple functions, pathologies, and potential applications. Front Biosci 12:2588–2600

Mackay D S, Andley U P and Shiels A (2003) Cell death triggered by a novel mutation in the alphaA-crystallin gene underlies autosomal dominant cataract linked to chromosome 21q. Eur J Hum Genet 11:784–793

Magen D, Georgopoulos C, Bross P et al (2008) Mitochondrial hsp60 chaperonopathy causes an autosomal-recessive neurodegenerative disorder linked to brain hypomyelination and leukodystrophy. Am J Hum Genet 83:30–42

Maloyan A, Sanbe A, Osinska H et al (2005) Mitochondrial dysfunction and apoptosis underlie the pathogenic process in alpha-B-crystallin desmin-related cardiomyopathy. Circulation 112:3451–3461

Mehlen P, Mehlen A, Godet J et al (1997) Hsp27 as a switch between differentiation and apoptosis in murine embryonic stem cells. J Biol Chem 272:31657–31665

Mersiyanova I V, Perepelov A V, Polyakov A V et al (2000) A new variant of Charcot-Marie-Tooth disease type 2 is probably the result of a mutation in the neurofilament-light gene. Am J Hum Genet 67:37–46

Michels A A, Kanon B, Konings A W et al (1997) Hsp70 and Hsp40 chaperone activities in the cytoplasm and the nucleus of mammalian cells. J Biol Chem 272:33283–9

Morimoto R I, Kline M P, Bimston D N et al (1997) The heat-shock response: regulation and function of heat-shock proteins and molecular chaperones. Essays Biochem 32:17–29

Mounier N and Arrigo A P (2002) Actin cytoskeleton and small heat shock proteins: how do they interact? Cell Stress Chaperones 7:167–176

Muchowski P J and Clark J I (1998) ATP-enhanced molecular chaperone functions of the small heat shock protein human alphaB crystallin. Proc Natl Acad Sci U S A 95:1004–1009

Muchowski P J, Wu G J, Liang J J, et al (1999) Site-directed mutations within the core "alpha-crystallin" domain of the small heat-shock protein, human alphaB-crystallin, decrease molecular chaperone functions. J Mol Biol 289:397–411

Nicholl I D, Quinlan R A (1994) Chaperone activity of alpha-crystallins modulates intermediate filament assembly. EMBO J 13:945–53

Nollen E A, Salomons F A, Brunsting J F et al (2001) Dynamic changes in the localization of thermally unfolded nuclear proteins associated with chaperone-dependent protection. Proc Natl Acad Sci U S A 98:12038–43

Outeiro T F, Klucken J, Strathearn K E et al (2006) Small heat shock proteins protect against alpha-synuclein-induced toxicity and aggregation. Biochem Biophys Res Commun 351:631–638

Parcellier A, Schmitt E, Gurbuxani S et al (2003) HSP27 is a ubiquitin-binding protein involved in I-kappaBalpha proteasomal degradation. Mol Cell Biol 23:5790–5802

Pasinelli P and Brown R H (2006) Molecular biology of amyotrophic lateral sclerosis: insights from genetics. Nat Rev Neurosci 7:710–723

Pasta S Y, Raman B, Ramakrishna T et al (2004) The IXI/V motif in the C-terminal extension of alpha-crystallins: alternative interactions and oligomeric assemblies. Mol Vis 10: 655–662

Patel Y J, Payne Smith M D, de Belleroche J et al (2005) Hsp27 and Hsp70 administered in combination have a potent protective effect against FALS-associated SOD1-mutant-induced cell death in mammalian neuronal cells. Brain Res Mol Brain Res 134:256–274

Perng M D, Cairns L, van den I P et al (1999) Intermediate filament interactions can be altered by HSP27 and alphaB-crystallin. J Cell Sci 112 (Pt 13):2099–2112

Perng M D, Wen S F, van den I P et al (2004) Desmin aggregate formation by R120G alphaB-crystallin is caused by altered filament interactions and is dependent upon network status in cells. Mol Biol Cell 15:2335–2346

Perrin V, Régulier E, Abbas-Terki T et al (2007) Neuroprotection by Hsp104 and Hsp27 in lentiviral-based rat models of Huntington's disease. Mol Ther 15:903–911

Perez-Olle R, Leung C L, Liem R K (2002) Effects of Charcot-Marie-Tooth-linked mutations of the neurofilament light subunit on intermediate filament formation.J Cell Sci 115:4937–46

Perez-Olle R, Jones S T, Liem R K (2004) Phenotypic analysis of neurofilament light gene mutations linked to Charcot-Marie-Tooth disease in cell culture models. Hum Mol Genet 13:2207-20

Pérez-Ollé R, López-Toledano M A, Goryunov D et al (2005) Mutations in the neurofilament light gene linked to Charcot-Marie-Tooth disease cause defects in transport. J Neurochem 93:861-74

Pivovarova A V, Chebotareva N A, Chernik I S et al (2007) Small heat shock protein Hsp27 prevents heat-induced aggregation of F-actin by forming soluble complexes with denatured actin. FEBS J 274:5937–5948

Pinz I, Robbins J, Rajasekaran N S et al (2008) Unmasking different mechanical and energetic roles for the small heat shock proteins CryAB and HSPB2 using genetically modified mouse hearts. FASEB J 22:84–92

Pou Serradell A, Lloreta Trull J, Corominas Torres J et al (2001) Familial myopathy with desmin storage seen as a granulo-filamentar, electron-dense material without mutation of the alpha-beta-crystallin gene. Neurologia 16:195–203

Preville X, Gaestel M and Arrigo A P (1998a) Phosphorylation is not essential for protection of L929 cells by Hsp25 against H2O2-mediated disruption actin cytoskeleton, a protection which appears related to the redox change mediated by Hsp25. Cell Stress Chaperones 3:177–187

Preville X, Schultz H, Knauf U et al (1998b) Analysis of the role of Hsp25 phosphorylation reveals the importance of the oligomerization state of this small heat shock protein in its protective function against TNFalpha- and hydrogen peroxide-induced cell death. J Cell Biochem 69:436–452

Quraishe S, Asuni A, Boelens W C et al (2008) Expression of the small heat shock protein family in the mouse CNS: Differential anatomical and biochemical compartmentalization. Neuroscience 153:483–491

Rajasekaran N S, Connell P, Christians E S et al (2007) Human alpha B-crystallin mutation causes oxido-reductive stress and protein aggregation cardiomyopathy in mice. Cell 130:427–39

Raman B, Ban T, Sakai M et al (2005) AlphaB-crystallin, a small heat-shock protein, prevents the amyloid fibril growth of an amyloid beta-peptide and beta2-microglobulin. Biochem J 392:573–581

Richter L, Flodman P, Barria von-Bischhoffshausen F et al (2008) Clinical variability of autosomal dominant cataract, microcornea and corneal opacity and novel mutation in the alpha A crystallin gene (CRYAA). Am J Med Genet A 146:833–842

Ritossa F M (1962) A new puffing pattern induced by a temperature shock and DNP in Drosophila. Experientia 18:571–573

Ross C A and Poirier M A (2005) Opinion: What is the role of protein aggregation in neurodegeneration? Nat Rev Mol Cell Biol 6:891–898

Rubinsztein D C (2006) The roles of intracellular protein-degradation pathways in neurodegeneration. Nature 443:780–786

Saji H, Iizuka R, Yoshida T (2008) Role of the IXI/V motif in oligomer assembly and function of StHsp14.0, a small heat shock protein from the acidothermophilic archaeon, Sulfolobus tokodaii strain 7. Proteins 71:771–82

Sanbe A, Osinska H, Saffitz J E et al (2004) Desmin-related cardiomyopathy in transgenic mice: a cardiac amyloidosis. Proc Natl Acad Sci U S A 101:10132–6

Sanbe A, Osinska H, Villa C et al (2005) Reversal of amyloid-induced heart disease in desmin-related cardiomyopathy. Proc Natl Acad Sci U S A 102:13592–13597

Schepers H, Geugien M, van der Toorn M et al (2005) HSP27 protects AML cells against VP-16-induced apoptosis through modulation of p38 and c-Jun. Exp Hematol 33:660–670

Selcen D and Engel A G (2003) Myofibrillar myopathy caused by novel dominant negative alpha B-crystallin mutations. Ann Neurol 54:804–810

Sharp P S, Akbar M T, Bouri S et al (2008) Protective effects of heat shock protein 27 in a model of ALS occur in the early stages of disease progression. Neurobiol Dis 30:42–55

Simon S, Fontaine J M, Martin J L et al (2007) Myopathy-associated alphaB-crystallin mutants: abnormal phosphorylation, intracellular location, and interactions with other small heat shock proteins. J Biol Chem 282:34276–34287

Singh B N, Rao K S, Ramakrishna T et al (2007) Association of alphaB-crystallin, a small heat shock protein, with actin: role in modulating actin filament dynamics in vivo. J Mol Biol 366:756–767

Snoeckx L H E H, Cornelussen R N, Van Nieuwenhoven R N et al (2001) Heat shock proteins and cardiovascular pathophysiology. Physiol Rev 81:1461–1497

Stege G J, Renkawek K, Overkamp P S et al (1999) The molecular chaperone alphaB-crystallin enhances amyloid beta neurotoxicity. Biochem Biophys Res Commun 262:152–156

Stone D L, Slavotinek A, Bouffard G G et al (2000) Mutation of a gene encoding a putative chaperonin causes McKusick-Kaufman syndrome. Nat Genet 25:79–82

Studer S, Obrist M, Lentze N et al (2002) A critical motif for oligomerization and chaperone activity of bacterial alpha-heat shock proteins. Eur J Biochem 269:3578–3586

Sun X, Fontaine J M, Rest J S et al (2004) Interaction of human HSP22 (HSPB8) with other small heat shock proteins. J Biol Chem 279:2394–402

Tang B S, Zhao G H, Luo W et al (2005) Small heat-shock protein 22 mutated in autosomal dominant Charcot-Marie-Tooth disease type 2L. Hum Genet 116:222–4

Tannous P, Zhu H, Johnstone J L et al (2008a) Autophagy is an adaptive response in desmin-related cardiomyopathy. Proc Natl Acad Sci U S A 105:9745–9750

Tannous P, Zhu H, Nemchenko A et al (2008b) Intracellular protein aggregation is a proximal trigger of cardiomyocyte autophagy. Circulation 117:3070–3078

Tissières A, Mitchell H K and Tracy U M (1974) Protein synthesis in salivary glands of Drosophila melanogaster. Relation to chromosomal puffs. J Mol Biol 84:389–398

Treweek T M, Ecroyd H, Williams D M et al (2007) Site-directed mutations in the C-terminal extension of human alphaB-crystallin affect chaperone function and block amyloid fibril formation. PLoS ONE 2:e1046

Treweek T M, Rekas A, Lindner R A et al (2005) R120G alphaB-crystallin promotes the unfolding of reduced alpha-lactalbumin and is inherently unstable. FEBS J 272:711–724

Van Rijk A F, Sweers M A, Merkx G F et al (2003) Pathogenesis of axonal dystrophy and demyelination in alphaA-crystallin-expressing transgenic mice. Int J Exp Pathol 84:91–9

Van Den Bosch L and Timmerman V (2006) Genetics of motor neuron disease. Curr Neurol Neurosci Rep 6:423–431

Vicart P, Caron A, Guicheney P et al (1998) A missense mutation in the alphaB-crystallin chaperone gene causes a desmin-related myopathy. Nat Genet 20:92–95

Vleminckx V, Van Damme P, Goffin K et al (2002) Upregulation of HSP27 in a transgenic model of ALS. J Neuropathol Exp Neurol 61:968–74

Wang K, Spector A (1996) alpha-crystallin stabilizes actin filaments and prevents cytochalasin-induced depolymerization in a phosphorylation-dependent manner. Eur J Biochem 242:56–66

Wang X, Klevitsky R, Huang W et al (2003) AlphaB-crystallin modulates protein aggregation of abnormal desmin. Circ Res 93:998–1005

Wang X, Osinska H, Klevitsky R et al (2001) Expression of R120G-alphaB-crystallin causes aberrant desmin and alphaB-crystallin aggregation and cardiomyopathy in mice. Circ Res 89:84–91

Wilhelmus M M, Boelens W C, Otte-Höller I et al (2006a) Small heat shock proteins inhibit amyloid-beta protein aggregation and cerebrovascular amyloid-beta protein toxicity. Brain Res 1089:67–78

Wilhelmus M M, Otte-Höller I, Wesseling P et al (2006b) Specific association of small heat shock proteins with the pathological hallmarks of Alzheimer's disease brains. Neuropathol Appl Neurobiol 32:119–130

Wood J D, Beaujeux T P and Shaw P J (2003) Protein aggregation in motor neurone disorders. Neuropathol Appl Neurobiol 29:529–545

Wyttenbach A, Sauvageot O, Carmichael J et al (2002) Heat shock protein 27 prevents cellular polyglutamine toxicity and suppresses the increase of reactive oxygen species caused by huntingtin. Hum Mol Genet 11:1137–1151

Xi J H, Bai F, Gross J et al (2008) Mechanism of small heat shock protein function in vivo: a knock-in mouse model demonstrates that the R49C mutation in alpha A-crystallin enhances protein insolubility and cell death. J Biol Chem 283:5801–14

Zhai J, Lin H, Julien J P et al (2007) Disruption of neurofilament network with aggregation of light neurofilament protein: a common pathway leading to motor neuron degeneration due to Charcot-Marie-Tooth disease-linked mutations in NFL and HSPB1. Hum Mol Genet 16:3103-16

Zourlidou A, Gidalevitz T, Kristiansen M et al (2007) Hsp27 overexpression in the R6/2 mouse model of Huntington's disease: chronic neurodegeneration does not induce Hsp27 activation. Hum Mol Genet 16:1078–1090

Zuchner S and Vance J M (2006) Mechanisms of disease: a molecular genetic update on hereditary axonal neuropathies. Nat Clin Pract Neurol 2:45–53

Part II
The Making of the Synapse: Transport of Proteins, Vesicles and Organelles

Chapter 5
Keeping it Together. Axonal Transport to the Synapse and the Effects of Molecular Chaperones in Health and Disease

Christopher Sinadinos and Amrit Mudher

Abstract Robust intracellular transport along the axon facilitates the delivery and replacement of somally synthesised macromolecules to and from the synapse, and is thus essential for the maintenance of neuronal function. In contact with an intricate cytoskeletal network, multi-subunit motor complexes drive axonal transport during journeys through an axonal compartment that is relatively devoid of machineries for de novo protein synthesis and turnover. Very little is known about how these complex transport machineries are assembled and maintained within the axon. In this review, possible roles for molecular chaperone proteins in the maintenance of processive, appropriately regulated axonal transport are considered. When such transport quality control is compromised, as in the case of neurodegenerative proteinopathies, such as Parkinson's, Alzheimer's and Huntington's diseases, pathological axonal transport disruption amidst aberrant protein folding and aggregation can result, severing an essential life-line and jeopardising synaptic function. How the stress-induced protein folding and aggregation-blocking functions of molecular chaperones are overcome in a range of pathological circumstances is also considered.

5.1 Introduction

Long-distance intracellular movement of vesicular organelles and their cargoes is mediated primarily by the kinesin and dynein motor superfamily proteins, multi-component mechanochemical enzymes that utilise the energy of ATP hydrolysis to power the movement of cellular components along polymeric microtubule tracks (Hirokawa and Takemura 2005). Although transport between spatially distant compartments is an essential housekeeping function for any eukaryotic cell, the role of microtubule-based transport has added significance within neurons. This is partly because the distal axonal and synaptic compartments subsist at a particularly great

C. Sinadinos (✉)
Southampton Neuroscience Group, School of Biological Sciences, University of Southampton, Highfield Campus, SO17 1BJ, UK
e-mail: cas1@soton.ac.uk

A. Wyttenbach and V. O'Connor (eds.), *Folding for the Synapse*,
DOI 10.1007/978-1-4419-7061-9_5,

distance from the predominant sites of mRNA and protein synthesis within the soma. The axoplasm can comprise over 99% of the total cell volume of a lower human motor neuron, the axon of which can exceed a metre in length (Brown 2003). Anterograde delivery of structural synaptic components, synaptic vesicle precursors and their contents, and of mitochondria to supply the demands of synaptic vesicle release and recycling (see Chap. 6), must regularly span such distances. Little protein turnover by de novo protein synthesis in the axon also means that local protein folding and quality control mechanisms in this neuronal compartment are likely to be of particular importance. When molecular chaperones that regulate protein folding and quality control mechanisms (as discussed in the chapters by J. Ellis and J. Höhfeld and colleagues) are compromised during neurodegenerative disease, axonal protein misfolding, aggregation and cargo accumulation could result, severing the essential life-line of axonal transport routes from primary sites of protein turnover in the soma. Here, we consider the potential roles of molecular chaperones in the maintenance of normal axonal transport, both under physiological conditions and during disease when defects in the axonal transport and protein folding machineries may synergise to compromise neuronal function and viability.

5.2 Do Molecular Chaperones Play a Role in Normal Axonal Transport?

Whereas the roles of molecular chaperones in protecting various neuronal processes during neurodegenerative disease have been intensely studied in recent years (Der Perng and Quinlan 2004; Muchowski and Wacker 2005), relatively little attention has been paid to basic housekeeping roles that chaperone proteins may play within the healthy axonal transport system. In the first part of this section, the basic molecular components of the axonal transport system are evaluated as potential sites of chaperone protein demand. Then, the emerging evidence for chaperone roles in axonal transport within healthy neurons is discussed.

5.2.1 Assembly of the Core Motor Complex

The kinesin, and particularly dynein motor complexes function as large, multi-subunit molecular complexes (Fig. 5.1) that are likely to represent a considerable challenge to the neuron in terms of nascent protein folding, axonal trafficking, assembly and maintenance. How these complex structures are assembled in neurons is currently unknown. Some clues may lie in the recent characterisation of mutant Medaka fish and Chlamydomonas strains showing defective assembly of the dynein outer arm complex found in the ciliary axoneme (Omran et al. 2008). In this system, the beating motion of the cilium is driven by ciliary dyneins that produce sliding

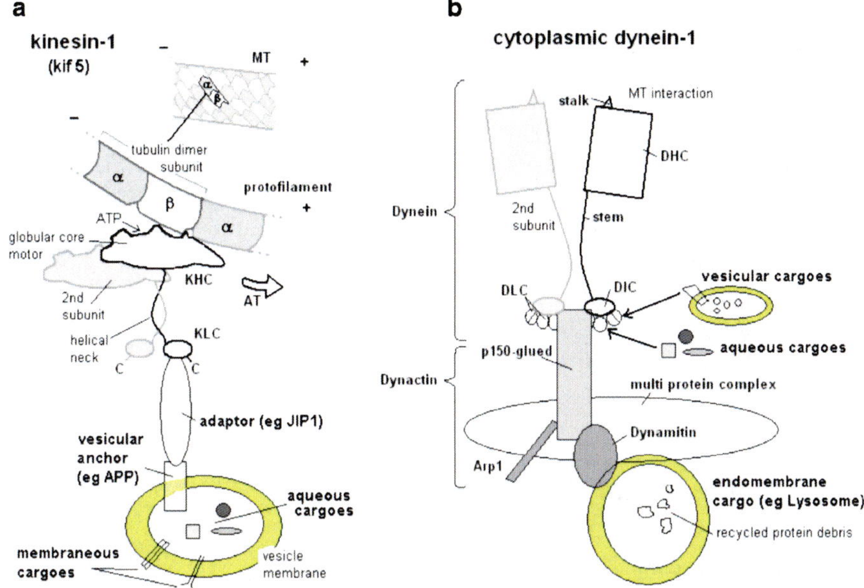

Fig. 5.1 Kinesin and dynein axonal transport motor complexes and their cargoes. (**a**) A current model for the interaction between kinesin-1 (kif5) and a single microtubule (MT) protofilament. The globular core motor within the kinesin heavy chain (KHC) comprises an α–β–α fold, with recognition helices making contact predominantly with the β-tubulin subunit. A central β-sheet provides inter-strand loops comprising a nucleotide-binding pocket adjacent to the MT-recognition interface, and successive rounds of ATP hydrolysis are thought to trigger conformational change and kinesin monomer exchange for a processive "hand-over-hand" model of polarised movement along the MT. An α-helical coiled-coil neck and tail region is involved in this exchange, and also connects the KHC to the kinesin light chain (KLC) and carboxy terminus, regions implicated in cargo receptor binding. The KLC interacts with JNK interacting protein 1 (JIP-1), for example, which recruits amyloid precursor protein as an adaptor to transport vesicles in the axon. At least 45 members constitute the kinesin superfamily in mice and humans, with extensive variation in KHC tail length, motor position and quarternary structure, although the core motor region remains relatively conserved (Hirokawa and Takemura 2005; Marx et al. 2006; Koushika 2008); (**b**) The dynein-1 multi-component complex. Cytoplasmic dynein-1, one of 15 vertebrate dynein complexes identified to date, is involved in retrograde transport of axonal cargoes to the neuronal cell body. Dynein heavy chain (DHC) possesses a 350 KDa globular core that comprises hexameric nucleotide binding rings with fold similarity to the AAA+ family of ATPases. Dynein intermediate chains (DIC) and light chains (DLC) connect the DHC stem with the rest of the complex. Instances of cargo association with both DLCs and the dynamitin component of the dynein-associated dynactin complex have been documented, such as binding of rhodopsin vesicles to the DLC component Tctex-1, soluble neuronal nitric oxide synthase and nuclear p53 binding protein to DLC8, and the association of endosomes and lysosomes to the dynactin complex. Despite growing knowledge of such cargo interactions, the structure, mechanics, and in particular any special requirements for long distance retrograde axonal transport of the dynein-1 complex as a whole, is currently at a relatively nascent level of understanding (Burkhardt et al. 1997; Tai et al. 1999; Crosby 2003; Vallee et al. 2004; Mohan and Hosur 2008)

forces between outer doublet microtubules (Gibbons 1996). The *kintoun* (*ktu*) Medaka mutant exhibits altered ciliary motility due to defective dynein outer arm assembly, and the homologous *Chlamydomonas* gene, PF13, is required for the attachment of the axonemal dynein heavy chain subunits to the nascent motor complex during cytoplasmic assembly prior to intraflagellar transport (Omran et al. 2008). The mouse *Ktu* homologue interacts specifically with Hsp70 (Omran et al. 2008), implicating molecular chaperone activity in the stabilisation of dynein heavy chains prior to the insertion and/or their assembly into the entire complex. Although axonemal dyneins are a divergent family of molecular motors from the single cytoplasmic dynein utilised for intracellular transport which is expressed in neurons (Gibbons 1996), it will be interesting to establish whether axonal transport motor complexes are pre-assembled in a similar fashion prior to axonal delivery, and to dissect possible roles for molecular chaperone proteins in this process.

5.2.2 Other Axonal Transport Components

In addition to the core subunits of the force-generating motor complexes, numerous molecular components comprise the intracellular transport machinery of the neuron. These include the neuronal cytoskeleton and microtubule network as the structural substrate for motor transport, a plethora of kinesin superfamily proteins with conserved motor domains but highly divergent cargo-interacting regions, and an emerging list of adaptor proteins involved in the attachment of diverse cargoes to these motor connection points. Cargoes destined for dendrites and post-synaptic regions are sorted from those targeted to the axolemma or pre-synaptic regions as they are attached in a regulated fashion to particular kinesins. Many of these kinesins exhibit unique structural characteristics, modes of transport processivity and sub-cellular destinations (Hirokawa and Takemura 2005). In theory, molecular chaperone function could serve to maintain the fidelity of the numerous molecular interactions involved in this transport process, and aid in the transit of some immature folding intermediates and individual subunits of protein complexes that have yet to attain their final conformation for reasons of energy efficiency during transport (Sousa and Lafer 2006). In some cases, pre-assembled motor complexes may alternatively transport intact organelles containing folded and functional protein cargoes, such as mitochondria and synaptic vesicle precursors, in the absence of a dedicated axonal molecular chaperone function. Evidence discussed below, however, suggests that such dedicated chaperone functions for normal axonal transport do exist. Molecular chaperones thus contribute to the structural integrity of transport components and the regulation of cargo attachment and/or release.

5.2.3 Neuronal Cytoskeleton

An intact stable neuronal cytoskeleton is an important prerequisite for efficient normal axonal transport to occur (Crosby 2003). This includes the microtubular transport tracks and associated microtubule stabilising proteins (called microtubule associated proteins, MAPs) within the axoplasm (Hirokawa 1998), actin filaments at the axonal peripheries that integrate axonal protein transport with other regions of cytoplasmic traffic (Hirokawa 1982; Brown et al. 2004; Lewis et al. 2009), and neurofilaments as stabilising cross junctions between parallel microtubules of the axonal fascicle (Hirokawa 1998). Chaperone proteins play an important role in the homeostasis of various components of the neuronal cytoskeleton. Members of the small heat shock protein (sHSP) family, for example, interact with and protect intermediate filament and microtubule networks from damage under stress conditions (Williams and Nelsen 1997; Koyama and Goldman 1999; Day et al. 2003). In the absence of external stress, Hsp27 (HspB1) and αβ-crystallin (HspB5) interact with various types of intermediate filament, and their attachment may prevent network collapse and aggregation as adjacent parallel filaments tend to adhere to each other by non-covalent interactions (Perng et al. 1999). A role for Hsp70 and Hsp40 homologues in the steady state stabilisation of microtubules has also been described (Oka et al. 1998), and Hsp70 associates with polymerised tubulin in vitro (Sanchez et al. 1994). Although this suggests that various HSPs are important in maintaining a normal cytoskeletal network, it remains to be investigated whether this is an important function in healthy neurons to facilitate axonal transport in vivo.

5.2.4 *Transport Vesicle Formation and Trafficking*

Intracellular vesicles for macromolecular transport require external membrane-associated protein structures, such as the clathrin vesicle coat, for their formation in the cytoplasm (Spang 2008). Hsc70, in association with the co-chaperone auxilin, plays a role in the uncoating of clathrin-coated vesicles that is important for recycling of the coat protein and for subsequent trafficking and fusion of the vesicle membrane (Ungewickell et al. 1995). Uncoating must also occur prior to vesicular axonal transport because fast moving vesicles appear to be devoid of a surrounding assembled clathrin coat while transiting through the axon (Black et al. 1991; Nakagawa et al. 2000). Even in a case whereby a component of the clathrin coat formation machinery, the AP1 complex, acts as adaptor between the transport vesicle and a kinesin family motor protein (Kif13A), the clathrin coat is absent from the vesicle (Nakagawa et al. 2000). Clathrin itself, required at the synapse for synaptic vesicle cycling, is predominantly transported along axon as soluble triskelions at a slower rate than that of membraneous vesicles, at 0.02–0.09 μm/s in slow component b (Black et al. 1991; Brown 2003). In this case, the soluble transported clathrin is associated with Hsc70 (de Waegh and Brady 1989; Black et al. 1991), possibly to prevent the spurious formation of clathrin cages and other tertiary structures during transit. Empty clathrin cages appear to form upon the expression of dominant negative Hsc70 mutants in cell culture (Newmyer and Schmid 2001), and such mutants perturb vesicular sorting and trafficking in the endosomal compartment to an extent that cannot be accounted for by defective uncoating of clathrin vesicles alone (Newmyer and Schmid 2001).

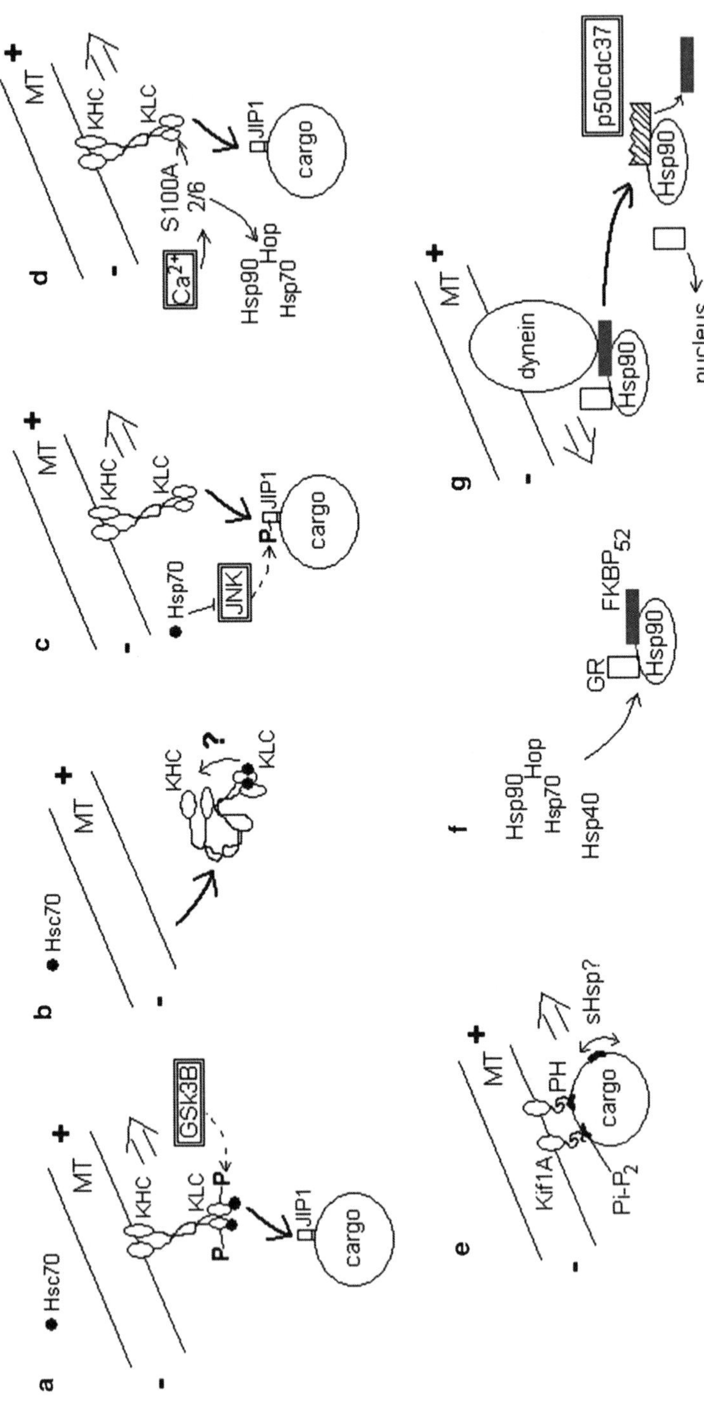
a
Hsc70
MT
KHC
KLC
GSK3B
P
JIP1
cargo
b
Hsc70
MT
KHC
KLC
?
c
Hsp70
JNK
MT
KHC
KLC
P
JIP1
cargo
d
MT
KHC
KLC
Ca^{2+}
S100A
2/6
Hsp90
Hop
Hsp70
JIP1
cargo
e
MT
Kif1A
PH
cargo
sHsp?
$Pi\text{-}P_2$
f
Hsp90
Hop
Hsp70
Hsp40
GR
Hsp90
$FKBP_{52}$
g
MT
dynein
Hsp90
p50cdc37
Hsp90
nucleus

Fig. 5.2 Chaperone roles in the regulation of axonal transport cargo release. Molecular chaperones are potentially involved in several mechanisms for the regulated attachment and release of molecular cargoes to axonal transport motor complexes. Motor proteins are represented as dimeric kinesin-1 (Kif5) unless otherwise stated. Double borders represent regulatory signal transduction inputs into the system. (**a**) Hsc70 interaction with the tandem repeat TPR/J-domain region of kinesin light chain (KLC) stimulates the release of JIP1-associated transport vesicles (Tsai et al. 2000) from the kinesin heavy chain (KHC). This general mechanism receives local/temporal specificity via phosphorylation of KLC by glycogen synthase kinase 3β (GSK3B), which may serve to uncover the KLC tandem repeat region for recognition by Hsc70 (Morfini et al. 2002). MT microtubule. (**b**) Following cargo release, kinesin-1 assumes a closed, compact conformation in which the motor domain is released from the MT and inactivated through contact with its own C-terminus (Coy et al. 1999) and/or KLC (Verhey et al. 1998). Though yet to be demonstrated, Hsc70 is well positioned to facilitate this major structural change, preventing accumulation of cargo-less transport complexes on the MT. (**c**) Phosphorylation of JIP1 by Jun N-terminal kinase (JNK) and JNK-associated kinases may also contribute to local regulation of cargo release (Horiuchi et al. 2007). As JNK signalling is sensitive to Hsp70 chaperone activity (Yaglom et al. 1999), this may represent an indirect influence of local chaperone levels on the release of axonal transport cargoes. (**d**) S100A2 and S100A6 are calcium sensitive signal proteins that are expressed in the nervous system (Zimmer et al. 2005) and that play a part in transport vesicle release from KLC (Shimamoto et al. 2008). S100A2 and S100A6 also regulate the formation of the Hsp70-Hsp90 cytoplasmic foldosome complex via interaction with a TPR motif on Hsp70 organising protein (Hop) (Shimamoto et al. 2008). S100A2/6 may thus contribute to co-regulated synchronisation of cargo release with Hsp activity in the axon. (**e**) The monomeric Kif1A kinesin family motor, involved in axonal transport of synaptic vesicle precursors, is devoid of a light chain and attaches directly to phosphatidyl inositol 4,5 bisphosphate (Pi-P_2) and associated lipids in the cargo vesicle membrane via its C-terminal pleckstrin homology domain (PH) (Klopfenstein et al. 2002). Axonal transport of vesicles by Kif1A is sensitive to Pi-P2 concentration and lipid bilayer composition/physical state (Klopfenstein et al. 2002), and sHsp may engage with membranes to modify bilayer physical characterisitics and microdomain organisation (Tsvetkova et al. 2002), although no direct role for sHsp in axonal transport has been described to date. (**f**) The Hsp90-Hop-Hsp70 foldosome, in tandem with Hsp40 cochaperone activity, is involved in assembly of Hsp90 heterocomplexes with retrograde signal components, such as the glucocorticoid steroid hormone receptor (GR) (Pratt et al. 1999). Via the immunophillin FKBP52, the heterocomplex attaches to dynein for MT-dependent GR transport (Galigniana et al. 2001). (**g**) When FKBP52 is replaced by a cell signal component such as p50cdc37, the complex is released from the motor and the GR can be freed for entry into the nucleus (Pratt et al. 1999). Although a role for the Hsp90 heterocomplex in long distance retrograde axonal transport is yet to be demonstrated, this mechanism reveals how several chaperones could synergise for transport complex formation, motor attachment and regulated-cargo release

5.2.5 Regulation of Cargo Attachment

As it has become clear that numerous modes of axonal transport exist, which utilise multiple motor protein complexes that target a plethora of diverse cargoes to different regions within the neuron, focus has centred upon how the axonal transport of particular cargoes is regulated in vivo. One mechanism of regulation that has received particular attention has been the controlled attachment and detachment of cargoes to their respective axonal motor complexes (Hirokawa and Takemura 2005). Interestingly, several roles for HSP chaperones in the assembly of transporter-cargo complexes and the linkage or release of cargo have been described (Fig. 5.2a, c, d, f, g). Molecular chaperones may act as the "workhorse" in these cases, facilitating the active remodelling of motor complexes and adaptor proteins for the release of cargo when/where required. Signal transduction pathways are emerging that may, in turn, regulate this chaperone activity in a spatial and temporal manner (Fig. 5.2). In other cases, hypothetical points of chaperone activity can be envisaged based upon the presence of chaperones around the transport machinery, and/or the previously described "generic" functions that chaperones are proposed to play (see Fig. 5.2b, e). Further work is needed to more fully understand the full range of chaperone activities within this neuronal process.

5.3 Chaperone Links to Axonal Transport Dysfunction in Disease

Neurodegeneration, involving a progressive, site-specific decline in neuronal function and cell death, exhibits the progressive accumulation of protein aggregates in cells of the nervous system. Protein misfolding in neurodegenerative "proteinopathies" such as Alzheimer's, Parkinson's and Huntington's disease (AD, PD, HD), is considered indicative of protein quality control defects (Ross and Poirier 2004), and has stimulated interest in the activity of molecular chaperones that may serve to protect against cellular dysfunction and toxicity resulting from protein folding stress. Despite this, our understanding of how site-specific neuronal stress may exceed the capacity of local chaperone function is limited. Axonal transport has received much recent attention as a neuronal process that is highly sensitive to neurodegenerative stress (Crosby 2003). Toxicity associated with protein misfolding and aggregation may be particularly acute in axoplasmic regions that are not specialised for de novo protein synthesis and turnover, particularly in cases where, as described below, disease-specific axonal transport dysfunction may serve to isolate aggregates from somal regions for sequestration and/or removal. In turn, cargo accumulates caused by defective axonal traffic may promote aggregation of misfolded disease protein in the axoplasm, instigating a vicious cycle towards "strangulation of the synapse", due to insufficient delivery of synaptic components and defective retrograde transport of senescent organelles. These processes will contribute or directly lead to synaptic dysfunction, axonal retraction and ensuing cell death. The following section considers how protein misfolding and chaperone function may interact with axonal pathology in several neurodegenerative diseases to illustrate how disruption of axonal transport may contribute key pathological events.

5.3.1 Parkinson's Disease

In PD, synaptic loss and cell death of dopaminergic neurons in the substantia nigra is accompanied by the formation of ubiquitin-positive cytoplasmic protein aggregates called Lewy bodies (LBs) (Robinson 2008), and molecular chaperone modulation alleviates aggregate load and toxicity in animal models of PD (Auluck et al. 2002; Klucken et al. 2004; Shen et al. 2005). It is unclear, however, how protein misfolding contributes to pathology in PD. LBs, which occur in several neurodegenerative disorders, are composed primarily of the pre-synaptic vesicle-interacting protein, α-synuclein (Spillantini et al. 1997), and mutations in the human α-synuclein gene that may destabilise the protein and promote its aggregation have been linked to rare familial forms of PD (Polymeropoulos et al. 1997) (see Chap. 13 for more details). This mutant α-synuclein is less able to interact with membraneous vesicles (Jensen et al. 1998) and its axonal transport, normally in slow component b, is perturbed in cultured primary rat cortical neurons, resulting in somal accumulation of the protein (Saha et al. 2004). Although conflicting results in spinal cord axons from transgenic PD-model mice suggest that α-synuclein axonal transport is not perturbed prior to its incorporation into cytoplasmic aggregates in all model systems (Li et al. 2004), axonal swellings rich in α-synuclein aggregates and accumulated transport vesicles and organelles have been identified in PD sufferer's brain tissue (Galvin et al. 1999). This would suggest that protein aggregation involving α-synuclein may disrupt axonal transport of other neuronal cargoes in PD, in addition to α-synuclein itself. The detection of bulging, dystrophic neurites in the striatum of pre-symptomatic rats expressing mutated human α-synuclein (Chung et al. 2009) confirms that α-synuclein abnormalities can potently disrupt axonal processes in vivo. In this rat animal model, early changes in the levels of cytoskeletal and axonal motor protein components, encompassing an apparent gross shift from anterograde to retrograde function, could reflect respective compensatory and pathological changes within a stressed axonal transport system (Chung et al. 2009).

Metabolic dysfunction and dopamine-processing abnormalities have also received much attention in relation to PD pathology (Miller et al. 2009; Leong et al. 2009). The neurotoxin, MPTP, which may target mitochondria and apoptotic caspases (Viswanath et al. 2001) and induces PD-like pathology in several model systems (Jenner 2008), disrupts fast axonal transport in a squid giant axon model, resulting in a reduced complement of synaptic vesicles at the axon terminal (Morfini et al. 2007b). It has also been shown that pharmacological induction of Hsp70 can alleviate MPTP neurotoxicity in a mouse model of neurotoxin-associated PD (Shen et al. 2005). It is currently unclear how chaperone modulation may protect from MPTP toxicity, especially given that LB pathology has not been identified in mice and primates exhibiting chronic MPTP-induced neurotoxicity (Shimoji et al. 2005; Halliday et al. 2009). The above findings nevertheless suggest that multiple stressors converging on the axonal transport system may be amenable to molecular chaperone modulation in PD.

5.3.2 Torsional Dystonia

TorsinA (TorA) is a non-HSP, AAA+ family..' chaperone that has been genetically linked to site-specific neurodegeneration and locomotor deficits in the early-onset torsional dystonia (Ozelius et al. 1997). Torsin proteins function as oligomers to regulate assembly, disassembly and operation of protein complexes within the cell (Ogura and Wilkinson 2001). A yeast-two-hybrid screen for binding partners of the chaperone revealed the cargo-interacting tandem repeat region of kinesin light chain as a key interactor (Kamm et al. 2006), suggesting that TorA may play a role in regulating cargo attachment similar to that described for Hsp70 (see Fig. 5.2). Consistent with this hypothesis, genetic reduction in the levels of the *Drosophila* TorA homologue, Torp4a, in fly retinal neurons results in an abnormal pigment distribution indicative of a defect in cellular transport (Muraro and Moffat 2006). This phenotype was enhanced in an unbiased modulator screen via a reduction in the levels of components involved in vesicular traffic (Muraro and Moffat 2006). TorA is also involved in intermediate filament transport during neuronal growth cone path finding, and cultured neurons transfected with mutant TorsinA, or primary fibroblasts isolated from human dystonia sufferers, reveal defects in cytoskeletal architecture and neurite extention (Hewett et al. 2006). TorA may thus play an essential role in maintaining functional transport within vulnerable neurons of the substantia nigra, a process that, if perturbed in torsional dystonia sufferers, could contribute to the synaptic defects observed in cell and animal models of the disease (Granata et al. 2008; Lee et al. 2009). Interestingly, TorA may additionally regulate both α-synuclein (McLean et al. 2002) and polyglutamine (polyQ) (Caldwell et al. 2003) aggregation and associates with nuclear inclusions from HD- and related polyQ disorder brain tissue (Walker et al. 2003). TorA may thus modulate other neurodegenerative pathologies involving aberrant protein folding and perturbed cellular transport, but whether this relates to its axonal transport function or other neuronal roles for the protein remains to be elucidated.

5.3.3 Alzheimer's Disease

AD is characterised by progressive cortical neurodegeneration and the deposition of extracellular β-amyloid (Aβ) protein aggregates, called neuritic plaques, and intracellular filaments that comprise the MAP tau, called neurofibrillary tangles (Mattson 2004) (see Chap. 14 for details on Aβ pathology). Evidence for early axonal degeneration in sporadic AD includes the detection of cortical white matter loss and dystrophic neurites in AD cortex analysed post-mortem (Gibson 1987; Rose et al. 2000), and an advanced age-dependent loss of brain microtubule density in AD (Cash et al. 2003). As discussed below, disease genes linked to rare familial AD point to early axonal transport disruption during pathology, and a kinesin light chain mutation has been linked to the early onset AD (Dhaenens et al. 2004). Toxic Aβ species are liberated through proteolysis of the amyloid precursor protein (APP), and over-expression of APP in *Drosophila* causes axonal accumulates of cargoes destined for the synapse (Torroja et al. 1999; Gunawardena and Goldstein 2001). Expression of mutant APP alleles linked to familial AD in mice results in accumulation of stalled motor proteins, vesicles and organelles within axonal swellings of cholinergic cortical inputs similar to those that are preferentially afflicted in AD (Stokin et al. 2005). Application of toxic Aβ to cultured hippocampal neurons perturbs axonal transport and results in the disease-specific hyper-phosphorylation of tau (Rui et al. 2006), and the latter directly affects the axonal transport system (see below). Any involvement of molecular chaperones in APP processing and/or toxic Aβ generation, such as association with and modulation of Aβ plaques by sHSPs (Renkawek et al. 1994; Wilhelmus et al. 2006), may thus impinge upon axonal transport disruption caused by this pathological process in AD.

Tau, which is hyperphosphorylated and misfolded in AD, forms intracellular tangles (Wischik et al. 1988) which are closely correlated with the clinical symptoms of dementia (Braak and Braak 1991; Ball and Murdoch 1997). Tau is intimately involved in the maintenance of an intact axonal transport system and has therefore received much attention with regard to axonal transport perturbations that may occur in this disease (Garcia and Cleveland 2001). This is also the case in various neurodegenerative tauopathies, some of which are caused by mutations in the human tau gene (Hernandez and Avila 2007). Over-expression of human tau causes retention of transport cargo within axonal swellings, microtubule disruption, axonal loss and synaptic dysfunction prior to cell loss (Ebneth et al. 1998; Ishihara et al. 1999; Lewis et al. 2000; Stamer et al. 2002), and tau effects in disturbing axonal transport in vivo are phosphorylation dependent (Ishihara et al. 1999; Mudher et al. 2004). The tau microtubule hypothesis posits that hyperphosphorylated tau detatches from microtubules and sequesters into aggregates, allowing microtubular tracks to destabilise and jeopardising axonal transport. Hsp70 modulates tau aggregation (Dou et al. 2003; Shimura et al. 2004b) and turnover (Dickey et al. 2006), increasing tau-microtubule interactions and reducing tau-associated neurotoxicity. In association with an Hsp70 binding partner, the E3 ubiquitin ligase CHIP, molecular chaperones may be involved in the control of NFT formation in human AD brain (Sahara et al. 2005), and a pathological reduction in CHIP levels downstream of toxic Aβ pathology may further excacerbate tau pathology in AD (Oddo et al. 2008). Hsp70 may bind to the same part of the microtubule as tau, and could even directly regulate tau binding (Liang and MacRae 1997). If this were the case, Hsp70 is well positioned to modulate a documented, albeit controversial, direct inhibition of axonal transport motor complex activity by microtubule-bound tau (Ebneth et al. 1998; Morfini et al. 2007a; Dixit et al. 2008). Tau interacts with the kinesin light chain and is itself transported to distal axonal regions to accumulate in a phosphorylation dependent manner in cell culture (Cuchillo-Ibanez et al. 2008).

Hsp40 co-chaperones may also play a role in tau turnover via ubiquitin-dependent pathways (Petrucelli et al. 2004; Dickey et al. 2006), whereas the co-chaperone BAG-2 may assist in the clearance of hyperphosphorylated tau via an alternative route (Carrettiero et al. 2009). Misfolded, aggregating tau may additionally disrupt the axonal transport of kinesin-1 dependent cargoes via a pathological sequestration of the JNK cargo adaptor protein JIP1 away from motor complex (Ittner et al. 2009).

This mechanism, hypothetically amenable to molecular chaperone modulation via tau folding/solubilisation effects, could result in preferential stalling of APP-containing vesicles and excess Aβ production at axonal swellings and degenerative neurites in AD (Roher et al. 2002; Stamer et al. 2002).

sHSPs may also be involved in tau-related and AD pathology. 'Hsp27 (HspB1) associates with hyperphosphorylated tau from AD brain and clears it via the proteasome in HCN2a cells (Shimura et al. 2004a). sHsps are elevated in glia in AD brain tissue (Renkawek et al. 1994; Dabir et al. 2004), and microglial nitric oxide release adjacent to axons reduces axonal transport via JNK signal pathway activation under pro-inflammatory conditions in cell culture (Stagi et al. 2005). sHsps are also of interest in light of their ability to stabilise actin polymer (Sun and MacRae 2005), as abnormalities in the axonal actin cytoskeleton have been described for both Aβ (Maloney et al. 2005) and tau (Fulga et al. 2007) induced neuropathology. It is thus possible that various HSPs are important with regard to Aβ aggregation, tau homeostasis, microtubule integrity and the intimately associated axonal cytoskeleton for the maintenance of robust axonal transport amidst neurodegenerative stress in tauopathies, such as AD. Sites of molecular chaperone interaction with axonal pathology in AD are summarised in Fig. 5.3a.

5.3.4 Huntington's Disease

HD is an autosomal dominant condition caused by a polyQ expansion within the Huntingtin protein (Htt) (The Huntington's Disease Collaborative Research Group 1993). It involves progressive striatal and neocortical neurodegeneration amidst the formation of nuclear and cytoplasmic insoluble Htt aggregates within afflicted cells (Reiner et al. 1988; Hoffner et al. 2007), and post-mortem brain tissue from pre-symptomatic HD sufferers exhibits axonal degeneration of striatal projection neurons (Albin et al. 1992) and cortico-striatal afferents (Sapp et al. 1999), indicative of early axonal degeneration. Dystrophic neurites, and swellings of accumulated cargoes that point to axonal transport disruption, have been localised to Htt aggregates from human brain (DiFiglia et al. 1997). Pathogenic human Htt-polyQ disrupts axonal transport in vitro (Szebenyi et al. 2003) and in HD animal models (Trushina et al. 2004; Sinadinos et al. 2009), where its toxicity has been associated with the formation of visible somal and axonal Htt aggregates. Aggregation of Htt-polyQ is also amenable to modulation by molecular chaperones. Hsp70 alters visible Htt aggregates and/or their biochemical characteristics in HD models (Jana et al. 2000; Wyttenbach et al. 2001; Fayazi et al. 2006). Similarly, Hsp40 co-chaperones (Wyttenbach et al. 2000; Chuang et al. 2002; Fayazi et al. 2006) and sHsps (Wyttenbach et al. 2002; Carra et al. 2005) modulate Htt aggregation and toxicity.

How Htt aggregation affects axonal transport has been considered in light of analysis of several different HD cell and animal models. One possibility is that physical axonal obstructions, such as that caused by a large insoluble axonal Htt aggregate, and later excacerbated by the accumulation of axonal transport cargoes

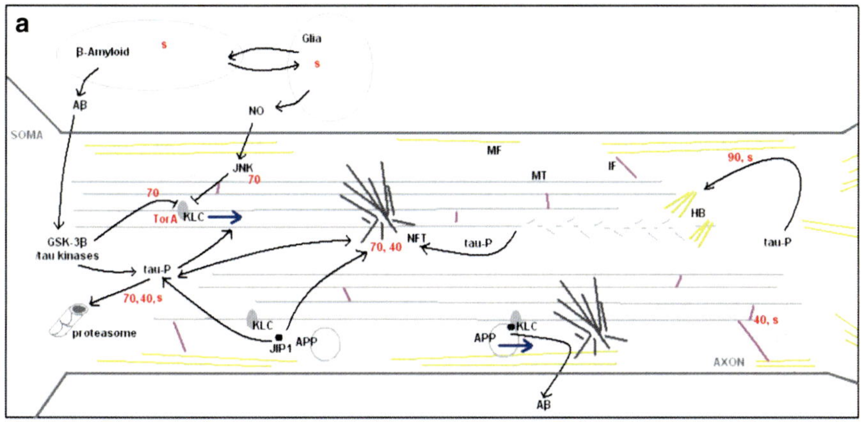

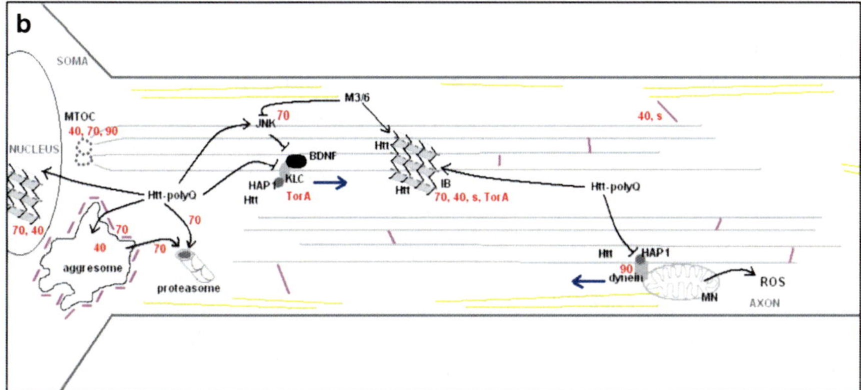

Fig. 5.3 Chaperone interactions with pathologic axonal events during chronic neurodegenerative stress in Alzheimer's and Huntington's disease. Disruption of axonal transport is being increasingly recognised as an early and important pathological event during neurodegeneration in Alzheimer's (**a**) and Huntington's (**b**) diseases. Several documented pathologic events are shown schematically above. Transport (*blue arrows*) is perturbed by abnormal interactions involving the microtubule-associated tau protein (**a**) and Huntingtin (**b**). Sites of potential molecular chaperone interaction with these events are shown in *red*. Mechanistic details and cited references are provided in the main text. *MF* microfilaments; *MT* microtubules; *IF* intermediate filaments; *90, 70, 40* heat shock protein 90, 70, 40; *s* small heat shock protein; *TorA* torsin A; *NO* nitric oxide; *JNK* Jun N-terminal kinase; *JIP1* JNK interacting protein 1; *KLC* kinesin light chain; *tau-P* hyperphosphorylated tau; *NFT* neurofibrillary tangle; *HB* Hirano body (filamentous actin aggregate); *GSK-3β* glycogen synthase kinase-3β; *APP* amyloid precursor protein; *MTOC* microtubule organising centre; *Htt* Huntingtin; *Htt-polyQ* polyglutamine-expanded Huntingtin; *HAP1* Huntingtin-associated protein 1; *BDNF* brain-derived neurotrophic factor; *IB* inclusion body; *MN* mitochondrion; *ROS* reactive oxygen species

at the aggregate, could physically impede the traffic of further materials along the narrow axon. Secondly, "sticky" Htt-polyQ aggregates, which interact strongly with many cellular components (Lee et al. 2004; Qin et al. 2004; Yamanaka et al. 2008), may directly sequester parts of the axonal transport motor complex machinery away from the microtubule, a hypothesis bolstered by genetic and biochemical evidence of kinesin-1 sequestration in animal models (Gunawardena et al. 2003; Trushina et al. 2004). Thirdly, axonal polyQ aggregates may sequester components involved in the indirect regulation of axonal transport, such as the phosphatase M3/6, leading to activation of the JNK signalling pathway (Merienne et al. 2003). Activation of JNK3 in Htt-polyQ knock-in mice leads to phosphorylation of kinesin heavy chain, detachment of the motor complex from the microtubule, and inhibition of fast axonal transport (Morfini et al. 2009). Regarding this pathological mechanism, molecular chaperones could thus modulate either phosphatase sequestration via aggregate interactions, or the regulation of JNK signalling (see above). Finally, transcription deficits, which have been linked to transcription factor sequestration into nuclear Htt aggregates (Yamanaka et al. 2008) and the abnormal expression of many genes, including those involved in vesicular transport (Chan et al. 2002; Sipione et al. 2002), could indirectly place further strain on the axonal transport system in HD, although this remains to be tested directly.

In *Drosophila* larval motor neurons, the anterograde axonal transport cargo syntaxin, bound for the synapse, forms axonal accumulates that co-localise with human Htt-polyQ aggregates (Lee et al. 2004), indicating that synaptic components may not reach their intended destination amidst Htt-polyQ aggregation. It remains unclear whether molecular chaperone modulation may alleviate axonal transport cargo accumulation in this system. Another synaptic cargo, cystein string protein (CSP), similarly accumulates into visible clumps in fly motor neuron axons expressing Htt-polyQ (Gunawardena et al. 2003). Although genetic elevation of Hsp70 clears such axonal CSP accumulates (Gunawardena et al. 2003), it is unclear whether this is due to correction of an underlying axonal transport deficit by the chaperone, or to sequestration of the cargo to other cellular compartments via its documented direct interaction with Hsp70 (Tobaben et al. 2001).

Htt is a pleiotropic protein with axonal transport function that is compromised by pathogenic polyQ expansion (Gunawardena et al. 2003; Gauthier et al. 2004; Zala et al. 2008). In this context, the loss of wildtype Htt function has received increasing attention in the context of HD pathology (Cattaneo et al. 2005). Normal Htt indirectly interacts with dynein as part of a multi-component complex (Tukamoto et al. 1997), and this interaction is altered when pathogenic Htt-polyQ is present (Block-Galarza et al. 1997). If, as shown in a HD mouse model (Trushina et al. 2004), retrograde axonal transport is defective in HD, this may in turn disrupt the formation of potentially protective perinuclear aggregates, or "aggresomes", that could serve as protective sinks for toxic misfolding species in the neuron (Johnston et al. 1998). Aggresome structures, caused by generic or neurodegenerative protein folding stress (Garcia-Mata et al. 1999; Taylor et al. 2003; Webb et al. 2004), rely on active dynein transport for their formation. Although Hsp70 and Hsp40 have respectively been found to associate with an outer intermediate filament "shell" and inner protease resistant "core" of somal inclusions (Fig. 5.3b) in a HD in vitro model (Qin et al. 2004), their exact role at this location, whether they promote or discourage aggresome formation, and its consequences for axonal transport and neuronal function are unknown.

Another consequence of perturbed retrograde transport may be inefficient recycling of senescent mitochondria and reactive oxygen species (ROS) generation in HD (Chang et al. 2006; Pilling et al. 2006). Indeed, the expression of expanded htt fragments cause an increase in ROS, as shown in HD cell models (Wyttenbach et al. 2002), and coexpression of several HSPs rescues htt-induced ROS and toxicity (Firdaus et al. 2006). Htt, as for the dynein complex, interacts with kinesin light chain via its binding partner HAP-1 (McGuire et al. 2006), and whether these interactions are altered due to Htt-polyQ misfolding defects that are amenable to molecular chaperone intervention remains to be investigated. If Htt itself serves important functions in the nervous system that are reliant upon robust axonal transport, such as delivery of brain-derived neurotophic factor to the striatum (Cattaneo et al. 2005), axonal aggregates that further disrupt transport (above) and reduce the available cytoplasmic pool of active molecular chaperones (Hands et al. 2008), could excacerbate Htt misfolding and lost transport function, leading to defective BDNF delivery, site-specific neuronal neglect, neuronal dysfunction and cell death in HD (Gauthier et al. 2004). Sites of molecular chaperone interaction with axonal pathology in HD are summarised in Fig. 5.3b.

5.4 Conclusions

Molecular chaperones act at numerous points in the axonal transport system and are likely involved in as yet undefined roles in the assembly and maintenance of transport apparatus. There is, however, an emerging involvement of chaperones in the regulation of axonal cargo attachment and release. In neurodegenerative proteinopathies, misfolding and aggregation can disrupt the axonal transport of a disease protein, block or sequester axonal motor complexes and signal components that regulate their activity, and perturb the disease protein's own transport function at the mircrotubule, motor complex or cargo vesicle. Disruption of axonal transport, and thus neuronal function, is therefore a pathological mechanism that is likely to be amenable to modulation by molecular chaperones. Consideration of local sites of misfolding, aggregation and chaperone recruitment within cell compartments, such as the axon, may prove crucial if we are to unravel the multitude of chaperone activities and their interacting, often synergistic effects in health and disease, thus paving the way for more controlled attempts at exogenous modulation of the chaperone system for therapeutic benefit.

References

Albin, R.L., A. Reiner, K.D. Anderson, L.S. Dure, B. Handelin, R. Balfour, W.O. Whetsell Jr., J.B. Penney, and A.B. Young (1992). Preferential loss of striato-external pallidal projection neurons in presymptomatic Huntington's disease. Ann. Neurol. 31:425–430.

Auluck, P.K., H.Y. Chan, J.Q. Trojanowski, V.M. Lee, and N.M. Bonini (2002). Chaperone suppression of alpha-synuclein toxicity in a Drosophila model for Parkinson's disease. Science 295:865–868.

Ball, M.J. and G.H. Murdoch (1997). Neuropathological criteria for the diagnosis of Alzheimer's disease: are we really ready yet? Neurobiol. Aging 18:S3–S12.

Black, M.M., M.H. Chestnut, I.T. Pleasure, and J.H. Keen (1991). Stable clathrin: uncoating protein (hsc70) complexes in intact neurons and their axonal transport. J. Neurosci. 11:1163–1172.

Block-Galarza, J., K.O. Chase, E. Sapp, K.T. Vaughn, R.B. Vallee, M. DiFiglia, and N. Aronin (1997). Fast transport and retrograde movement of huntingtin and HAP 1 in axons. Neuroreport 8:2247–2251.

Braak, H. and E. Braak (1991). Neuropathological stageing of Alzheimer-related changes. Acta Neuropathol. (Berl) 82:239–259.

Brown, A. (2003). Axonal transport of membranous and nonmembranous cargoes: a unified perspective. J. Cell Biol. 160:817–821.

Brown, J.R., P. Stafford, and G.M. Langford (2004). Short-range axonal/dendritic transport by myosin-V: a model for vesicle delivery to the synapse. J. Neurobiol. 58:175–188.

Burkhardt, J.K., C.J. Echeverri, T. Nilsson, and R.B. Vallee (1997). Overexpression of the dynamitin (p50) subunit of the dynactin complex disrupts dynein-dependent maintenance of membrane organelle distribution. J. Cell Biol. 139:469–484.

Carrettiero, D.C., I. Hernandez, P. Neveu, T. Papaginnakopouler, and K.S. Kosik (2009).

Caldwell, G.A., S. Cao, E.G. Sexton, C.C. Gelwix, J.P. Bevel, and K.A. Caldwell (2003). Suppression of polyglutamine-induced protein aggregation in *Caenorhabditis elegans* by torsin proteins. Hum. Mol. Genet. 12:307–319.

Carra, S., M. Sivilotti, A.T. Chavez Zobel, H. Lambert, and J. Landry (2005). HspB8, a small heat shock protein mutated in human neuromuscular disorders, has in vivo chaperone activity in cultured cells. Hum. Mol. Genet. 14:1659–1669.

Carrettiero, D.C., I. Hernandez, P. Neven, T. Papaginnakopoulos, and K.S. Kosik (2009). The cochaperone BAG2 sweeps paired helical filament- insoluble tau from the microtubule. J. Neurosci. 29:2151–2161.

Cash, A.D., G. Aliev, S.L. Siedlak, A. Nunomura, H. Fujioka, X. Zhu, A.K. Raina, H.V. Vinters, M. Tabaton, A.B. Johnson, M. Paula-Barbosa, J. Avila, P.K. Jones, R.J. Castellani, M.A. Smith, and G. Perry (2003). Microtubule reduction in Alzheimer's disease and aging is independent of tau filament formation. Am. J. Pathol. 162:1623–1627.

Cattaneo, E., C. Zuccato, and M. Tartari (2005). Normal huntingtin function: an alternative approach to Huntington's disease. Nat. Rev. Neurosci. 6:919–930.

Chan, E.Y., R. Luthi-Carter, A. Strand, S.M. Solano, S.A. Hanson, M.M. DeJohn, C. Kooperberg, K.O. Chase, M. DiFiglia, A.B. Young, B.R. Leavitt, J.H. Cha, N. Aronin, M.R. Hayden, and J.M. Olson (2002). Increased huntingtin protein length reduces the number of polyglutamine-induced gene expression changes in mouse models of Huntington's disease. Hum. Mol. Genet. 11:1939–1951.

Chang, D.T., G.L. Rintoul, S. Pandipati, and I.J. Reynolds (2006). Mutant huntingtin aggregates impair mitochondrial movement and trafficking in cortical neurons. Neurobiol. Dis. 22:388–400.

Chuang, J.Z., H. Zhou, M. Zhu, S.H. Li, X.J. Li, and C.H. Sung (2002). Characterization of a brain-enriched chaperone, MRJ, that inhibits Huntingtin aggregation and toxicity independently. J. Biol. Chem. 277:19831–19838.

Chung, C.Y., J.B. Koprich, H. Siddiqi, and O. Isacson (2009). Dynamic changes in presynaptic and axonal transport proteins combined with striatal neuroinflammation precede dopaminergic neuronal loss in a rat model of AAV alpha-synucleinopathy. J. Neurosci. 29:3365–3373.

Coy, D.L., W.O. Hancock, M. Wagenbach, and J. Howard (1999). Kinesin's tail domain is an inhibitory regulator of the motor domain. Nat. Cell Biol. 1:288–292.

Crosby, A.H. (2003). Disruption of cellular transport: a common cause of neurodegeneration? Lancet Neurol. 2:311–316.

Cuchillo-Ibanez, I., A. Seereeram, H.L. Byers, K.Y. Leung, M.A. Ward, B.H. Anderton, and D.P. Hanger (2008). Phosphorylation of tau regulates its axonal transport by controlling its binding to kinesin. FASEB J. 22:3186–3195.

Dabir, D.V., J.Q. Trojanowski, C. Richter-Landsberg, V.M. Lee, and M.S. Forman (2004). Expression of the small heat-shock protein alphaB-crystallin in tauopathies with glial pathology. Am. J. Pathol. 164:155–166.

Day, R.M., J.S. Gupta, and T.H. MacRae (2003). A small heat shock/alpha-crystallin protein from encysted Artemia embryos suppresses tubulin denaturation. Cell Stress. Chaperones. 8:183–193.

de, W.S. and S.T. Brady (1989). Axonal transport of a clathrin uncoating ATPase (HSC70): a role for HSC70 in the modulation of coated vesicle assembly in vivo. J. Neurosci. Res. 23: 433–440.

Der, P.M. and R.A. Quinlan (2004). Neuronal diseases: small heat shock proteins calm your nerves. Curr. Biol. 14:R625–R626.

Dhaenens, C.M., B.E. Van, S. Schraen-Maschke, F. Pasquier, A. Delacourte, and B. Sablonniere (2004). Association study of three polymorphisms of kinesin light-chain 1 gene with Alzheimer's disease. Neurosci. Lett. 368:290–292.

Dickey, C.A., J. Dunmore, B. Lu, J.W. Wang, W.C. Lee, A. Kamal, F. Burrows, C. Eckman, M. Hutton, and L. Petrucelli (2006). HSP induction mediates selective clearance of tau phosphorylated at proline-directed Ser/Thr sites but not KXGS (MARK) sites. FASEB J. 20:753–755.

DiFiglia, M., E. Sapp, K.O. Chase, S.W. Davies, G.P. Bates, J.P. Vonsattel, and N. Aronin (1997). Aggregation of huntingtin in neuronal intranuclear inclusions and dystrophic neurites in brain. Science 277:1990–1993.

Dixit, R., J.L. Ross, Y.E. Goldman, and E.L. Holzbaur (2008). Differential regulation of dynein and kinesin motor proteins by tau. Science 319:1086–1089.

Dou, F., W.J. Netzer, K. Tanemura, F. Li, F.U. Hartl, A. Takashima, G.K. Gouras, P. Greengard, and H. Xu (2003). Chaperones increase association of tau protein with microtubules. Proc. Natl. Acad. Sci. U. S. A. 100:721–726.

Ebneth, A., R. Godemann, K. Stamer, S. Illenberger, B. Trinczek, and E. Mandelkow (1998). Overexpression of tau protein inhibits kinesin-dependent trafficking of vesicles, mitochondria, and endoplasmic reticulum: implications for Alzheimer's disease. J. Cell Biol. 143:777–794.

Fayazi, Z., S. Ghosh, S. Marion, X. Bao, M. Shero, and P. Kazemi-Esfarjani (2006). A Drosophila ortholog of the human MRJ modulates polyglutamine toxicity and aggregation. Neurobiol. Dis. 24:226–244.

Firdaus, W.J., A. Wyttenbach, C. Diaz-Latoud, R.W. Currie, and A.P. Arrigo (2006). Analysis of oxidative events induced by expanded polyglutamine huntingtin exon 1 that are differentially restored by expression of heat shock proteins or treatment with an antioxidant. FEBS J. 273:3076–3093.

Fulga, T.A., I. Elson-Schwab, V. Khurana, M.L. Steinhilb, T.L. Spires, B.T. Hyman, and M.B. Feany (2007). Abnormal bundling and accumulation of F-actin mediates tau-induced neuronal degeneration in vivo. Nat. Cell Biol. 9:139–148.

Galigniana, M.D., C. Radanyi, J.M. Renoir, P.R. Housley, and W.B. Pratt (2001). Evidence that the peptidylprolyl isomerase domain of the hsp90-binding immunophilin FKBP52 is involved in both dynein interaction and glucocorticoid receptor movement to the nucleus. J. Biol. Chem. 276:14884–14889.

Galvin, J.E., K. Uryu, V.M. Lee, and J.Q. Trojanowski (1999). Axon pathology in Parkinson's disease and Lewy body dementia hippocampus contains alpha-, beta-, and gamma-synuclein. Proc. Natl. Acad. Sci. U. S. A. 96:13450–13455.

Garcia, M.L. and D.W. Cleveland (2001). Going new places using an old MAP: tau, microtubules and human neurodegenerative disease. Curr. Opin. Cell Biol. 13:41–48.

Garcia-Mata, R., Z. Bebok, E.J. Sorscher, and E.S. Sztul (1999). Characterization and dynamics of aggresome formation by a cytosolic GFP-chimera. J. Cell Biol. 146:1239–1254.

Gauthier, L.R., B.C. Charrin, M. Borrell-Pages, J.P. Dompierre, H. Rangone, F.P. Cordelieres, M.J. De, M.E. MacDonald, V. Lessmann, S. Humbert, and F. Saudou (2004). Huntingtin controls neurotrophic support and survival of neurons by enhancing BDNF vesicular transport along microtubules. Cell 118:127–138.

Gibbons, I.R. (1996). The role of dynein in microtubule-based motility. Cell Struct. Funct. 21:331–342.

Gibson, P.H. (1987). Ultrastructural abnormalities in the cerebral neocortex and hippocampus associated with Alzheimer's disease and aging. Acta Neuropathol. (Berl) 73:86–91.

Granata, A., R. Watson, L.M. Collinson, G. Schiavo, and T.T. Warner (2008). The dystonia-associated protein torsinA modulates synaptic vesicle recycling. J. Biol. Chem. 283:7568–7579.

Gunawardena, S. and L.S. Goldstein (2001). Disruption of axonal transport and neuronal viability by amyloid precursor protein mutations in Drosophila. Neuron 32:389–401.

Gunawardena, S., L.S. Her, R.G. Brusch, R.A. Laymon, I.R. Niesman, B. Gordesky-Gold, L. Sintasath, N.M. Bonini, and L.S. Goldstein (2003). Disruption of axonal transport by loss of huntingtin or expression of pathogenic polyQ proteins in Drosophila. Neuron 40:25–40.

Halliday, G., M.T. Herrero, K. Murphy, H. McCann, F. Ros-Bernal, C. Barcia, H. Mori, F.J. Blesa, and J.A. Obeso (2009). No Lewy pathology in monkeys with over 10 years of severe MPTP Parkinsonism. Mov Disord. 24:1519–1523.

Hands, S., C. Sinadinos, and A. Wyttenbach (2008). Polyglutamine gene function and dysfunction in the ageing brain. Biochim. Biophys. Acta 1779:507–521.

Hernandez, F. and J. Avila (2007). Tauopathies. Cell Mol. Life Sci. 64:2219–2233.

Hewett, J.W., J. Zeng, B.P. Niland, D.C. Bragg, and X.O. Breakefield (2006). Dystonia-causing mutant torsinA inhibits cell adhesion and neurite extension through interference with cytoskeletal dynamics. Neurobiol. Dis. 22:98–111.

Hirokawa, N. (1982). Cross-linker system between neurofilaments, microtubules, and membranous organelles in frog axons revealed by the quick-freeze, deep-etching method. J. Cell Biol. 94:129–142.

Hirokawa, N. (1998). Kinesin and dynein superfamily proteins and the mechanism of organelle transport. Science 279:519–526.

Hirokawa, N. and R. Takemura (2005). Molecular motors and mechanisms of directional transport in neurons. Nat. Rev. Neurosci. 6:201–214.

Hoffner, G., S. Soues, and P. Djian (2007). Aggregation of expanded huntingtin in the brains of patients with Huntington disease. Prion 1:26–31.

Horiuchi, D., C.A. Collins, P. Bhat, R.V. Barkus, A. Diantonio, and W.M. Saxton (2007). Control of a kinesin-cargo linkage mechanism by JNK pathway kinases. Curr. Biol. 17:1313–1317.

Ishihara, T., M. Hong, B. Zhang, Y. Nakagawa, M.K. Lee, J.Q. Trojanowski, and V.M. Lee (1999). Age-dependent emergence and progression of a tauopathy in transgenic mice overexpressing the shortest human tau isoform. Neuron 24:751–762.

Ittner, L.M., Y.D. Ke, and J. Gotz (2009). Phosphorylated tau interacts with c-JUN N-terminal kinase (JNK) interacting protein 1 (JIP1) in Alzheimer's disease. J. Biol. Chem. 284:28909–28916.

Jana, N.R., M. Tanaka, G. Wang, and N. Nukina (2000). Polyglutamine length-dependent interaction of Hsp40 and Hsp70 family chaperones with truncated N-terminal huntingtin: their role in suppression of aggregation and cellular toxicity. Hum. Mol. Genet. 9:2009–2018.

Jenner, P. (2008). Functional models of Parkinson's disease: a valuable tool in the development of novel therapies. Ann. Neurol. 64 Suppl 2:S16–S29.

Jensen, P.H., M.S. Nielsen, R. Jakes, C.G. Dotti, and M. Goedert (1998). Binding of alpha-synuclein to brain vesicles is abolished by familial Parkinson's disease mutation. J. Biol. Chem. 273:26292–26294.

Johnston, J.A., C.L. Ward, and R.R. Kopito (1998). Aggresomes: a cellular response to misfolded proteins. J. Cell Biol. 143:1883–1898.

Kamm, C., H. Boston, J. Hewett, J. Wilbur, D.P. Corey, P.I. Hanson, V. Ramesh, and X.O. Breakefield (2006). The early onset dystonia protein torsinA interacts with kinesin light chain 1. J.Biol.Chem. 279:19882–19892.

Klopfenstein, D.R., M. Tomishige, N. Stuurman, and R.D. Vale (2002). Role of phosphatidylinositol(4,5)bisphosphate organization in membrane transport by the Unc104 kinesin motor. Cell 109:347–358.

Klucken, J., Y. Shin, E. Masliah, B.T. Hyman, and P.J. McLean (2004). Hsp70 reduces alpha-synuclein aggregation and toxicity. J. Biol. Chem. 279:25497–25502.

Koushika, S.P. (2008). "JIP"ing along the axon: the complex roles of JIPs in axonal transport. Bioessays 30:10–14.

Koyama, Y. and J.E. Goldman (1999). Formation of GFAP cytoplasmic inclusions in astrocytes and their disaggregation by alphaB-crystallin. Am. J. Pathol. 154:1563–1572.

Lee, D.W., J.B. Seo, B. Ganetzky, and Y.H. Koh (2009). DeltaFY mutation in human torsin A [corrected] induces locomotor disability and abberant synaptic structures in Drosophila. Mol. Cells 27:89–97.

Lee, W.C., M. Yoshihara, and J.T. Littleton (2004). Cytoplasmic aggregates trap polyglutamine-containing proteins and block axonal transport in a Drosophila model of Huntington's disease. Proc. Natl. Acad. Sci. U. S. A. 101:3224–3229.

Leong, S.L., R. Cappai, K.J. Barnham, and C.L. Pham (2009). Modulation of alpha-synuclein aggregation by dopamine: a review. Neurochem. Res. 34:1838–1846.

Lewis, J., E. McGowan, J. Rockwood, H. Melrose, P. Nacharaju, S.M. Van, K. Gwinn-Hardy, M.M. Paul, M. Baker, X. Yu, K. Duff, J. Hardy, A. Corral, W.L. Lin, S.H. Yen, D.W. Dickson, P. Davies, and M. Hutton (2000). Neurofibrillary tangles, amyotrophy and progressive motor disturbance in mice expressing mutant (P301L) tau protein. Nat. Genet. 25:402–405.

Lewis, T.L. Jr., T. Mao, K. Svoboda, and D.B. Arnold (2009). Myosin-dependent targeting of transmembrane proteins to neuronal dendrites. Nat. Neurosci. 12:568–576.

Li, W., P.N. Hoffman, W. Stirling, D.L. Price, and M.K. Lee (2004). Axonal transport of human alpha-synuclein slows with aging but is not affected by familial Parkinson's disease-linked mutations. J. Neurochem. 88:401–410.

Liang, P. and T.H. MacRae (1997). Molecular chaperones and the cytoskeleton. J. Cell Sci. 110 (Pt 13):1431–1440.

Maloney, M.T., L.S. Minamide, A.W. Kinley, J.A. Boyle, and J.R. Bamburg (2005). Beta-secretase-cleaved amyloid precursor protein accumulates at actin inclusions induced in neurons by stress or amyloid beta: a feedforward mechanism for Alzheimer's disease. J. Neurosci. 25:11313–11321.

Marx, A., J. Muller, E.M. Mandelkow, A. Hoenger, and E. Mandelkow (2006). Interaction of kinesin motors, microtubules, and MAPs. J. Muscle Res. Cell Motil. 27:125–137.

Mattson, M.P. (2004). Pathways towards and away from Alzheimer's disease. Nature 430:631–639.

McGuire, J.R., J. Rong, S.H. Li, and X.J. Li (2006). Interaction of Huntingtin-associated protein-1 with kinesin light chain: implications in intracellular trafficking in neurons. J. Biol. Chem. 281:3552–3559.

McLean, P.J., H. Kawamata, S. Shariff, J. Hewett, N. Sharma, K. Ueda, X.O. Breakefield, and B.T. Hyman (2002). TorsinA and heat shock proteins act as molecular chaperones: suppression of alpha-synuclein aggregation. J. Neurochem. 83:846–854.

Merienne, K., D. Helmlinger, G.R. Perkin, D. Devys, and Y. Trottier (2003). Polyglutamine expansion induces a protein-damaging stress connecting heat shock protein 70 to the JNK pathway. J. Biol. Chem. 278:16957–16967.

Miller, R.L., M. James-Kracke, G.Y. Sun and A.Y. Sun (2009). Oxidative and inflammatory pathways in Parkinson's disease. Neurochem. Res. 34:55–65.

Mohan, P.M. and R.V. Hosur (2008). NMR characterization of structural and dynamics perturbations due to a single point mutation in Drosophila DLC8 dimer: functional implications. Biochemistry 47:6251–6259.

Morfini, G., G. Pigino, N. Mizuno, M. Kikkawa, and S.T. Brady (2007a). Tau binding to microtubules does not directly affect microtubule-based vesicle motility. J. Neurosci. Res. 85:2620–2630.

Morfini, G., G. Pigino, K. Opalach, Y. Serulle, J.E. Moreira, M. Sugimori, R.R. Llinas, and S.T. Brady (2007b). 1-Methyl-4-phenylpyridinium affects fast axonal transport by activation of caspase and protein kinase C. Proc. Natl. Acad. Sci. U. S. A. 104:2442–2447.

Morfini, G., G. Szebenyi, R. Elluru, N. Ratner, and S.T. Brady (2002). Glycogen synthase kinase 3 phosphorylates kinesin light chains and negatively regulates kinesin-based motility. EMBO J. 21:281–293.

Morfini, G.A., Y.M. You, S.L. Pollema, A. Kaminska, K. Liu, K. Yoshioka, B. Bjorkblom, E.T. Coffey, C. Bagnato, D. Han, C.F. Huang, G. Banker, G. Pigino, and S.T. Brady (2009).

Pathogenic huntingtin inhibits fast axonal transport by activating JNK3 and phosphorylating kinesin. Nat. Neurosci. 12:864–871.

Muchowski, P.J. and J.L. Wacker (2005). Modulation of neurodegeneration by molecular chaperones. Nat. Rev. Neurosci. 6:11–22.

Mudher, A., D. Shepherd, T.A. Newman, P. Mildren, J.P. Jukes, A. Squire, A. Mears, J.A. Drummond, S. Berg, D. MacKay, A.A. Asuni, R. Bhat, and S. Lovestone (2004). GSK-3beta inhibition reverses axonal transport defects and behavioural phenotypes in Drosophila. Mol. Psychiatry 9:522–530.

Muraro, N.I. and K.G. Moffat (2006). Down-regulation of torp4a, encoding the Drosophila homologue of torsinA, results in increased neuronal degeneration. J. Neurobiol. 66:1338–1353.

Nakagawa, T., M. Setou, D. Seog, K. Ogasawara, N. Dohmae, K. Takio, and N. Hirokawa (2000). A novel motor, KIF13A, transports mannose-6-phosphate receptor to plasma membrane through direct interaction with AP-1 complex. Cell 103:569–581.

Newmyer, S.L. and S.L. Schmid (2001). Dominant-interfering Hsc70 mutants disrupt multiple stages of the clathrin-coated vesicle cycle in vivo. J. Cell Biol. 152:607–620.

Oddo, S., A. Caccamo, B. Tseng, D. Cheng, V. Vasilevko, D.H. Cribbs, and F.M. LaFerla (2008). Blocking Abeta42 accumulation delays the onset and progression of tau pathology via the C terminus of heat shock protein70-interacting protein: a mechanistic link between Abeta and tau pathology. J. Neurosci. 28:12163–12175.

Ogura, T. and A.J. Wilkinson (2001). AAA+ superfamily ATPases: common structure–diverse function. Genes Cells 6:575–597.

Oka, M., M. Nakai, T. Endo, C.R. Lim, Y. Kimata, and K. Kohno (1998). Loss of Hsp70-Hsp40 chaperone activity causes abnormal nuclear distribution and aberrant microtubule formation in M-phase of *Saccharomyces cerevisiae*. J. Biol. Chem. 273:29727–29737.

Omran, H., D. Kobayashi, H. Olbrich, T. Tsukahara, N.T. Loges, H. Hagiwara, Q. Zhang, G. Leblond, E. O'Toole, C. Hara, H. Mizuno, H. Kawano, M. Fliegauf, T. Yagi, S. Koshida, A. Miyawaki, H. Zentgraf, H. Seithe, R. Reinhardt, Y. Watanabe, R. Kamiya, D.R. Mitchell, and H. Takeda (2008). Ktu/PF13 is required for cytoplasmic pre-assembly of axonemal dyneins. Nature 456:611–616.

Ozelius, L.J., J.W. Hewett, C.E. Page, S.B. Bressman, P.L. Kramer, C. Shalish, L.D. de, M.F. Brin, D. Raymond, D.P. Corey, S. Fahn, N.J. Risch, A.J. Buckler, J.F. Gusella, and X.O. Breakefield (1997). The early-onset torsion dystonia gene (DYT1) encodes an ATP-binding protein. Nat. Genet. 17:40–48.

Perng, M.D., L. Cairns, P. van de Ijssel, A. Prescott, A.M. Hutcheson, and R.A. Quinlan (1999). Intermediate filament interactions can be altered by HSP27 and alphaB-crystallin. J. Cell Sci. 112 (Pt 13):2099–2112.

Petrucelli, L., D. Dickson, K. Kehoe, J. Taylor, H. Snyder, A. Grover, M. De Lucia, E. McGowan, J. Lewis, G. Prihar, J. Kim, W.H. Dillmann, S.E. Browne, A. Hall, R. Voellmy, Y. Tsuboi, T.M. Dawson, B. Wolozin, J. Hardy, and M. Hutton (2004). CHIP and Hsp70 regulate tau ubiquitination, degradation and aggregation. Hum. Mol. Genet. 13:703–714.

Pilling, A.D., D. Horiuchi, C.M. Lively, and W.M. Saxton (2006). Kinesin-1 and Dynein are the primary motors for fast transport of mitochondria in Drosophila motor axons. Mol. Biol. Cell 17:2057–2068.

Polymeropoulos, M.H., C. Lavedan, E. Leroy, S.E. Ide, A. Dehejia, A. Dutra, B. Pike, H. Root, J. Rubenstein, R. Boyer, E.S. Stenroos, S. Chandrasekharappa, A. Athanassiadou, T. Papapetropoulos, W.G. Johnson, A.M. Lazzarini, R.C. Duvoisin, I.G. Di, L.I. Golbe, and R.L. Nussbaum (1997). Mutation in the alpha-synuclein gene identified in families with Parkinson's disease. Science 276:2045–2047.

Pratt, W.B., A.M. Silverstein, and M.D. Galigniana (1999). A model for the cytoplasmic trafficking of signalling proteins involving the hsp90-binding immunophilins and p50cdc37. Cell Signal. 11:839–851.

Qin, Z.H., Y. Wang, E. Sapp, B. Cuiffo, E. Wanker, M.R. Hayden, K.B. Kegel, N. Aronin, and M. DiFiglia (2004). Huntingtin bodies sequester vesicle-associated proteins by a polyproline-dependent interaction. J. Neurosci. 24:269–281.

Reiner, A., R.L. Albin, K.D. Anderson, C.J. D'Amato, J.B. Penney, and A.B. Young (1988). Differential loss of striatal projection neurons in Huntington disease. Proc. Natl. Acad. Sci. U. S. A. 85:5733–5737.

Renkawek, K., G.J. Bosman, and W.W. de Jong (1994). Expression of small heat-shock protein hsp 27 in reactive gliosis in Alzheimer disease and other types of dementia. Acta Neuropathol. (Berl) 87:511–519.

Robinson, P.A. (2008). Protein stability and aggregation in Parkinson's disease. Biochem. J. 413:1–13.

Roher, A.E., N. Weiss, T.A. Kokjohn, Y.M. Kuo, W. Kalback, J. Anthony, D. Watson, D.C. Luehrs, L. Sue, D. Walker, M. Emmerling, W. Goux, and T. Beach (2002). Increased A beta peptides and reduced cholesterol and myelin proteins characterize white matter degeneration in Alzheimer's disease. Biochemistry 41:11080–11090.

Rose, S.E., F. Chen, J.B. Chalk, F.O. Zelaya, W.E. Strugnell, M. Benson, J. Semple, and D.M. Doddrell (2000). Loss of connectivity in Alzheimer's disease: an evaluation of white matter tract integrity with colour coded MR diffusion tensor imaging. J. Neurol. Neurosurg. Psychiatry 69:528–530.

Ross, C.A. and M.A. Poirier (2004). Protein aggregation and neurodegenerative disease. Nat. Med. 10 Suppl:S10–S17.

Rui, Y., P. Tiwari, Z. Xie, and J.Q. Zheng (2006). Acute impairment of mitochondrial trafficking by beta-amyloid peptides in hippocampal neurons. J. Neurosci. 26:10480–10487.

Saha, A.R., J. Hill, M.A. Utton, A.A. Asuni, S. Ackerley, A.J. Grierson, C.C. Miller, A.M. Davies, V.L. Buchman, B.H. Anderton, and D.P. Hanger (2004). Parkinson's disease alpha-synuclein mutations exhibit defective axonal transport in cultured neurons. J. Cell Sci. 117:1017–1024.

Sahara, N., M. Murayama, T. Mizoroki, M. Urushitani, Y. Imai, R. Takahashi, S. Murata, K. Tanaka, and A. Takashima (2005). In vivo evidence of CHIP up-regulation attenuating tau aggregation. J. Neurochem. 94:1254–1263.

Sanchez, C., R. Padilla, R. Paciucci, J.C. Zabala, and J. Avila (1994). Binding of heat-shock protein 70 (hsp70) to tubulin. Arch. Biochem. Biophys. 310:428–432.

Sapp, E., J. Penney, A. Young, N. Aronin, J.P. Vonsattel, and M. DiFiglia (1999). Axonal transport of N-terminal huntingtin suggests early pathology of corticostriatal projections in Huntington disease. J. Neuropathol. Exp. Neurol. 58:165–173.

Shen, H.Y., J.C. He, Y. Wang, Q.Y. Huang, and J.F. Chen (2005). Geldanamycin induces heat shock protein 70 and protects against MPTP-induced dopaminergic neurotoxicity in mice. J. Biol. Chem. 280:39962–39969.

Shimamoto, S., M. Takata, M. Tokuda, F. Oohira, H. Tokumitsu, and R. Kobayashi (2008). Interactions of S100A2 and S100A6 with the tetratricopeptide repeat proteins, Hsp90/Hsp70-organizing protein and kinesin light chain. J. Biol. Chem. 283:28246–28258.

Shimoji, M., L. Zhang, A.S. Mandir, V.L. Dawson, and T.M. Dawson (2005). Absence of inclusion body formation in the MPTP mouse model of Parkinson's disease. Brain Res. Mol. Brain Res. 134:103–108.

Shimura, H., Y. Miura-Shimura, and K.S. Kosik (2004a). Binding of tau to heat shock protein 27 leads to decreased concentration of hyperphosphorylated tau and enhanced cell survival. J. Biol. Chem. 279:17957–17962.

Shimura, H., D. Schwartz, S.P. Gygi, and K.S. Kosik (2004b). CHIP-Hsc70 complex ubiquitinates phosphorylated tau and enhances cell survival. J. Biol. Chem. 279:4869–4876.

Sinadinos, C., T. Burbidge-King, D. Soh, L. Thompson, L. Marsh, A. Wyttenbach, and A.K. Mudher (2009). Live axonal transport disruption by mutant huntingtin fragments in Drosophila motor neuron axons. Neurobiol. Dis. 34:389–395.

Sipione, S., D. Rigamonti, M. Valenza, C. Zuccato, L. Conti, J. Pritchard, C. Kooperberg, J.M. Olson, and E. Cattaneo (2002). Early transcriptional profiles in huntingtin-inducible striatal cells by microarray analyses. Hum. Mol. Genet. 11:1953–1965.

Sousa, R. and E.M. Lafer (2006). Keep the traffic moving: mechanism of the Hsp70 motor. Traffic. 7:1596–1603.

Spang, A. (2008). The life cycle of a transport vesicle. Cell Mol. Life Sci. 65:2781–2789.

Spillantini, M.G., M.L. Schmidt, V.M. Lee, J.Q. Trojanowski, R. Jakes, and M. Goedert (1997). Alpha-synuclein in Lewy bodies. Nature 388:839–840.

Stagi, M., P.S. Dittrich, N. Frank, A.I. Iliev, P. Schwille, and H. Neumann (2005). Breakdown of axonal synaptic vesicle precursor transport by microglial nitric oxide. J. Neurosci. 25:352–362.

Stamer, K., R. Vogel, E. Thies, E. Mandelkow, and E.M. Mandelkow (2002). Tau blocks traffic of organelles, neurofilaments, and APP vesicles in neurons and enhances oxidative stress. J. Cell Biol. 156:1051–1063.

Stokin, G.B., C. Lillo, T.L. Falzone, R.G. Brusch, E. Rockenstein, S.L. Mount, R. Raman, P. Davies, E. Masliah, D.S. Williams, and L.S. Goldstein (2005). Axonopathy and transport deficits early in the pathogenesis of Alzheimer's disease. Science 307:1282–1288.

Sun, Y. and T.H. MacRae (2005). Small heat shock proteins: molecular structure and chaperone function. Cell Mol. Life Sci. 62:2460–2476.

Szebenyi, G., G.A. Morfini, A. Babcock, M. Gould, K. Selkoe, D.L. Stenoien, M. Young, P.W. Faber, M.E. MacDonald, M.J. McPhaul, and S.T. Brady (2003). Neuropathogenic forms of huntingtin and androgen receptor inhibit fast axonal transport. Neuron 40:41–52.

Tai, A.W., J.Z. Chuang, C. Bode, U. Wolfrum, and C.H. Sung (1999). Rhodopsin's carboxy-terminal cytoplasmic tail acts as a membrane receptor for cytoplasmic dynein by binding to the dynein light chain Tctex-1. Cell 97:877–887.

Taylor, J.P., F. Tanaka, J. Robitschek, C.M. Sandoval, A. Taye, S. Markovic-Plese, and K.H. Fischbeck (2003). Aggresomes protect cells by enhancing the degradation of toxic polyglutamine-containing protein. Hum. Mol. Genet. 12:749–757.

The Huntington's Disease Collaborative Research Group (1993). A novel gene containing a trinucleotide repeat that is expanded and unstable on Huntington's disease chromosomes. Cell 72:971–983.

Tobaben, S., P. Thakur, R. Fernandez-Chacon, T.C. Sudhof, J. Rettig, and B. Stahl (2001). A trimeric protein complex functions as a synaptic chaperone machine. Neuron 31:987–999.

Torroja, L., H. Chu, I. Kotovsky, and K. White (1999). Neuronal overexpression of APPL, the Drosophila homologue of the amyloid precursor protein (APP), disrupts axonal transport. Curr. Biol. 9:489–492.

Trushina, E., R.B. Dyer, J.D. Badger, D. Ure, L. Eide, D.D. Tran, B.T. Vrieze, V. Legendre-Guillemin, P.S. McPherson, B.S. Mandavilli, H.B. Van, S. Zeitlin, M. McNiven, R. Aebersold, M. Hayden, J.E. Parisi, E. Seeberg, I. Dragatsis, K. Doyle, A. Bender, C. Chacko, and C.T. McMurray (2004). Mutant huntingtin impairs axonal trafficking in mammalian neurons in vivo and in vitro. Mol. Cell Biol. 24:8195–8209.

Tsai M.Y., G. Morfini, G. Szebenyi, and S.T. Brady (2000). Release of kinesin from vesicles by hsc70 and regulation of fast axonal transport. Mol Biol Cell. 11:2161–2173.

Tsvetkova, N.M., I. Horvath, Z. Torok, W.F. Wolkers, Z. Balogi, N. Shigapova, L.M. Crowe, F. Tablin, E. Vierling, J.H. Crowe, and L. Vigh (2002). Small heat-shock proteins regulate membrane lipid polymorphism. Proc. Natl. Acad. Sci. U. S. A. 99:13504–13509.

Tukamoto, T., N. Nukina, K. Ide, and I. Kanazawa (1997). Huntington's disease gene product, huntingtin, associates with microtubules in vitro. Brain Res. Mol. Brain Res. 51:8–14.

Ungewickell, E., H. Ungewickell, S.E. Holstein, R. Lindner, K. Prasad, W. Barouch, B. Martin, L.E. Greene, and E. Eisenberg (1995). Role of auxilin in uncoating clathrin-coated vesicles. Nature 378:632–635.

Vallee, R.B., J.C. Williams, D. Varma, and L.E. Barnhart (2004). Dynein: an ancient motor protein involved in multiple modes of transport. J. Neurobiol. 58:189–200.

Verhey, K.J., D.L. Lizotte, T. Abramson, L. Barenboim, B.J. Schnapp, and T.A. Rapoport (1998). Light chain-dependent regulation of Kinesin's interaction with microtubules. J. Cell Biol. 143:1053–1066.

Viswanath, V., Y. Wu, R. Boonplueang, S. Chen, F.F. Stevenson, F. Yantiri, L. Yang, M.F. Beal, and J.K. Andersen (2001). Caspase-9 activation results in downstream caspase-8 activation and bid cleavage in 1-methyl-4-phenyl-1,2,3,6-tetrahydropyridine-induced Parkinson's disease. J. Neurosci. 21:9519–9528.

Walker, R.H., P.F. Good, and P. Shashidharan (2003). TorsinA immunoreactivity in inclusion bodies in trinucleotide repeat diseases. Mov Disord. 18:1041–1044.

Webb, J.L., B. Ravikumar, and D.C. Rubinsztein (2004). Microtubule disruption inhibits autophagosome-lysosome fusion: implications for studying the roles of aggresomes in polyglutamine diseases. Int. J. Biochem. Cell Biol. 36:2541–2550.

Wilhelmus, M.M., I. Otte-Holler, P. Wesseling, R.M. de Waal, W.C. Boelens, and M.M. Verbeek (2006). Specific association of small heat shock proteins with the pathological hallmarks of Alzheimer's disease brains. Neuropathol. Appl. Neurobiol. 32:119–130.

Williams, N.E. and E.M. Nelsen (1997). HSP70 and HSP90 homologs are associated with tubulin in hetero-oligomeric complexes, cilia and the cortex of Tetrahymena. J. Cell Sci. 110 (Pt 14):1665–1672.

Wischik, C.M., M. Novak, H.C. Thogersen, P.C. Edwards, M.J. Runswick, R. Jakes, J.E. Walker, C. Milstein, M. Roth, and A. Klug (1988). Isolation of a fragment of tau derived from the core of the paired helical filament of Alzheimer disease. Proc. Natl. Acad. Sci. U. S. A. 85:4506–4510.

Wyttenbach, A., J. Carmichael, J. Swartz, R.A. Furlong, Y. Narain, J. Rankin, and D.C. Rubinsztein (2000). Effects of heat shock, heat shock protein 40 (HDJ-2), and proteasome inhibition on protein aggregation in cellular models of Huntington's disease. Proc. Natl. Acad. Sci. U. S. A. 97:2898–2903.

Wyttenbach, A., O. Sauvageot, J. Carmichael, C. az-Latoud, A.P. Arrigo, and D.C. Rubinsztein (2002). Heat shock protein 27 prevents cellular polyglutamine toxicity and suppresses the increase of reactive oxygen species caused by huntingtin. Hum. Mol. Genet. 11:1137–1151.

Wyttenbach, A., J. Swartz, H. Kita, T. Thykjaer, J. Carmichael, J. Bradley, R. Brown, M. Maxwell, A. Schapira, T.F. Orntoft, K. Kato, and D.C. Rubinsztein (2001). Polyglutamine expansions cause decreased CRE-mediated transcription and early gene expression changes prior to cell death in an inducible cell model of Huntington's disease. Hum. Mol. Genet. 10:1829–1845.

Yaglom, J.A., V.L. Gabai, A.B. Meriin, D.D. Mosser, and M.Y. Sherman (1999). The function of HSP72 in suppression of c-Jun N-terminal kinase activation can be dissociated from its role in prevention of protein damage. J. Biol. Chem. 274:20223–20228.

Yamanaka, T., H. Miyazaki, F. Oyama, M. Kurosawa, C. Washizu, H. Doi, and N. Nukina (2008). Mutant Huntingtin reduces HSP70 expression through the sequestration of NF-Y transcription factor. EMBO J. 27:827–839.

Zala, D., E. Colin, H. Rangone, G. Liot, S. Humbert, and F. Saudou (2008). Phosphorylation of mutant huntingtin at S421 restores anterograde and retrograde transport in neurons. Hum. Mol. Genet. 17:3837–3846.

Zimmer, D.B., J. Chaplin, A. Baldwin, and M. Rast (2005). S100-mediated signal transduction in the nervous system and neurological diseases. Cell Mol. Biol. (Noisy-le-grand) 51:201–214.

Chapter 6
Mechanisms of Neuronal Mitochondrial Transport

F. Anne Stephenson and Kieran Brickley

Abstract Mitochondria are transported in cells to meet local demands for energy and for the buffering of Ca^{2+} ions. In neurones, the regulated trafficking of mitochondria to specific sites is particularly important to satisfy the requirements demanded by neurotransmission and ATP-dependent molecular chaperone machineries. Aberrant distribution of mitochondria is a feature of certain neurodegenerative diseases. Mitochondria are transported in anterograde and retrograde directions along microtubules using the molecular motors kinesin and dynein. Specific adaptor proteins which link the mitochondria to the motors have recently been identified. This chapter will summarize these new studies which resulted in the identification of these adaptor proteins, the identification of mitochondrial acceptors for the adaptors and anchoring mechanisms that ensure the arrest of mitochondria at required sites. Also discussed are the signalling mechanisms that regulate mitochondrial trafficking which are now beginning to be unravelled following the identification of the major players in the transport of this organelle.

6.1 Introduction

Mitochondria serve several functions within cells. These include the generation of energy in the form of ATP via the citric acid cycle and oxidative phosphorylation, the buffering of calcium ions and the regulation of apoptosis. Thus within cells, mitochondria have to be mobile so that they can respond to local needs. In the nervous system, mitochondria and mitochondrial transport are particularly important because of the requirement for high energy and calcium buffering during neurotransmission; indeed mitochondria are known to be accumulated at synapses to serve these needs (reviewed in Mattson et al. 2008). It is pertinent to note that patients with mutations in mitochondrial DNA (mitochondrial encephalopathies) and those with mutations in nuclear genes encoding mitochondrial proteins present with neurological symptoms

F.A. Stephenson (✉)
School of Pharmacy, University of London, 29/39 Brunswick Square, London WC1N 1AX, UK
e-mail: anne.stephenson@pharmacy.ac.uk

A. Wyttenbach and V. O'Connor (eds.), *Folding for the Synapse*,
DOI 10.1007/978-1-4419-7061-9_6,

(reviewed in e.g. Mattson et al. 2008, Chen and Chan 2006). In most neurones, the synapse is distant from the neuronal cell body. Mechanisms must therefore exist to transport mitochondria from their sites of biogenesis along either, axons to nerve terminals or, dendrites to deliver mitochondria to sites adjacent to post-synaptic membranes. This chapter will review the current understanding of motor-dependent mitochondrial trafficking mechanisms focusing particularly on proteins that mediate their regulated movement. Other reviews that consider other aspects of mitochondrial motility not described in detail here include Boldogh and Pon (2007), Cerveny et al., (2007) and Chen and Chan (2006).

6.2 General Features of Neuronal Mitochondrial Distribution and Mobility

Mitochondrial distribution and mobility in neurones are generally studied in primary cultures of defined neuronal cell types where they are visualized by mitochondrial-selective stains such as the Mitotracker dyes (Invitrogen). Alternatively, neurones may be transfected with vectors that express mitochondrial-targeted fluorescent proteins. One example is the vector pDsRed1-Mito (Clontech) that contains the mitochondrial targeting sequence from subunit VIII of human cytochrome c oxidase fused to a red fluorescent protein from *Discosoma sp.* Imaging of neuronal mitochondria in living mice was recently made possible by the development of the "Mitomouse" transgenic mice lines in which mitochondrially targeted cyan or yellow fluorescent proteins are selectively expressed in neurones under the control of either the *Thy1* or *NSE* regulator elements (Misgeld et al. 2007).

In general, mitochondria are distributed throughout neurones but they are enriched at sites that require high energy such as the synapse, active growth cones and branches, the nodes of Ranvier, distal initial segments, myelination boundaries and sites of axonal protein synthesis (reviewed in Hollenbeck and Saxton 2005). They are also found in dendrites but not in spines due to size constraints (Li et al. 2004). Figure 6.1 is a typical example of the distribution of mitochondria in hippocampal pyramidal neurones in primary culture.

Under basal conditions, it is evident that in neurones mitochondria are mobile. This movement is developmentally regulated with higher mitochondrial mobility being found early in development (Chang and Reynolds 2006). Mitochondrial movement appears not to be synchronized. The majority of mitochondria (in the range 60–87%) are immobile, some seem to hover backwards and forwards in a defined location, others move in an anterograde direction away from the cell body and others in a retrograde direction back to the cell nucleus. Axonal mitochondria with a normal membrane potential have a higher level of anterograde transport compared to mitochondria that have a compromised membrane potential and ATP synthetic activity and exhibit a higher level of retrograde transport (Miller and Sheetz 2004). This empirical observation suggests that anterograde transport

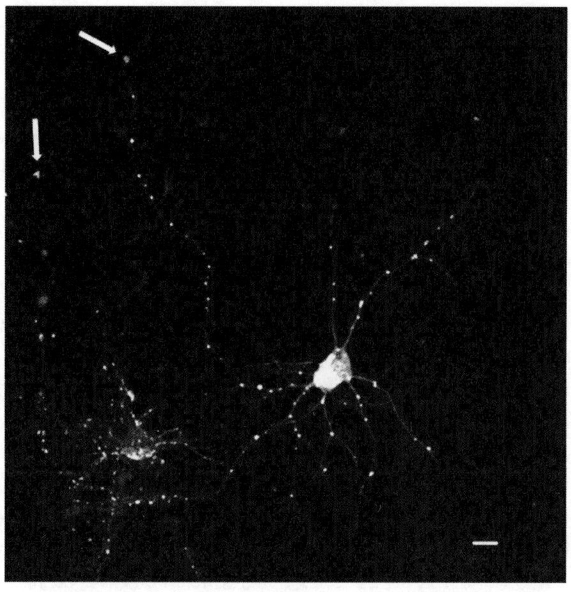

Fig. 6.1 The distribution of mitochondria in processes of primary cultures of hippocampal pyramidal neurones. The figure shows hippocampal pyramidal neurones in primary culture stained at day 14 in vitro with the mitochondrial stain, MitoTracker Red CMXRos. Each puncta is a mitochondrion. Note the distribution of mitochondria with different morphologies throughout all the processes but being enriched at their tips (*white arrows*). The scale bar is 10 μm

moves mitochondria to their sites of action, whereas retrograde transport returns mitochondria to the cell body where they may be degraded and recycled. The rates at which mitochondria move are of the order, in axons of ~0.1–0.3 μm/s, a speed characteristic of relatively fast axonal transport. In a study where anterograde and retrograde velocities were compared, retrograde transport was ~37% faster (Misgeld et al. 2007). Comparisons of mitochondrial transport between axons and dendrites in hippocampal neurons found that mitochondria in axons travelled further than those localized in dendrites (Overly et al. 1996; Ligon and Steward 2000). One study reported that axonal mitochondria were more motile compared to those localized in dendrites (Overly et al. 1996). A later analysis, however, found a similar percentage of mobile mitochondria (~30%) for axons and dendrites (Ligon and Steward 2000). Interestingly, although mitochondria may move further in axons, they were found to be more metabolically active in dendrites (Overly et al. 1996).

In addition to moving, mitochondria are constantly undergoing fusion and fission. This process contributes to an observed pleomorphism of mitochondria. In one report studying transport of mitochondria in peripheral nerves, mitochondria moving in both anterograde and retrograde directions were shorter (1.7 μm and 1.5 μm, respectively) versus a length of 3 μm for immobile mitochondria (Misgeld et al. 2007). Fusion and fission of mitochondria are mediated via a complex series of protein–protein interactions and post-translational modifications. The dynamin-related, mitochondrial GTPases, mitofusin and OPA1 (fusion) and the mitochondrial protein Fis1 and dynamin-related peptide (Drp1, fission) are the major players; mitochondrial fusion and fission in health and disease are reviewed in Mattson et al. (2008) and Cerveny et al. (2007).

6.3 Modes of Mitochondrial Movement

Since in neurones mitochondria are known to translocate to relevant sites in response to particular needs, it is probable that there are intracellular signalling mechanisms that initiate this movement and that appropriate trafficking mechanisms exist to facilitate transport. Furthermore, it is likely that there are docking mechanisms that ensure that mitochondria are retained at the required sites. Each of these components is considered below.

The majority of neuronal (axonal) mitochondrial transport occurs via motor-driven translocation along the microtubule tracks. This movement is driven in an anterograde direction to the plus ends of microtubules by kinesin motor proteins. Kinesins (KIFs) are a large gene family with 45 KIF members in the human genome, 38 of which are expressed in the brain (Miki et al. 2001). Two kinesins have been implicated in mitochondrial trafficking, the conventional kinesin-1 (alternative nomenclature KIF5, often referred to as simply kinesin heavy chain; there are three KIF5 genes, KIF5A, KIF5B and KIF5C) and kinesin-3 (also termed KIF1B; Hirokawa and Takemura 2005). It has been estimated that between 2 and 200 motor proteins associate with mitochondria at any one time (Gross et al. 2007). The kinesins, however, are not thought to interact directly with mitochondrial cargoes when moving, but they do so indirectly via kinesin adaptor proteins. Kinesin adaptors include the trafficking kinesin (TRAK) family of proteins, syntabulin and possibly the c-Jun N-terminal kinase scaffolding protein, JIP1 and GTP-binding nuclear protein, Ran, binding protein 2 (RanBP2). The TRAK proteins are also alternatively known as Milton following their discovery in *Drosophila*. These protein families are discussed in more detail below. Dynein has been implicated in retrograde transport of mitochondria in *Drosophila*. Mutations in fly cytoplasmic dynein heavy chain result in axonal swellings containing mitochondria suggesting impaired trafficking processes (Pilling et al. 2006).

In addition to microtubule-based transport of mitochondria, it has also been shown that mitochondria can move along actin filaments. This implies that here, myosin motors are utilized for the movement of the organelle. Candidate myosins, I, II, V and VI, are expressed in neuronal tissue; Myosins I, II and V transport cargoes to the (+) end of actin filaments whereas myosin VI moves in the opposite direction. Thus, there is potential for actin/myosin anterograde and retrograde transport of mitochondria. There is, however, no evidence for their direct or adaptor-mediated role in mitochondrial transport (reviewed in Hollenbeck and Saxton 2005).

6.4 Proteins Directly Involved in Mitochondrial Transport

6.4.1 The TRAK Family of Kinesin Adaptor Proteins

The most compelling candidates to function as kinesin adaptor proteins for the anterograde trafficking of mitochondria are the TRAK/Milton family of proteins. The *Drosophila* protein, Milton, was first discovered in a genetic screen searching for mutant flies which had phenotypes suggestive of an impairment of axonal or nerve terminal function (Stowers et al. 2002). One mutant strain was blind, and in the photoreceptor cells of these flies, the mitochondria were found accumulated in neuronal cell bodies, whereas synaptic terminals and axons were depleted of mitochondria. The same anomalous distribution of mitochondria was later observed in central and peripheral neurones of homozygous null milton *Drosophila* (Gorska-Andrzejak et al. 2003; Glater et al. 2006). The mutated gene was cloned and aptly named Milton after the blind English poet, John Milton. Milton is the *Drosophila* orthologue of a 913 amino acid protein that was cloned and characterized earlier in the same year. That gene was γ-aminobutyric acid$_A$ (GABA$_A$) receptor interacting factor-1 (GRIF-1 now termed TRAK2) due to the fact that it had been identified in a yeast two-hybrid screen searching for GABA$_A$ receptor interacting proteins (Beck et al. 2002). In *Drosophila*, there is one *milton* gene but in mammals, there is a gene family containing the *TRAK1* and *TRAK2* genes. TRAK1 and TRAK2 are homologous sharing 48% amino acid identity; 58% homology with each other and ~40% homology with Milton. The gene family is known as a coiled-coil family of proteins due to the presence of these predicted structural domains within their N-termini. The gene encoding TRAK2 (human chromosome 2q33) has 16 exons. There are three proven TRAK2 splice variants; TRAK2a = the full length protein of 913 amino acids; TRAK2b has a truncated C-terminus and TRAK2c has an internal deletion. There are no known functional differences between the splice variants. TRAK1 is localized on chromosome 3p25.3–p24.13. TRAK1 and TRAK2 also have amino acid similarity with Huntingtin-associated proteins (HAP1A and HAP1B) within their predicted coiled-coil regions. The HAP1 proteins bind kinesin light chain (McGuire et al. 2006). Evidence supports a role for HAPs in the mediation of intracellular trafficking in neurones (e.g. Gauthier et al. 2004; Rong et al. 2006). There is no evidence, however, to link them with mitochondrial trafficking and it is unclear whether they are members of the TRAK gene family.

Endogenous TRAK2 and Milton have all been shown to co-immunoprecipitate with kinesin heavy chain from detergent-solubilized extracts of neuronal tissue (Stowers et al. 2002; Brickley et al. 2005). In mammalian brain, TRAK2 co-immunoprecipitates predominantly with the kinesin-1 heavy chain, KIF5A (Brickley et al. 2005). No data are currently available for TRAK1 due to a lack of specific antibodies. Exogenous TRAKs1/2 and Milton will also co-immunoprecipitate with endogenous kinesin in mammalian cells (Stowers et al. 2002; Brickley et al. 2005). Fluorescent resonance energy transfer (FRET) measurements using fluorescently tagged TRAKs and another member of the kinesin-1 heavy chain family, KIF5C, showed that the association between TRAKs1/2 is direct (Smith et al. 2006). Together with yeast two-hybrid interaction assays and co-immunoprecipitations, FRET studies mapped the TRAK binding site on the kinesin heavy chain to the C-terminal cargo binding domain, KIF5C (827–957) (Smith et al. 2006).

Over-expression in mammalian cells of Milton or the TRAKs results in aggregation of mitochondria in the cell cytoplasm adjacent to cell nuclei. When kinesin heavy chain is expressed together with either TRAK1, TRAK2 or Milton, the distribution of mitochondria is changed such that each is now enriched at the ends of cellular processes where they co-localize with kinesin heavy chain and either

TRAK1, TRAK2 or Milton, respectively (Smith et al. 2006; Glater et al. 2006). In relatively mature rat hippocampal pyramidal neurones (day 14 in culture), endogenous TRAK2 partially co-distributed with mitochondria (MacAskill et al. 2009a). Approximately 40% of TRAK2 co-localized with labelled mitochondria under basal conditions. The co-distributed TRAK2 and mitochondria were observed in both axons and dendritic processes (MacAskill et al. 2009a; Kenny et al. unpublished results).

It is now known that the association and trafficking of mitochondria are mediated via the interaction of the TRAK family of proteins with the mitochondrial proteins, Miro1 and Miro2 (however, it should be noted that MacAskill et al. (2009b) recently reported that in in vitro pull down assays, in the absence of Ca^{2+}, Miro1 bound directly to kinesin-1 motor proteins). Miro1 and Miro2 are atypical RhoGTPases that are found in eukaryotic organisms from yeast, to *Drosophila* to mammals. The human genes were characterized in detail by Fransson and colleagues (2003). Early clues to their function were the demonstration that over-expression of a constitutively active variant resulted in aggregation of the mitochondrial network and the appearance of thread-like mitochondria suggesting abnormal mitochondrial biogenesis and/or trafficking processes. In *Drosophila* dMiro mutant flies, mitochondria are abnormally distributed being accumulated in parallel rows in neuronal cell bodies (Guo et al. 2005). At the neuromuscular junction of these flies, presynaptic termini lack mitochondria and their larvae have impaired locomotion (Guo et al. 2005). Miro1 and Miro2 are integrated into the outer mitochondrial membrane. They share a similar domain structure and they have 59% amino acid sequence identity and 75% homology. Each has a short C-terminal domain that is located inside mitochondria; one transmembrane domain and then in their N-terminal regions which protrude from the mitochondria into the cell cytoplasm, each has two GTPase domains, I and II, which are separated by two EF-hand calcium-binding domains, i.e. EF-hands I and II. Transfection of non-neuronal cells with fluorescent Miro constructs results in the targeting of Miro to mitochondria (Fransson et al. 2006; Glater et al. 2006). Similarly, expression in neuronal cells in primary culture results in targeting to mitochondria where Miro is localized to both axonal and dendritic mitochondria (Wang and Schwarz 2009; MacAskill et al. 2009a, b). Miro mutants lacking the transmembrane domain are not targeted to mitochondria (Fransson et al. 2006). A functional difference between Miro1 and Miro2 has been reported. Whereas after expression in cell lines, Miro1 induced aggregation and the appearance of thread-like mitochondria, Miro2 induced aggregation only (Fransson et al. 2006). In *Drosophila*, Miro is required for both anterograde and retrograde mitochondrial transport (Russo et al. 2009).

Miro was first shown to be a binding partner of Milton in a large-scale study that generated a protein interaction map of *Drosophila* (Giot et al. 2003). Miro was later shown to co-distribute and co-immunoprecipitate with Milton, TRAK1 or TRAK2 following over-expression in mammalian cells (Glater et al. 2006; Fransson et al. 2006). Over-expression of Miro1 in hippocampal pyramidal neurones was found to increase the percentage of TRAK2 associated with mitochondria (from ~40 to >90%). This concomitantly resulted in an enhanced distribution of mitochondria to the distal end of processes (MacAskill et al. 2009a).

In addition to co-associating with kinesin heavy chain and Miro1, TRAK1 and TRAK2 interact with the post-translational modification enzyme, *N*-acetylglucosamine transferase (OGT; Iyer et al. 2003). Indeed both TRAK1 and TRAK2 were identified from a yeast two-hybrid screen as OGT interacting proteins (OIPs), hence the alternative nomenclature in the literature for TRAK1 and TRAK2 of OIP106 and OIP98, respectively (Iyer et al. 2003). OGT catalyses the covalent modification of serine or threonine residues in proteins by *N*-acetylglucosamine (GlcNAc). This post-translational modification differs from conventional *N*-glycosylation of proteins in that it occurs in the cell cytoplasm. *O*-glycosylation by GlcNAc can compete with phosphorylation for these serine/threonine residues. Unsurprisingly therefore, it has been shown to modulate a variety of cellular processes including signalling, protein expression, protein degradation and of particular relevance to TRAK1 and TRAK2, intracellular cellular trafficking events (reviewed in Hart et al. 2007). It was recently demonstrated that OGT is an integral component of a KIF/TRAK/mitochondrial trafficking complex (Brickley et al. 2010). TRAK1 and TRAK2 are both substrates for OGT and are glycosylated in vivo (Iyer et al. 2003), but neither the sites of *O*-glycosylation nor their functional significances are known.

Key features of the proteins implicated in mitochondrial trafficking are shown in Fig. 6.2. A schematic diagram of the kinesin/TRAK/Miro mitochondrial trafficking complex is shown in Fig. 6.3.

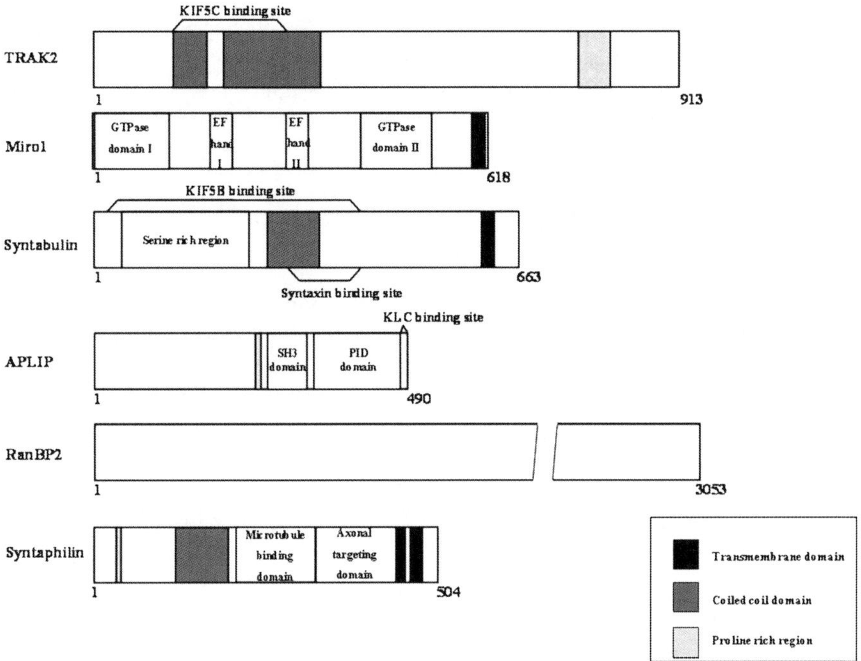

Fig. 6.2 The predicted domain structures and protein binding domains of proteins implicated in kinesin-mediated mitochondrial trafficking in neurones

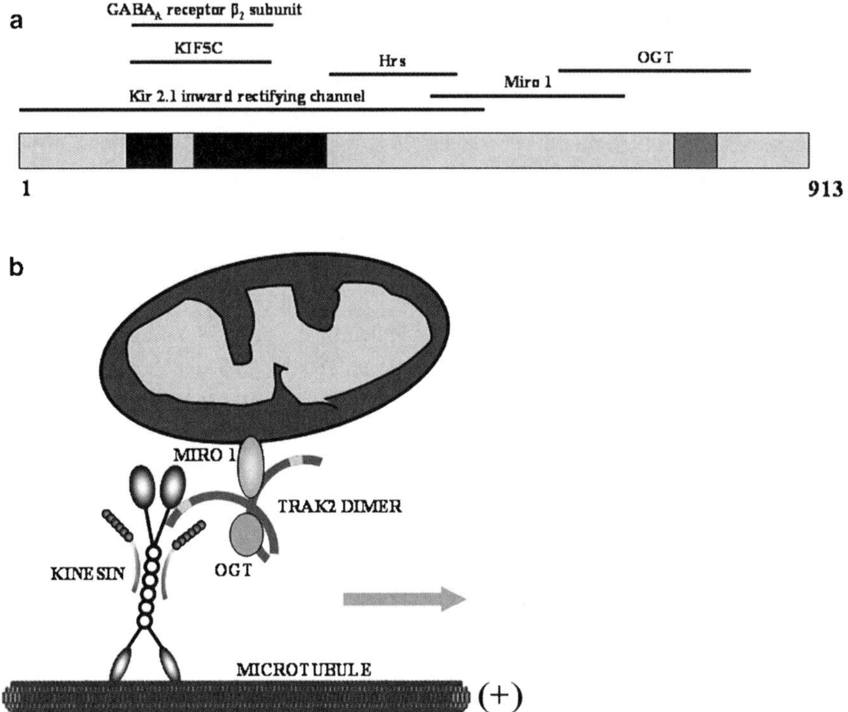

Fig. 6.3 Schematic diagram of the kinesin/TRAK/Miro mitochondrial trafficking complex and the identified protein-binding domains of TRAK 2. **a** is a schematic showing the different protein-binding domains of TRAK2; **b** is a schematic diagram depicting the kinesin/TRAK/Miro multi-component mitochondrial trafficking complex. *Dark-shaded area* – coiled-coil domain; *grey-shaded area* – proline rich region

6.4.2 The Syntaxin-Binding Protein, Syntabulin

Syntabulin is another example of a kinesin adaptor protein that has a postulated role in anterograde mitochondrial trafficking mechanisms in neurones. In hippocampal neurones in culture, syntabulin co-localizes with mitochondria, co-migrates with mitochondria along processes and siRNA knockdown of syntabulin resulted in a significant reduction in the mitochondrial density in processes (Cai et al. 2005). It is known to co-associate in vitro with the cargo binding domain of the ubiquitous conventional kinesin heavy chain, KIF5B (Su et al. 2004). Syntabulin is a membrane-associated protein with an internal coiled-coil domain and a predicted C-terminal transmembrane domain that was shown to target it to mitochondria (Fig. 6.2). It is not known, however, how syntabulin interacts with mitochondria. It was speculated that it could be either via association of the hydrophobic C-terminal tail with mitochondrial phospholipids or alternatively, with an as yet unidentified outer mitochondrial membrane protein (Cai et al. 2005).

It is important to note that both syntabulin and the TRAK proteins may be multi-functional in terms of the intracellular transport of organelles. Syntabulin is implicated in the anterograde transport of syntaxin-containing vesicles (Su et al. 2004) and active zone components essential for presynaptic assembly (Cai et al. 2007). There is evidence to support a role for TRAKs in endosome to lysosome trafficking by virtue of its association with hepatocyte growth factor-regulated tyrosine kinase substrate (hrs) (Kirk et al. 2006; Webber et al. 2008; MacAskill et al. 2009a) and in the cell surface trafficking of Kir2.1 inwardly rectifying potassium channels (Grishin et al. 2006). In support of the original discovery of TRAK2 as a $GABA_A$ receptor trafficking molecule, *Trak1* is the mutated gene in *hyrt* mice, a mouse model of hypertonia (Gilbert et al. 2006). These *hyrt* mice have reduced levels of $GABA_A$ receptors particularly in lower motor neurones. Their hypertonic symptoms can be ameliorated by the administration of benzodiazepines, drugs that potentiate GABAergic neurotransmission and compensate for the reduced expression of $GABA_A$ receptors thus implying a role for TRAK1 in receptor transport mechanisms.

6.4.3 Ran-Binding Protein 2

Ran-binding protein 2 (RanBP2), a putative E3 SUMO ligase, is another candidate mitochondrial trafficking protein. Its down-regulation results in, among other defects, delocalization of mitochondria in photosensory neurones (Cho et al. 2007). RanBP2 binds directly to the kinesins, KIF5B and KIF5C but not to KIF5A. There is no information about how RanBP2 may associate with mitochondria at the molecule level.

6.4.4 APP-Like Interacting Protein 1

APP-like interacting protein 1 (APLIP1) is the *Drosophila* homologue of c-Jun N-terminal kinase (JNK)-interacting protein 1 (JIP-1). JIP-1 is a scaffolding protein that has been shown to bind kinesin light chain; a reelin receptor (apolipoprotein E receptor 2; ApoER2) and amyloid precursor protein. It also serves as a scaffold molecule for components of the JNK signalling pathway, mitogen-activated protein kinase kinase kinase (MAPKKK), mitogen-activated protein kinase kinase (MAPKK) and mitogen-activated protein kinase (MAPK) (Horiuchi et al. 2005, 2007). JIP-1 and APLIP1 have both been proposed to have a functional role in the axonal transport of synaptic vesicles. Mutation of the *Aplip-1* gene resulted in traits associated with transport deficits, i.e. larval paralysis, axonal swelling and reduced anterograde and retrograde vesicle transport (Horiuchi et al. 2005). Interestingly, the mutation also resulted in a decrease in the retrograde transport of mitochondria. This suggested an association between APLIP1 and dynein and, thereby, a role for

APLIP1 in regulated retrograde mitochondrial transport. There is no evidence, however, to link JIP-1 with the trafficking of mitochondria in mammalian cells.

6.5 Mitochondrial Docking Mechanisms

As mentioned above, at any one time, 60–70% of mitochondria are immobile. These stationary mitochondria may be considered as "ready and waiting" or alternatively they have been transported then anchored to sites to meet particular high energy requirements of neurones. Two mechanisms that result in the docking of mitochondria have recently been elucidated. The first involves the axonal docking protein, syntaphilin (Kang et al. 2008). Syntaphilin knock-out mice have a substantially higher proportion of axonal mitochondria in the mobile state (~75% compared to ~30% for wild-type animals). They also have a decreased density of axonal mitochondria. Syntaphilin is highly enriched in axonal tracts. Expression of fluorescently tagged syntaphilin in hippocampal neurones results in its co-distribution with mitochondria and all syntaphilin labelled mitochondria were observed to be stationary. The domain structure of syntaphilin includes a microtubule binding domain, an axonal targeting domain and close to its C-terminus, reminiscent of syntabulin, a short transmembrane domain that targets it to outer mitochondrial membranes (Fig. 6.2). Deletion of the microtubule-binding domain resulted in restoration in mobility to syntaphilin-labelled mitochondria. These observations are consistent with a role for syntaphilin as a mitochondrial docking protein that anchors mitochondria via interaction with microtubules. Like syntabulin, it is as yet unclear whether syntaphilin is integrated into the mitochondrial outer membrane or if it interacts with mitochondrial phospholipids or a mitochondrial membrane acceptor protein. Furthermore, it is not known what controls the formation or the dissociation of microtubule/syntaphilin/mitochondrial complexes.

The second mitochondrial docking mechanism involves the TRAK/kinesin/Miro complex. In an elegant study, Wang and Schwarz (2009) confirmed what was already known from the work of Yi et al. (2004) and Chang et al. (2006) that increases in intracellular Ca^{2+} result in a decrease in axonal mitochondrial motility in hippocampal pyramidal neurones. They expressed a tagged Miro mutant that lacked a Ca^{2+} binding EF-hand, i.e. $Miro^{KK}$ in neurones. This resulted in a loss of the decrease in mitochondrial motility observed for wild-type Miro transfected neurones. Further studies showed that in the presence of elevated Ca^{2+}, the motor domain of kinesin heavy chain co-immunoprecipitated with Miro in the absence of Milton (the *Drosophila* homologue of the TRAKs). Thus, a scheme was proposed whereby in the presence of Ca^{2+}, kinesin binds to Miro rather than requiring the intermediary TRAK adaptor protein. It is the kinesin motor domain that interacts with Miro thus in the presence of elevated Ca^{2+}, kinesin is released from microtubules thus resulting in the arrest of mitochondrial movement (Wang and Schwarz 2009). It was implied from these studies that Miro and the kinesin motor domain associate directly, although this is not unequivocally proven. It was also unclear whether, in the presence of Ca^{2+}, Milton is still part of the kinesin/Miro mitochondrial complex. In complementary studies, Saotome et al. (2008) and MacAskill et al. (2009b) also demonstrated the importance of Miro EF-hands in Ca^{2+}-induced arrest of mitochondria in cortical and pyramidal neurones, respectively. Saotome et al. (2008) found that the Ca^{2+}-induced arrest of mitochondria was promoted by over-expression of Miro. But, when either Miro was depleted by siRNA or, when the EF- hand was mutated, the Ca^{2+}-induced arrest was suppressed (Saotome et al. 2008). Interestingly and in addition, they reported that over-expression of Miro had effects on mitochondrial morphology. At resting Ca^{2+} levels, mitochondrial fusion was promoted but at high Ca^{2+}, fission was more apparent (Saotome et al. 2008). The study by MacAskill et al. (2009b) focused on dendritic mitochondria in hippocampal pyramidal neurones. They elegantly showed that elevation of intra-cellular Ca^{2+} by glutamate receptor activation resulted in mitochondrial arrest. Mutation of the Miro1 EF hands to prevent the binding of Ca^{2+} resulted in the blocking of the glutamate-induced arrest of mitochondrial movement although the mutated Miro1 could still facilitate mitochondrial motility. As part of their study, the Ca^{2+} dependence of the Miro1/kinesin interaction was investigated. In direct contrast to Wang and Schwarz (2009), they found that in the presence of 2 mM Ca^{2+}, Miro1 and KIF5 did not co-immunoprecipitate from brain extracts, and using pull down assays, the direct binding of Miro1 to kinesin-1 was inhibited. The binding of TRAK2 to Miro1 was Ca^{2+} independent (MacAskill et al. 2009b). Clearly there are issues that need to be reconciled.

6.6 Molecular Mechanisms of the Regulation of Mitochondrial Trafficking

Mitochondrial trafficking in neurones occurs to meet the demands of normal physiological processes as discussed earlier, but also transport may be required in response to pathophysiological insults. For example, it is known that elevation of intracellular Ca^{2+} results in a decrease in motility (Yi et al. 2004; Chang et al. 2006); glutamate treatment (induction of excitotoxicity) decreases mitochondrial movement probably via Ca^{2+} influx (Rintoul et al. 2003); elevation of intracellular Zn^{2+} also halts mitochondria (Malaiyandi et al. 2005) and focal stimulation of axon shafts with nerve growth factor causes accumulation of mitochondria (Chada and Hollenbeck 2003, 2004). The transport processes that respond to these events may be regulated by the formation of the kinesin/adaptor, the kinesin/adaptor/cargo mitochondrial complex, the dissociation of the kinesin/adaptor/mitochondrial complex, i.e. the delivery of cargo and the docking of the mitochondria. Since several kinesins, several adaptors and several mitochondrial or mitochondrial-associated proteins form multi-protein complexes that are implicated in transport, there is high potential for regulation by intracellular signalling pathways to control this multiplicity of protein–protein interactions.

Several signalling pathways have been implicated in the control of mitochondrial movement. These include the phosphoinositide kinase-3 (PI3K) pathway in the NGF mediated-docking of axonal mitochondria and the Zn^{2+} immobilization of mitochondria (Chada and Hollenbeck 2003, 2004; Malaiyandi et al. 2005) and Abl tyrosine kinase signalling pathways in *Drosophila* (Martin et al. 2005). Furthermore, different signalling pathways have been implicated in the formation of kinesin/adaptor complexes. Examples include phosphorylation of the kinesin, KIF17, by calcium-calmodulin-dependent protein kinase II (CaMKII) identified as the molecular switch that controlled the interaction between the KIF17 molecular motor and the scaffold protein, Mint1 (Guillaud et al. 2008) and an activated JNK pathway that functions as a kinesin-cargo dissociation factor (Horiuchi et al. 2007).

Interestingly, axonal organelle transport can be blocked by inhibition of the GTPase cycle of monomeric G proteins (Bloom et al. 1993). Miro1, the mitochondrial acceptor for the TRAKs, is a GTPase. It was reported recently that co-expression of a constitutively active Miro1 mutant with TRAK2 in a mammalian cell line resulted in impairing the recruitment of TRAK2 to mitochondria (MacAskill et al. 2009a). Live cell imaging of mitochondrial movement in hippocampal neurones showed that transfection with the constitutively active Miro1 mutant yielded aggregation of mitochondria in the cell soma when compared to transfections with wild-type Miro1 (MacAskill et al. 2009a). This suggests that the TRAK-mediated mitochondrial transport is dependent on the GTPase activity of Miro1 leading to the question, what controls the GTPase activity of Miro1? A further potential regulatory mechanism for the formation or the dissociation of the kinesin/TRAK/Miro mitochondrial trafficking complex is the *O*-glycosylation post-translational modification of each constituent by the enzyme OGT. OGT activity has been linked to glucose availability (Hart et al. 2007). Since it is a known integral component of the TRAK mitochondrial transport complex and that TRAKs are O-glycoyslated in vivo, it seems reasonable to propose that they may be important regulators of mitochondrial mobility.

6.7 Concluding Remarks

The early studies of mitochondria in neurones were empirical observing their morphology, their fusion and fission, their movements and describing paradigms in which these traits were altered. In this chapter, we have reviewed recent advances in the elucidation of mitochondrial trafficking processes in neurones, whereby a number of the proteins that mediate mitochondrial motility and docking have now been identified. Furthermore, progress has been made in understanding the mechanisms that control protein–protein interactions in the formation and consequent activation of multi-protein trafficking complexes and the anchoring of mitochondria. From the available information, however, it is evident that there are different modes of transport using different molecular motors and protein players. It is not clear how or indeed if the alternative mechanisms interact. The functional significances of the different splice forms (TRAK2) or related gene variants, i.e. TRAK1 and TRAK2, Miro1 and Miro2, within the well-characterized TRAK/Miro multi-protein complex is not known. Possible differences in the transport mechanisms between different types of neurones and also between axons and dendrites within one cell type have yet to be explored in depth. Furthermore, much remains to be learnt regarding the signalling mechanisms that intitiate anterograde and retrograde movement of mitochondria. The challenge will then be to combine all the information to yield an integrated view of mitochondrial movement in cells. It is known that mitochondrial distribution (i.e. trafficking) and function are compromised in protein aggregation neurodegenerative diseases such as amyotrophic lateral sclerosis (ALS) and Alzheimer's disease (reviewed in Mattson et al. 2008). Impaired mitochondrial transport would be predicted to impact detrimentally upon ATP-dependent processes on which neurones are especially dependent. These would include the maintenance or necessitated enhanced activity of ATP-dependent chaperone proteins to combat protein aggregation, the maintenance of ATP-dependent molecular motors for appropriate axonal transport and, crucially, the preservation of energy homeostasis at the synapse. Thus, unraveling the mechanisms of mitochondrial transport will surely impact on understanding the pathogenesis of neurodegenerative diseases in which defective trafficking is implicated. Furthermore, it may provide therapeutic opportunities such as gene delivery or enhanced/reduced targeted intracellular signalling regulation to correct mitochondrial trafficking deficits.

Acknowledgements Work in the author's laboratory is funded by the BBSRC (UK). We would like to thank Dr Anna Kenny for Fig. 6.1 and Dr Sarah Cousins for assistance with Figs. 6.2 and 6.3.

References

Beck M, Brickley K, Wilkinson H, Sharma S, Smith M, Chazot PL, Pollard S, Stephenson FA (2002) Identification, molecular cloning and characterization of a novel $GABA_A$ receptor associated protein, GRIF-1 J. Biol. Chem. 277: 30079–30090.

Bloom GS, Richards BW, Leopold PL, Ritchey DM, Brady ST (1993) GTP gamma S inhibits organelle transport along axonal microtubules. J. Cell Biol. 120: 467–476.

Boldogh IR, Pon LA (2007) Mitochondria on the move. Trends Cell Biol. 17: 502–510.

Brickley K, Smith MJ, Beck M, Stephenson FA (2005) GRIF-1 and OIP106, members of a novel gene family of coiled-coil domain proteins: Association in vivo and in vitro with kinesin. J. Biol. Chem. 280: 14723–14732.

Brickley K, Pozo K, Stephenson FA (2010) *N*-Acetylglucosamine transferase is an integral component of a kinesin-directed mitochondrial trafficking complex. J. Biol. Chem. submitted.

Cai Q, Gerwin C, Sheng Z-H (2005) Syntabulin-mediated anterograde transport of mitochondria along neuronal processes. J. Cell Biol. 170: 959–969.

Cai Q, Pan, PY, Sheng Z-H (2007) Syntabulin-kinesin-1 family member 5B-mediated axonal transport contributes to activity-dependent presynaptic assembly. J. Neurosci. 27: 7284–7296.

Cerveny KL, Tamura Y, Zhang Z, Jensen RE, Sesaki H (2007) Regulation of mitochondrial fusion and division. Trends Cell Biol. 17: 563–569.

Chada SR, Hollenbeck PJ (2003) Mitochondrial movement and positioning in axons: the role of growth factor signalling. J. Exp. Biol. 206: 1985–1992.

Chada SR, Hollenbeck PJ (2004) Nerve growth factor signalling regulates motility and docking of axonal mitochondria. Curr. Biol. 14: 1272–1276.

Chang DTW, Reynolds IJ (2006) Differences in mitochondrial movement and morphology in young and mature primary cortical neurons in culture. Neuroscience 141: 727–736.

Chang DTW, Honick A, Reynolds IJ (2006) Mitochondrial trafficking to synapses in cultured primary cortical neurons. J. Neurosci. 26: 7035–7045.

Chen H, Chan DC (2006) Critical dependence of neurons on mitochondrial dynamics. Curr. Opin. Cell Biol. 18: 453–459.

Cho K-I, Cai Y, Yi H, Yeh A, Aslanukov A, Ferreira PA (2007) Association of the kinesin-binding domain of RanBP2 to KIF5B and KIF5C determines mitochondria localization and function. Traffic 8: 1722–1735.

Fransson A, Ruusala A, Aspenstrom P (2003) Atypical Rho GTPases have roles in mitochondrial homeostasis and apoptosis. J. Biol. Chem. 278: 6495–6502.

Fransson A, Ruusala A, Aspenstrom P (2006) The atypical rho GTPases Miro-1 and Miro-2 have essential roles in mitochondrial trafficking. Biochem. Biophys. Res. Commun. 344: 500–510.

Gauthier LR, Charrin BC, Borrell-Pages M, Dompierre JP, Rangone H, Cordellieres FP, De Mey J, MacDonald ME, Lebmann V, Humbert S, Saudou F (2004) Huntingtin controls neurotrophic support and survival of neurons by enhancing BDNF vesicular transport along microtubules. Cell 118: 127–138.

Gilbert SL, Zhang L, Forster ML, Anderson JR, Iwase T, Soliven B, Donahue LR, Sweet HO, Bronson RT, Davisson MT, Wollman RL, Lahn BT (2006) *Trak1* mutation disrupts $GABA_A$ receptor homeostasis in hypertonic mice. Nat. Genet. 38: 245–250.

Giot L et al. (2003) A protein interaction map of *Drosophila melanogaster*. Science 302: 1727–1736.

Glater EE, Megeath LJ, Stowers RS, Schwarz TL (2006) Axonal transport of mitochondria requires Milton to recruit kinesin heavy chain and is light chain independent. J. Cell Biol. 173: 545–557.

Gorska-Andrzejak J, Stowers RS, Borycz J, Kostyleva R, Schwarz TL, Meinertzhagen IA (2003) Mitochondria are redistributed in photoreceptors lacking Milton, a kinesin-associated protein. J. Comp. Neurol. 463: 372–388.

Grishin A, Li H, Levitan ES, Zaks-Makhina E (2006) Identification of γ-aminobutyric acid receptor-interacting factor 1 (TRAK2) as a trafficking factor for the K^+ channel Kir2.1. J. Biol. Chem. 281: 30104–30111.

Gross SP, Vershinin M, Shubeita GT (2007) Cargo transport: two motors are sometimes better than one. Curr. Biol. 17: 545–557.

Guillaud L, Wong R, Hirokawa N (2008) Disruption of KIF17–Mint1 interaction by CaMKIIdependent phosphorylation: a molecular model of kinesin–cargo release Nat Cell Biol. 10: 19–29.

Guo X, Macleod GT, Wellington A, Hu F, Panchumarthi S, Schoenfield M, Marin L, Charlton MP, Atwood HL, Zinsmaier KE (2005) The GTPase dMiro is required for axonal transport of mitochondria to *Drosophila* synapses. Neuron 47: 379–393.

Hart GW, Housley MP, Slawson C (2007) Cycling of O-linked beta-N- acetylglucosamine on nucleocytoplasmic proteins Nature 446: 1017–1022.

Hirokawa N, Takemura R (2005) Molecular motors and mechansims of directional transport in neurons. Nat. Rev. Neurosci. 6: 201–214.

Hollenbeck PJ, Saxton WM (2005) The axonal transport of mitochondria. J. Cell Sci. 118: 5411–5419.

Horiuchi D, Barkus RV, Pilling AD, Gassman A, Saxton WM (2005) APLIP1, a kinesin binding JIP-1/JNK scaffold protein, influences the axonal transport of both vesicles and mitochondria in *Drosophila*. Curr. Biol. 15: 2137–2141.

Horiuchi D, Collins CA, Bhat P, Barkus RV, DiAntonio A, sexton WM (2007) Control of a kinesin-cargo linkage mechanism by JNK pathway kinases. Curr. Biol. 17: 1313–1317.

Iyer SPN, Akimoto Y, Hart GW (2003) Identification and cloning of a novel family of coiled-coil domain proteins that interact with *O*-GlcNAc transferase. J. Biol. Chem. 278: 5399–5409.

Kang J-S, Tian J-H, Pan, P-Y, Zald P, Li C, Deng C, Sheng Z-H (2008) Docking of axonal mitochondria by syntaphilin controls their mobility and affects short-term facilitation. Cell 132: 137–148.

Kirk E, Chin L-S, Li L (2006) GRIF1 binds Hrs and is a new regulator of endosomal trafficking. J. Cell Sci. 119: 4689–4701.

Li Z, Okamoto K-I, Hayashi Y, Sheng M (2004) The importance of dendritic mitochondria in the morphogenesis and plasticity of spines and synapses. Cell 119: 873–887.

Ligon LA, Steward O (2000) Movement of mitochondria in the axons and dendrites of cultured hippocampal neurons. J. Comp. Neurol. 427: 340–350.

MacAskill AF, Brickley K, Stephenson FA, Kittler, JT (2009a) GTPase dependent recruitment of GRIF-1 by miro1 regulates mitochondrial trafficking in hippocampal neurons. Mol. Cell. Neurosci. 40: 301–312.

MacAskill AF, Rinholm JE, Twelvetrees AE, Arancibia-Carcamo L, Muir J, Fransson A, Aspenstrom P, Attwell D, Kittler JT (2009b) Miro1 is a calcium sensor fro glutamate receptor-dependent localization of mitochondria at synapses. Neuron 61: 541–555.

McGuire JR, Rong J, Li S-H, Li X-J (2006) Interaction of Huntingtin-associated protein-1 with kinesin light chain. J. Biol. Chem. 281: 3552–3559.

Malaiyandi L, Honick A, Rintoul G, Wang Q, Reynolds I (2005) Zn^{2+} inhibits mitochondrial movement in neurons by PI-3 kinase activation. J. Neurosci. 25: 9507–9514.

Martin MA, Ahern-Djamali SM, Hoffman FM, Saxton WM (2005) Abl tyrosine kinase and its substrate Ena/VASP have functional interactions with kinesin-1. Mol. Biol. Cell 16: 4225–4230.

Mattson MP, Gleichmann M, Cheng A (2008) Mitochondria in neuroplasticity and neurological disorders. Neuron 60: 748–766.

Miki H, Setou M, Kaneshiro K, Hirokawa N (2001) All kinesin superfamily protein, KIF, genes in mouse and human. Proc. Natl. Acad. Sci. U.S.A. 98: 7004–7011.

Miller KE, Sheetz MP (2004) Axonal mitochondrial transport and potential are correlated. J. Cell Sci. 117: 2791–2804.

Misgeld T, Kerschensteiner M, Bareyre FM, Burgess RW, Lichtman JW (2007) Imaging axonal mitochondria in vivo. Nat. Methods 4: 559–561.

Overly CC, Rieff HI, Hollenbeck PJ (1996) Organelle motility and metabolism in axons vs dendrites of cultured hippocampal neurons. J. Cell Sci. 109: 971–980.

Pilling AD, Horiuchi D, Lively CM, Saxton WM (2006) Kinesin-1 and dynein are the primary motors for fast transport of mitochondria in *Drosophila* motor axons. Mol. Biol. Cell 17: 2057–2068.

Rintoul GL, Filiano AJ, Brocard JB, Kress GJ, Reynolds IJ (2003) Glutamate decreases mitochondrial size and movement in primary forebrain neurons. J. Neurosci. 23: 7881–7888.

Rong J, McGuire JR, Fang Z-H, Sheng G, Shin J-Y, Li S-H, Li X-J (2006) Regulation of intracellular trafficking of Huntingtin-associated protein-1 is critical for TrkA protein levels and neurite outgrowth. J. Neurosci. 26: 6019–6030.

Russo GJ, Louie K, Wellington A, Macleod GT, Hu F, Panchumarti S, Zinsmaier KE (2009) Drosophila Miro is required for both anterograde and retrograde axonal mitochondrial transport. J. Neurosci. 29: 5443–5455.

Saotome M, Safiulina D, Szabadkai G, Das S, Fransson A, Aspenstrom P, Rizzuto R, hajnoczky G (2008) Bidirectional Ca^{2+}-dependent control of mitochondrial dynamics by the Miro GTPase. Proc. Natl. Acad. Sci. U.S.A. 105: 20728–20733.

Smith MJ, Pozo K, Brickley K, Stephenson FA. (2006). Mapping the GRIF-1 binding domain of the kinesin, KIF5C, substantiates a role for GRIF-1 as an adaptor protein in the anterograde trafficking of cargoes. J. Biol. Chem. 281: 27216–27228.

Stowers RS, Megeath LJ, Gorska-Andrzejak J, Meinertzhagen IA, Schwarz TL (2002) Axonal transport of mitochondria to synapses depends on Milton, a novel *Drosophila* protein. Neuron 36: 1063–1077.

Su Q, Cai Q, Gerwin C, Smith CL, Sheng Z-H (2004) Syntabulin is a microtubule-associated protein implicated in syntaxin transport in neurons. Nat. Cell Biol. 6: 941–953.

Yi M, Weaver D, Hajnoczky G (2004) Control of mitochondrial motility and distribution by the calcium signal: a homeostatic circuit. J. Cell. Biol. 167: 661–672.

Wang X, Schwarz TL (2009) The mechanism of Ca^{2+}-dependent regulation of kinesin-mediated mitochondrial motility. Cell 136: 163–174.

Webber E, Li L, Chin L-S (2008) Hypertonia-associated protein Trak1 is a novel regulator of endosome-to-lysosome trafficking. J. Mol. Biol. 382: 638–651.

Part III
Chaperone Modalities and Homeostatic Mechanisms in the Synaptic Compartment

Chapter 7
Molecular Chaperones in the Mammalian Brain: Regional Distribution, Cellular Compartmentalization and Synaptic Interactions

Andreas Wyttenbach, Shmma Quraishe, Joanne Bailey, and Vincent O'Connor

Abstract The expression of molecular chaperones in the human central nervous system (CNS) is not well explored although they regulate key aspects of neuronal functions and likely play crucial roles during chronic neurodegeneration associated with protein misfolding. Current data suggest a rather complex expression profile with a distinct distribution to brain regions and cell types that are further modulated under stress. Synapses represent a particular folding environment due to the many and highly regulated protein interactions that occur in relative isolation. Despite the synaptic localization of several chaperones and proteins with identifiable chaperone modality, we argue that additional synaptic chaperones and chaperone pathways exist that may regulate the synaptic protein homeostasis during protein folding and re-folding. Given the early synapse dysfunction and alterations of chaperone function that occur during CNS diseases associated with protein misfolding, we suggest that further efforts should be made to better define and understand the synaptic "chaperome".

7.1 Introduction

Molecular chaperones serve essential "house-keeping" functions in all cellular environments and help proteins attain their precise three-dimensional structures to fulfil their biological functions (Hartl and Hayer-Hartl 2009). They function by providing the substrates they bind to with a pathway that allows (re)folding of protein conformation. This is achieved by either "passive" chaperone binding to for example ameliorate aggregation (like the small heat shock proteins (HSPs) or by a more active process in which ATP hydrolysis, as used by some classes of chaperones, is employed to promote (re)folding (e.g. HSP70 family). These distinct chaperone modalities are sometimes coupled together and the efficacy of chaperone pathways often relies on

A. Wyttenbach (✉) and V. O'Connor (✉)
Southampton Neuroscience Group, School of Biological Sciences, Life Science Building (B85), University of Southampton, Southampton SO17 1BJ, UK
e-mail: aw3@soton.ac.uk; voconno@soton.ac.uk

A. Wyttenbach and V. O'Connor (eds.), *Folding for the Synapse*,
DOI 10.1007/978-1-4419-7061-9_7,

this (see Chap. 2). In the central nervous system (CNS), with its high-energy demand that underpins a sophisticated information-processing network, molecular chaperones need to integrate their activities in a highly complex environment. There are three levels of complexity: brain-to-brain regions, intercellular communication (cell to cell) and intracellular signalling (molecule to molecule). However, the major CNS cells; neurons, astrocytes and oligodendrocytes are extremely compartmentalized. Intracellular compartments themselves, perhaps similar to organelles (e.g. the nucleus or mitochondria), could therefore be considered as a "particular folding environment" with differences in macromolecular crowding impacting on the actions of molecular chaperones. The synapse may be such a sub-compartment where the demands for chaperones to regulate protein turnover, protein interactions and vesicle cycling are particularly challenging.

To date data exist to describe the organ-specific distribution and function of many chaperones, but less is known about their cell-type specificity and how this changes in response to different stressors. A systems view on the entire "chaperome" would be essential to understand chaperone function under basal and stress conditions. This lack of knowledge is particular for the CNS whether considering normal function or during conditions of chronic neurodegeneration. In a disease context, dysfunction associated with protein misfolding is particularly likely to engage or result from disruption of molecular chaperones (e.g. HSPs) as they play crucial roles in the regulation of protein homeostasis (Powers et al. 2009).

The aim here is to outline typical examples and summarize the current knowledge on the brain-regional and cell-type-specific distribution of molecular chaperones in the mammalian CNS. We will also describe situations that allow us to make the case that chaperones have a role to play in specific cellular compartments under physiological conditions and during protein misfolding stress. Finally, to add weight to the concept of a "synaptic chaperome" we will present mechanistic insight into fundamental processes of synaptic transmission to illustrate that chaperone modality lies at the heart of synaptic function.

7.2 The Presence and Distribution of Molecular Chaperones in the Mammalian Brain

7.2.1 From Cell Compartments to Brain Regions

Molecular chaperones such as the HSPs are primordial proteins and are involved in the regulation of homeostatic mechanisms and stress in order to adjust protein biogenesis to changing physiological demands in all cells. Therefore, many HSPs are continuously present in and/or translocate to several intracellular organelles (e.g. mitochondria, lysosomes or endoplasmic reticulum) and compartments such as the cytosol, the nucleus, the axon/synapses of neurons under basal- or stress conditions (e.g. Bender et al. 2010; Kirkegaard et al. 2010; Buck et al. 2010; Kabani and Martineau 2008; Terada et al. 2010; Chen and Brown 2007a, b). While some (co)-chaperones show a highly compartment-specific expression and function (Stevens and Argon 1999), others are constitutively expressed in most cells. For example, the constitutively expressed Hsp60 and its co-chaperone Hsp10, eukaryotic homologues of the bacterial proteins GroEL and GroES, are mainly localized to mitochondria under physiological conditions (Martin 1997). In the mitochondrial matrix Hsp60/10 are dominant stress proteins that regulate turnover of misfolded proteins, but other chaperones also participate in the complex chaperone network that is essential for mitochondrial protein biogenesis (reviewed in Voos and Rottgers 2002). Hsp60 is constitutively expressed in neurons, astrocytes and oligodendrocytes (e.g. Schwarz et al. 2010). On the other hand, the constitutively expressed form of Hsp70, Hsp73 (or also named Hsc70 and designated as such from herein) is expressed in all neural cells and functions in routine protein folding and transport in many compartments including the synapse where Hsc70 participates in selective functions by forming the specific trimeric chaperone complex at the synaptic vesicle surface (as detailed in the Chap. 8).

There are also major differences in the CNS distribution of individual chaperones within HSP families such as the small HSPs (sHSPs). For example, HspB5 (alphaBcrystallin) is expressed in the brain, particularly in glial cells (Iwaki et al. 1992; Klemenz et al. 1993). We have found a distinct white-matter-specific expression of HspB5 in the mouse CNS, but neither HspB1 (Hsp25) nor HspB8 (Hsp22) showed such a distribution (Quraishe et al. 2008) (also see below, Sect. 7.2.3). This suggested cell-specific transcriptional regulation of HSPs under physiological conditions in the CNS. Limited evidence also exist that HSPs are differentially expressed in particular brain regions, notably during development (e.g. e-Souza and Brown 1998, also see below). Intrinsically, the various groups of classic HSPs are soluble proteins and the mechanism that recruit them to specific membranes and discrete reactions are not well resolved, although specific protein interactions and the use of targeting of co-chaperones are becoming increasingly established.

Finally, of particular importance are the more recent observations that HSP chaperones not only act inside cells, but also serve extracellular roles that could have important consequences for mediating extracellular signals between cells and brain regions, especially during inflammatory processes (Noort 2008). Both release of Hsp70 and Hsp110 from neuronal cells in vitro has been known for some time (Hightower and Guidon 1989) and it has been suggested that HSPs released by glial cells may protect axons (Tytell 2005; Houenou et al. 1996). It is now clearly established that Hsp60, Hsp70 and HspB5 are present in the cerebrospinal fluid (CSF) and sometimes serum of humans during exercise (Lancaster and Febbraio 2005) and patients suffering from neurodegeneration (e.g. Ousman et al. 2007). HSP release occurs in the absence of cell death by an active, non-classical secretory pathway via exosomes or by an endo-lysosomal-dependent pathway that have mainly been studied for cancer cells and cells of the immune system (Hightower and Guidon 1989; Clayton et al. 2005; Mambula and Calderwood 2006; Vega et al. 2008).

7.2.2 *Expression of Molecular Chaperones Under Physiological Conditions and Their Induction During Protein Misfolding Stress*

7.2.2.1 Small Heat Shock Proteins

A systematic investigation of expression and distribution of the sHSP family in the rodent brain under physiological conditions found a differential anatomical and biochemical compartmentalization (Quraishe et al. 2008). Out of the 10 sHSPs present in the mammalian genome, expression of 5 sHSP mRNAs (HspB1, B5, B6, B7 and B8) was clarified, for which 4 sHSPs showed protein expression in the mouse CNS. The expression of sHSPs in the human brain has not been systematically studied, but limited information on mRNA and protein presence of HspB1, HspB5 and HspB6 is available in the Allen Brain Atlas (Lein et al. 2007; http://brain-map.org). Immunohistochemical analysis shows that while moderate neuronal expression of HspB1 is found in various regions, expression of HspB5 is mainly non-neuronal. These findings are consistent with our study in the rodent CNS showing a white-matter-specific expression of HspB5 (Quraishe et al. 2008). A comparative study of HspB1 neuronal expression in different brain regions in the rat showed that its constitutive expression in the entorhinal cortex, hippocampus and within dopaminergic neurons is basically undetectable and not different, but much higher in motorneurons (Chen and Brown 2007a, b). As outlined in Chap. 4 the family of sHSPs is involved in a number of diseases of the nervous system and hence information on their physiological expression in various human brain regions would be a critical starting point for a further understanding of their roles during neurodegeneration.

Some human data are available, mainly from comparative studies on naive tissue and tissue examined from sufferers of neurodegenerative conditions due to acute or chronic stress. While HspB1 is constitutively expressed in neuronal- and non-neuronal cells, in Alzheimer's disease (AD) its selective expression includes an up-regulation in glial cells (Perng and Quinlan 2004). Accumulation of HspB1 has been reported around extracellular Abeta plaques (Shinohara et al. 1993). Such observations are similar to reports detailing increased expression of HspB1 in astrocytes and microglia in the vicinity of the protease-resistant PrPsc plaques associated with prion disease (Renkawek et al. 1992). Furthermore, there is a selective expression in tissues of patients suffering from Parkinson's disease (PD) (Renkawek et al. 1999) and dementia with Lewy bodies (DLB) (Outeiro et al. 2006). HspB5 appears to show a similar distribution in diseased states compared with HspB1. HspB5 is also found up-regulated in glial cells (astrocytes and microglia) in AD, prion disease and PD in a fashion similar to HspB1 (Renkawek et al. 1992; 1999; Shinohara et al. 1993; Wilhelmus et al. 2006a, b). But additionally, HspB5 expression in the mouse and human brain is localized to oligodendrocytes (Quraishe et al. 2008; Shinohara et al. 1993). These findings do not rule out expression and up-regulation of HspB5 in neurons during misfolding diseases, but indicate that an up-regulation generally occurs in glial cells and is likely associated with an "activation" of microglia and astrocytes (gliosis).

Despite some studies on HspB1 in cell- and mouse models of Huntington's disease and spinocerebellar ataxia 3 (HD, SCA-3) (Wyttenbach et al. 2002; Hay et al. 2004; Chang et al. 2005) changes in HspB1 expression (and indeed other sHSPs) and the role of sHSPs in the human brain in polyglutamine (polyQ) disorders are unknown. The distribution and expression of HspB6 and HspB8 is even less well studied. While similar findings have been made in AD (up-regulation of HspB6/B8 in glial cells and their association with Abeta plaques) (Wilhelmus et al. 2006a, b), no studies were performed in prion disease, PD and the polyQ disorders. Under conditions of acute brain damage (such as traumatic brain injury and stroke) it seems that an immediate up-regulation of sHSPs occurs in neuronal cells (for neuroprotection) followed by a more sustained expression of several sHSPs in glia (reviewed in Wyttenbach et al. 2010), perhaps pointing to mechanisms of CNS repair by glial cells via sHSPs. The crucial and diverse roles of sHSPs such as redox regulation, interactions with the cytoskeleton and direct involvement in the regulation of cell death- and survival signalling pathways will make future studies on their expression- and intracellular distribution pattern a worthwhile effort. Furthermore, given that sHSPs appear to regulate inflammatory processes, most likely via both their intra- and extracellular presence (see above), they are well placed as therapeutic drug targets during chronic neurodegeneration (Sajjad et al. 2010).

7.2.2.2 Other Heat Shock Proteins

Not surprisingly, studies on the physiological expression of the Hsp60, Hsp70, Hsc70 and Hsp90 family of molecular chaperones in the CNS goes back some time. In the adult mammalian brain Hsp60, Hsp70, Hsc70 and Hsp90 are neuronally localized (Quraishi and Brown 1995; Foster and Brown 1996; Manzerra and Brown 1996). In the rat brain Hsc70 expression is high and maintained during post-natal development (D'Souza and Brown 1998). However, Hsp70 expression is low in the unstressed brain and shows an increase during brain development. While both Hsp60 and Hsp90 levels in the adult brain are high, no change during brain development is seen for Hsp90. However, a major increase is seen in several brain regions for Hsp60, reflecting mitochondrial content (D'Souza and Brown 1998).

Hsp40 and Hsp110 proteins (like members of the Hsp60, 70 and 90 families) are constitutively expressed in several neuronal populations of the rodent CNS (Chen and Brown 2007a). Given that neurons express many different HSPs, an interesting question is whether some are differentially expressed in neurons, a potential phenomenon that could inform the debate on why certain neuronal populations are more vulnerable to protein misfolding stress than others. A quantitative study in the rat brain showed no differences in neuronal expression of Hsp40, Hsp60, Hsp90 and Hsp110 in the entorhinal cortex, hippocampus, substantia nigra and motorneurons of the spinal cord. However, Hsc70 expression appeared increased in dopaminergic neurons and was further elevated in motorneurons (similar to HspB1, see above) (Chen and Brown 2007a).

Given the marked differences in baseline expression of various HSPs, a related question is which HSPs are induced in what type of brain cells under conditions of stress. Clearly, under heat stress, the classic HSP-inducing treatment, some HSPs in the CNS are induced, but not others. For example, after 24 h recovery from a whole body heat shock, HspB1 and Hsp70 expression are increased in the rat cerebral cortex, hippocampus, midbrain and spinal cord, but not Hsp40, Hsp90 and Hsp110 (Chen and Brown 2007a). Hyperthermic induction of these Hsps appears to occur in glial cells and not in neurons (Brown 2007). Under conditions of neuronal protein misfolding stress HSPs are expected to be induced in neurons to cope with the elevated levels of protein accumulation and defend against neuronal death. As outlined above, however, chronic sHSP induction occurs mainly in glial cells and not within neurons, similar to what is observed after hyperthermic stress. But, despite several reports to date, changes in CNS expression and distribution of members of the Hsp40/70, Hsp90 and Hsp110 family in the various proteinopathies have not been exhaustively studied. Rather than determining expression level changes within cell types, most studies performed biochemical analysis of the whole brain or CNS regions. For example, in a mouse model of the polyQ disease HD, Hdj1, Hdj2, Hsp70 and both alpha- and betaSGT expression was shown to be progressively lost in the CNS, but not HspB1, Hsc70, Grp78, Hsp84 and Hsp90 (Hay et al. 2004). Whether these findings translate to human polyQ disorders is unknown. In contrast, in the AD brain HSPs appear to be elevated in the affected brain regions (for a recent review see Koren et al. 2009), but to various degrees in different brain regions (e.g. Yoo et al. 2001; Di Domenico et al. 2010). Similarly, in prion disease HSPs appear to be induced, mainly in glial cells (Wyttenbach et al. 2010). Other than an induction of HSPs in glial cells in several of these CNS disorders, it is too early to draw firm conclusions about how the HSP response is regulated during chronic neurodegeneration. This is in contrast to genetic and drug studies convincingly showing that the modulation of HSP induction by drugs and over/under expression changes the misfolding and aggregation behaviours of polyQ- and other molecules such as tau, Abeta and alpha-synuclein with significant effects on their toxicity in vitro and in vivo models of disease (reviewed in Wyttenbach 2004; Muchowski and Wacker 2005; Wyttenbach and Arrigo 2006; Sajjad et al. 2010; Witt 2010). To translate such mechanistic studies to human pathology, investigation into the distribution of HSPs at the cellular and sub-cellular level in rodent models and indeed in the human brain are urgently needed. Given that the synaptic compartment is prone to early dysfunction in many proteinopathies (see above and Chaps. 11–15) it will be interesting to explore the level of expression and functional role of molecular chaperones in the axonal and synaptic compartments in the various CNS regions.

7.2.3 Synaptic Distribution of HSPs

The significance of synaptic chaperones is well illustrated by the trimeric chaperone complex, consisting of cysteine-string protein (CSP), Hsc70 and SGT (small glutamine-rich tetratricopeptide containing protein), which performs several house-keeping synaptic functions (discussed in the Chap. 8). However, a relevant question is how important chaperone modality is to the synapse and if there are indications of whether it is particularly important under basal/physiological conditions or under conditions of synaptic dysfunction.

Bechthold et al. suggested that apart from Hsc70 functioning in the pre- and post-synaptic elements, the constitutively expressed Hsp60 and Hsp90 are also associated with synapses in the unstressed mammalian brain (Bechthold et al. 2000). Further, hyperthermic-induced Hsp70 was also shown to localize to synapses. In a different study HspB1 and Hsp32 were examined after hyperthermic stress and found to localize to synaptic sites in the rat cerebellum (Bechhold and Brown 2000). Given that both proteins were induced in Bergmann glial cells and their radial fibres and also detected in synaptic structures, the authors proposed a stress-induced transfer from perisynaptic glial processes to synapses. Apart from sHSPs (Quraishe et al. 2008), we also found Hsp40, Hsc70 and Hsp90 to fractionate to synaptosomes, consistent with these earlier studies (Fig. 7.1). Additionally we detected Hsp70 in both the Triton X-100 insoluble and soluble fractions and clear differences in solubility of HspB1, HspB6 and Hsp90 (see Fig. 7.1). These results show a modest enrichment in synaptic fractions, but this and the association with the detergent-insoluble fractions that contain sub-synaptic specializations indicates that sub-pools of the total "chaperome" could be performing a selective activity. In keeping with this proposition, a recent study showed that in the rat cortex hyperthermic stress induced a relocation of Hsc70 and increased association of Hsc70 with Hsp40 in synapse-enriched areas (Chen and Brown 2007b). These events were transient, lasting up to 48 h and hence they likely occurred to preserve synapse function, for example to assist in the re-folding of synaptic proteins. An earlier study in medullary slice preparations showed that thermal pre-conditioning and exogenous Hsp72 preserve synaptic transmission during thermal stress, perhaps directly implicating Hsps during stress-induced release of neurotransmitters (Kelty et al. 2002). Furthermore, 2-deoxy-d-glucose (2DG) treatment in rats has been proposed to protect synaptic terminals via a pre-conditioning mechanism correlating with increased levels of synaptic Hsp70 and GRP78 (Guo and Mattson 2000). Together, these findings establish that HSPs other than the ones of the trimeric synaptic complex are associated with synapses and play a functional role under basal and stress conditions.

7.3 The Biochemical Organization of Synaptic Chaperones

The above indicates abundant, differential expression of classic chaperones, but provides limited evidence for discrete synaptic chaperones. However, a key feature of these classic chaperones, in particular those associated with an intrinsic ATPase activity, is the pleiotropic use of their re-folding activity that may not lend itself to readily defining local function through classic localization studies like those detailed in Fig. 7.1. In the following sections we introduce how biochemical investigations in vitro provide a resounding case for the role of chaperone modality as pivotal to

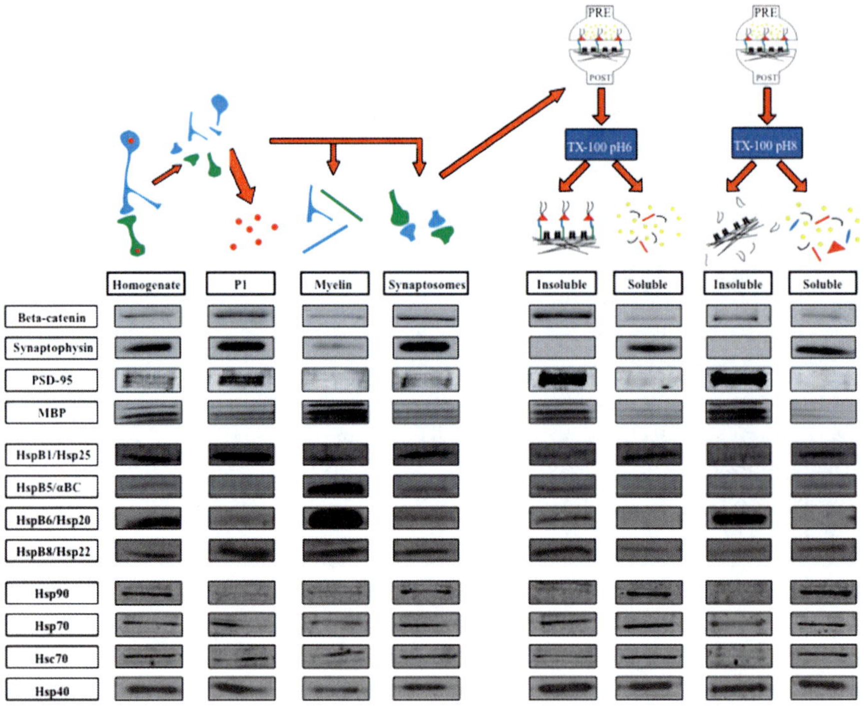

Fig. 7.1 Brain and synaptosomal compartmentalization of molecular chaperones. Brain fractionation and synaptosomal sub-fractionation based of differential Triton X-100 solubility was performed as described previously (Quraishe et al. 2008). The major components of the different fractions are depicted in cartoon form above the immunoreactivities. Well-defined biochemical markers of indicated fractions were used to authenticate the fractionation. Beta-catenin (control protein representing synaptic adhesion junction proteins) showed expression in all fractions, with some enrichment in the P1 and synaptosome-insoluble fractions. The synaptic vesicle membrane protein synaptophysin showed enrichment in the synaptosome fraction and localization to the soluble fraction consistent with its enrichment in the pre-synaptic compartment. PSD-95 (a post-synaptic protein) showed enrichment in the Triton X-100-insoluble fractions consistent with the enrichment of the post-synaptic density. MBP shows enrichment of the myelin fraction and highlights that some chaperones are localized in white matter. A range of classic chaperones were screened for their relative expression in each fraction. There was no strong enrichment in these proteins across brain fraction, but the data supports a role for sub-pools present in synaptic fractions. The distribution of the sHSPs has been detailed elsewhere (Quraishe et al. 2008). In the case of the larger ATP-dependent HSPs, Hsp90 was preferentially associated with both pH 6 and pH 8 soluble fractions supporting its presence in the synaptic compartment and a fractionation pattern characteristic of one enriched in the pre-synaptic compartment. Hsp70, Hsc70 and Hsp40 displayed similar expression patterns and were detected in both the Triton-insoluble and -soluble fractions. Antibodies other than anti-Hsp90 (SPA-836; 1:1,000); anti-Hsp70 (SPA-812; 1:100): anti-Hsc70 (Stressgen SPA-816; 1:2,000), and anti-Hsp40 (SPA-400; 1:1,000) were used as described in Quraishe et al. (2008)

synaptic function. The clear evidence provided by the *N*-ethylmaleimide sensitive fusion protein's (NSF) regulation of transmitter release and the pivotal role of Hsc70 in synaptic vesicle re-cycling is used here to support the contention of a wider significance for chaperones in this compartment (Whiteheart and Mateeva 2004; Eisenberg and Greene 2007).

7.3.1 The ATP-Dependent Basis of a Key Synaptic Chaperone Modality

The co-ordinated cycle of ATP binding, hydrolysis and ADP release by the classic chaperones allows them to sequentially bind and release misfolded substrate. This cycle promotes re-folding and provides the energy that allows the substrate protein to adopt a distinct conformation (Mayer and Bukau 2005). The classic chaperones are soluble and not intrinsically targeted. When they require more precise cellular localization they use co-chaperones (Craig et al. 2006). In Chap. 8 the importance of co-chaperones was outlined by discussing how CSP localizes Hsc70 to the synapse by recruiting it to synaptic vesicles. In other cases it may be that there is no explicit targeting, but discrete chaperone function appears to be still achievable.

7.3.2 A Fundamental Chaperone Folding Activity in Synaptic Transmission

The pre-synaptic component of transmission involves a vesicle-mediated process in which pre-stored neurotransmitters in synaptic vesicles are released by exocytosis following membrane fusion with the plasma membrane. At the most active synapses this may occur several times in 1 s and even where there are large numbers of preformed transmitter containing vesicles, this supply is limiting. To work against this requirement for continued exocytosis, local re-cycling of the fused synaptic vesicle ensures a rapid retrieval of the vesicle. Overall this requires endocytosis and the re-cycling of the lipid- and membrane protein components that make up the synaptic vesicle (Fig. 7.2). Thus, a highly coupled exo/endocytotic cycle of synaptic vesicle membranes is essential to prevent pools of stored transmitter becoming exhausted. Although the detailed requirements for this synaptic vesicle cycle will vary across different synapses, its execution is performed by a core set of protein–protein and protein–lipid interactions (Südhof 2004). Although not initially obvious, we now appreciate that chaperone modality is crucial to both arms of this process.

A critical juncture of this cycle is the point where synaptic vesicles undergo fusion with the plasma membrane. This is triggered by the local influx of Ca^{2+} via spatially co-localized ion channels within restricted domains of the pre-synaptic

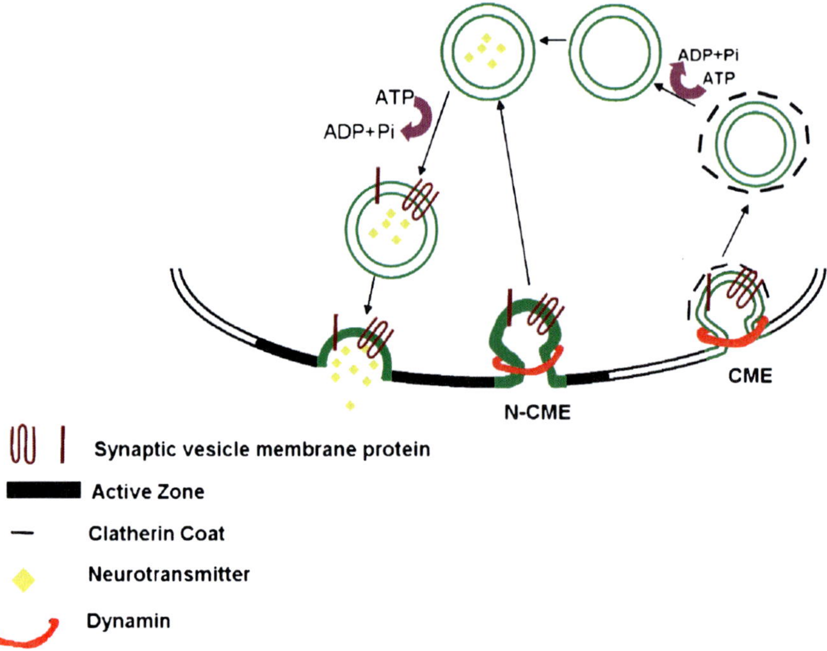

Fig. 7.2 Synaptic vesicle cycle. The key intermediates in synaptic vesicle cycle are illustrated to depict the essential steps in this exo/endocytotic coupling. This cycle ensures that the vesicle that fuses during exocytosis is efficiently retrieved by endocytotic mechanisms. Gross steps depicted include priming, calcium-dependent fusion and distinct modes of retrieval that are found in the nerve terminal. There are distinct steps that exhibit an ATP requirement, including the priming and the uncoating reactions that precede and follow fusion respectively. The context of this general overview is a full cycle that operates between 2 and 30 s depending on route use and the realization that many vesicles can fuse during periods of high neuronal activity. Thus, the cartoon *under represents* the extreme temporal and spatial modulation that underpins synaptic transmitter release

membrane called active zones (Südhof 2004). Subsequent to this the vesicle membrane is retrieved by nerve terminal endocytosis. The retrieved membrane will undergo local re-cycling in which the composition of proteins required for subsequent rounds of fusion is achieved. Additionally, during this dynamic process the vesicle will mature the conformation of re-cycling proteins to ensure that they are optimized by priming reactions for subsequent rounds of fusion. Put simply, there is an explicit requirement for certain proteins to undergo a transformation from a post-fusion to a pre-fusion conformation. This modulation of protein conformation and lipid composition involves a number of vesicle priming reactions that require ATP and would be conceptually consistent with a requirement for chaperone activity. The protein composition of the synaptic vesicle is well defined and the cohort of proteins found on this organelle have been attributed to several functions (Takamouri et al. 2006). These include vesicle/cytomatrix interactions, transmitter uptake, docking, Ca^{2+} sensing and fusion, which need to be integrated with high temporal and spatial resolution to drive efficient transmitter release (Südhof 2004).

The other side of this critical juncture is the endocytosis where membrane and associated proteins, which collapse into the plasma membrane after fusion, undergo selective retrieval. An important aspect of the re-cycling process is the resorting and reactivation of retrieved proteins into synaptic vesicles. A major route for this re-cycling is clathrin-mediated endocytosis. Here, the cytosolic protein clathrin, in the form of soluble triskelions, is recruited to form a coat around the retrieved membrane. The biogenesis of these coated vesicles involves coat recruitment, membrane invagination and the pinching off that leads to membrane vesicles surrounded by a clathrin coat (Eisenberg and Greene 2007). After successful internalization the clathrin coat is lost and generates a naked vesicle that is re-primed and capable of providing new transmitter for Ca^{2+}-triggered exocytosis. Clathrin-mediated events dominate the retrieval of synaptic vesicle membranes post-fusion, but there are other modes of membrane internalization including retrieval without clathrin (Dittman and Ryan 2009).

Figure 7.2 describes the key principles of this process of the synaptic vesicle cycle, but rather ignores the molecular complexity of the biochemical activities associated with it (see Fig. 12.1 in Chap. 12). The number of biochemical activities required to execute the full cycle is greater than 100 based on the most conservative estimates. Figure 7.2 makes no attempt to assimilate this complexity, but provides a framework for a more detailed discussion of the significance of NSF's and Hsc70's chaperone activity.

7.3.3 *Is NSF a Fusion Protein with Chaperone Modality?*

At the core of release processes is the fusion of the synaptic vesicle and the plasma membrane. Central to this is a pre-activation that involves the co-ordinated control of protein conformation of the synaptic vesicle and plasma membrane. A series of experiments using in vitro measures of intra Golgi membrane trafficking defined an activity known as NSF that appeared essential for membrane fusion (Rothman and Orci 1992). NSF was identified as the mammalian homologue of the yeast ATPase sec-18, which when deficient in yeast caused defective vesicle-mediated intracellular membrane trafficking. NSF is a multi-domain protein consisting of an N-terminal domain and two D domains (D1 and D2), which have intrinsic ATPase activity. This cytsolic protein's catalysis of membrane trafficking depends on additional soluble proteins called SNAPs (soluble NSF attachment protein). As their name suggests these recruit NSF to membranes (Whiteheart and Mateeva 2004). There is a broad requirement for SNAP and NSF at several membrane trafficking events in which lipids and proteins are transferred from one compartment to the next via membrane fusion (Rothman and Warren 1994).

Conventional biochemical purification using solubilised brain extracts identified the first of the so-called SNAP receptors (SNAREs) as a protein complex made up of the synaptic membrane proteins synaptobrevin, on the synaptic vesicle, and syntaxin and SNAP-25, on the plasma membrane, respectively. SNAP binding to membrane receptors was known to recruit NSF to form a high order NSF/SNAP/SNARE complex in which NSF's ATPase is activated to drive dissolution of the high order complex into its components. Initial thinking was that this NSF-dependent disassembly was the direct correlate of membrane fusion, which in the case of transmitter release involved synaptic vesicle (v) fusing with the target (t) plasma membrane (Sollner et al. 1993).

Although the dissolution of the SNARE complex with its requirement for ATP made a compelling case for a direct role in transmitter exocytosis, it was clear from several physiological measurements that ATP's requirement in transmitter release was not during fusion (O'Connor et al. 1994). Rather ATP seemed to be required in a series of reactions that preceded the Ca^{2+}-triggered event. This pre-fusion requirement of ATP during vesicle priming meant the precise role of NSF was revisited and suggestions quickly pointed to it acting as a chaperone-like molecule in which its ATPase activity modulated protein folding to indirectly catalyze fusion (Morgan and Burgoynne 1995).

This understanding has been "crystallized" by defining the structural determinants of the association that allows the SNAREs to form a complex. In the case of the synaptic proteins synaptobrevin, syntaxin and SNAP-25 provide 1, 1 and 2 helical coil–coil domains respectively that form a four-helix bundle that bind SNAP/NSF (Sutton et al. 1998). These structures are formed from individual SNAREs when the coil–coil domains align against each other in register in a parallel fashion. The formation of this complex is energetically very favourable and represents a state to which the v-SNAREs and t-SNARES gravitate. The topology of vesicular synaptobrevin in one membrane and syntaxin and SNAP-25 in the plasma membrane dictates that the favoured conformation within the four-helix bundle can take two forms. One so-called trans-conformation involves the four-helix bundle forming between the vesicle and plasma membrane and the cis-complex when the three proteins (and the four-helix bundle) are in the same membrane. The trans-configuration is achieved when the vesicle is very close to the plasma membrane, a situation that is supported by vesicle docking. The cis-configuration occurs after fusion when synaptobrevin is in the same membrane as syntaxin and SNAP-25. Both conformations exist and are thought to follow in sequence. As the individual SNARES are pulled into this association, they pull membranes together causing fusion and accumulating the SNARE complex in the plasma membrane. Thus, NSF binds to this post-fusion intermediate and triggers the dissociation of the SNARE complex (summarized in Fig. 7.3). The dissociation of the SNARE complex into individual components causes them to become relatively unstructured with a higher enthalpic potential, which is dissipated as the proteins enter into SNARE complexes. This provides a major contribution to the energy requirements that allows membrane fusion and the otherwise unfavourable mixing of two lipid bi-layers (Stein et al. 2009). Thus, in a kind of reverse of conventional chaperone action, out of these unfolded states proteins fall into a stable complex in which proteins are more folded, but used fruitfully to drive membrane fusion.

The dependence of NSF modulation of SNARE conformation makes it relatively easy to conceive of it acting in a classic chaperone function on SNARE proteins.

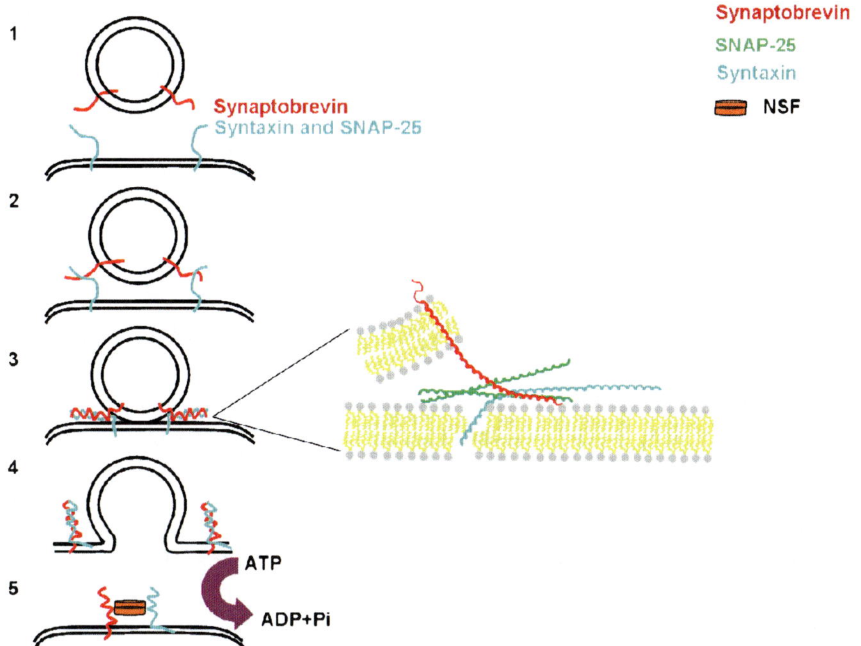

Fig. 7.3 SNARE-dependent fusion and reactivation. The synaptic vesicle SNARE, synaptobrevin and the plasma membrane SNAREs, Syntaxin and SNAP-25 are depicted. *1* Prior to fusion the unfolded SNAREs are independent of each other. *2* As the vesicle approaches the plasma membrane the SNAREs undergo a re-folding to *3* generate interacting coil–coil domains of the SNAREs that are initially in a trans-configuration. *4* These go onto to adopt a cis-configuration when the two defined lipid membranes that hold the SNARE complex in trans begin to fuse. Fusion is promoted by the stability of the "cis" complex. The inset shows an expanded view of the four coil–coil domains amalgamated from synaptobrevin (*red*) syntaxin (*blue*) and SNAP-25 (*green*). *5* After fusion the "cis" complex bind SNAP and NSF, which reactivate them via ATP hydrolysis driving the SNAREs back to their unfolded state. These unfolded states can again be used in trans to drive another round of fusion. The cartoon only crudely depicts the controlled change in SNARE protein conformation and this is modulated by additional protein interactions not depicted (see text for listing)

The fundamental role of SNAREs has been borne out by several experiments that highlight that they and the various free and complexed protein conformations are essential to transmitter release (Südhof 2004). However, it is clear that the gross switch in the protein conformation of the SNAREs mapped out by NSF dependence have further, more discrete regulation of various SNARE conformations by additional synaptic proteins. In some ways this modulation superimposes a refinement of the protein conformations that arise out of the NSF's chaperone modality. These include pre-SNARE complex binding of synaptophysin/synapotobrevin, munc-13/syntaxin and several modes of munc-18/syntaxin. Furthermore, there are post-SNARE complex binding of a distinct munc-18/syntaxin complex, complexins/SNARE complex binding and synaptotagmin/SNAP25 binding. Many of these interactions involve discrete exposure or restraint of the key SNARE-binding domains and these non-ATP-dependent interactions modulate protein folding (Wojcik and Brose 2007; Sudhof and Rothman 2009). This has some parallels with other non-ATP-dependent protein chaperone modalities (e.g. sHSPs), which act as co-factors in folding pathways and assist the more classic chaperones (e.g. Hsp70; Haslebeck 2002). Whether these additional modulators of SNARE complexes are operating with a chaperone modality is unclear, but the increased understanding of their structural basis might provide useful insight into such speculation.

7.3.4 Distinct NSF Modality at the Post-Synaptic Cell

There are several SNARE-dependent trafficking steps in which distinct combination of SNAREs drive heterotypic and homotypic membrane fusions. This involves NSF chaperoning a vast array of distinct SNARE complexes underpinned by a four-helix bundle. Nevertheless SNARE-independent use of NSF's chaperone activity has been identified in a synaptic context (Whiteheart and Matveeva 2004). This includes a modality in which NSF shows interaction with intracellular domains of synaptic glutamate receptors. These belong to the sub-class of glutamate receptors known as AMPA receptors. These are the major ligand-gated ion channels that mediate the activation of the post-synaptic neuron after pre-synaptic glutamate release. Whether this interaction is SNAP dependent or not remains unclear. However, functional studies indicate that recruiting NSF is essential in maintaining these glutamate receptors at the cell surface. This is thought to be dependent on NSF utilizing its ATPase activity to release an additional glutamate-receptor-binding partner namely PICK 1. PICK 1's recruitment to the receptors has implications for their membrane trafficking as its association with the receptor promotes their removal from the membrane. The result of this is that when NSF binding is inhibited, more PICK1 is recruited and receptors are internalized. This would lead to a decrease in glutamate signalling as post-synaptic AMPA receptors are lost from the membrane. Integration of this PICK1/NSF competition is thought to be a significant determinant of the synaptic plasticity called long-term depression. This modality reveals a potential synaptic chaperone function for NSF beyond SNARE complexes (Collingridge and Isaac 2003) and a capability that may extend to other synaptic protein complexes (Whiteheart and Mateeva 2004).

7.3.5 Mechanisms Contributing to Synapse Selectivity of NSF Chaperone Function

NSF-dependent SNARE conformations are ubiquitously used in intracellular membrane trafficking and their function is clearly not restricted to synaptic chaperoning. Nevertheless, the membrane trafficking associated with the synaptic function places an increased demand for NSF chaperone activity. This seems to be supported by the expression of NSF, which indicates its expression is much higher in brain relative to other tissue (Pueschel et al. 1994). These particularly high steady-state levels of NSF may contribute to increasing the efficiency of this chaperone in SNARE-mediated transmitter release. This view is supported by the high expression of alpha-SNAP and the evolution and selective expression of the brain homologue, beta-SNAP. Further membrane-binding motifs in the SNAPs pre-target it and facilitate the interaction with SNARE complexes. Thus, relatively high neuronal expression of NSF and SNAP, membrane association domains and their steady-state association with the abundant synaptic SNARES underpins the observed high proportion of the total pool of these soluble proteins on synaptic membranes (Winter et al. 2009).

7.3.6 Clathrin Coats and Hsc70 Chaperones

NSF is selectively expressed in neurons contrasting the situation with Hsc70, which appears uniformly expressed across tissues under resting conditions (see above). One of the critical functions of Hsc-70 function is in the clathrin-mediated endocytosis. In the case of transmitter release synaptic vesicle membrane proteins efficiently bind adaptor proteins (AP) that allow subsequent clathrin recruitment and coat formation. In nerve terminals where transmitter release is the most dynamic membrane trafficking event, AP-2 and AP-180 are particularly important adaptor proteins (Dittman and Ryan 2009). This requirement is aped in the post-synaptic compartment where adaptor-mediated clathrin endocytosis is pivotal to dynamically regulate the plasma membrane receptors and control synaptic plasticity that underpins learning and memory (Puthenveedu et al. 2007).

The initiation and subsequent retrieval of membrane through internalization makes use of the cytosolic clathrin triskelion that self-associates and polymerizes under the membrane. The adaptors act as a bridge and bind to cytosolic domains of the synaptic vesicle protein and recruit and nucleate individual triskelions. This builds up an increasing concentration of clathrin, which by virtue of intrinsic tendency to curve, accentuate or directly contribute to the invagination. From here the combined exchange of triskelions at the neck domain and the accumulation of dynamin's activated GTPase activity generate membrane fission and the production of a clathrin-coated vesicle (see Fig. 7.4; Slepnev and De Camilli 2000). Membrane fission is the opposite of fusion and involves an equally unfavourable removal of lipid from one bi-layer to generate a new vesicle bi-layer. The final gross aspect of this energetically intense process is the shedding of the clathrin coat and its dissolution into individual clathrin triskelions.

In vitro reconstitution of this process has revealed that it is made up of many independent steps and indicated a role for several important factors (Figs. 7.2 and 7.4). One important determinant of clathrin coat function is the ATP-dependent chaperone Hsc70 where its ATPase activity is assimilated to drive uncoating. The Hsc-70 undergoes an ATP loading that accumulates under the coat and produces the

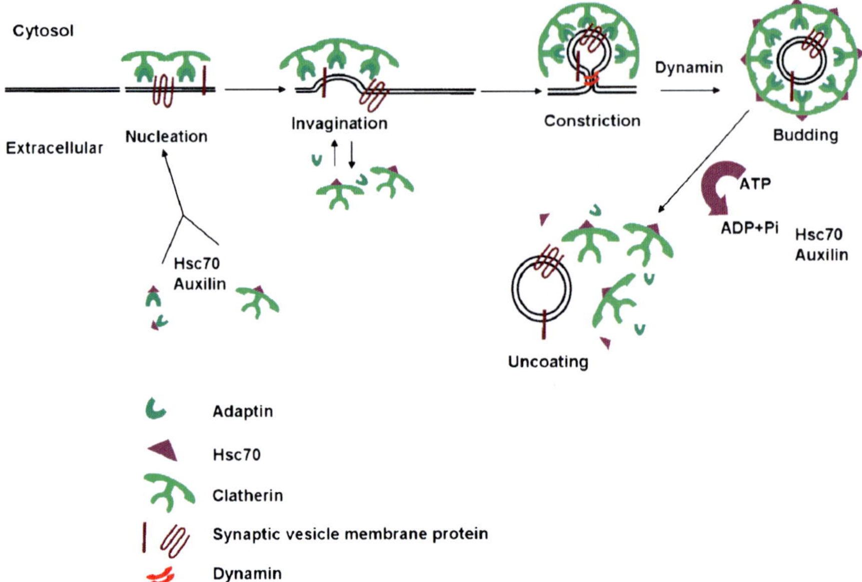

Fig. 7.4 The role of Hsc70 in clathrin-mediated endocytosis. After fusion synaptic vesicle proteins in the membrane bind adaptor proteins that recruit and nucleate clathrin on the intracellular side of the membrane. The recruitment of clathrin and the GTPase dynamin lead to clathrin-coated pits and later vesicle budding. The clathrin-coated vesicle is uncoated by auxilin and its associated Hsc70's ATPase activity. The vesicle coat proteins (adaptors and clathrin) are released and assisted in maintaining their solubility by the chaperone modality of Hsc70. These soluble proteins can then be recruited back to the membrane for subsequent rounds of membrane retrieval

controlled hydrolysis of ATP which is used in uncoating (Xing et al. 2010). Refined in vitro analysis identified auxilin as an important co-factor in uncoating. Auxilin is selectively expressed in neurons and mediates its co-activity by binding Hsc70 through its C-terminal DNA-J domain. This enables Hsc70 to be selectively recruited to auxilin-specific reactions. The simple interpretation is that auxilin acts as a co-chaperone to bring Hsc70's ATPase activity to the uncoating reaction. However, there are a number of observations that suggest that this chaperone/co-chaperone's contribution to endocytosis may extend beyond simply driving uncoating (Eisenberg and Greene 2007).

First, Hsc70 and auxilin appear to allow discrete exchange of clathrin during coat formation. Second, during the dissociation, Hsc70 modulates the conformation of both adaptors and clathrin and imparts chaperone activity towards these proteins. In the former case this is thought to facilitate rebinding of the adaptors at the plasma membrane. In the case of clathrin this action, which may include persistent Hsc70 binding after uncoating, prevents inappropriate self-association of clathrin triskelions in the cytoplasm. Finally, the hydrolysis of ATP by Hsc70 is thought to prime the clathrin triskelions with energy providing capacity to curve membranes and directly contribute to the otherwise thermodynamically unfavourable process of membrane bending. It is difficult to resolve whether these apparently independent molecular events are strictly sub-functions or arise out of the integrated use of Hsc70 and auxilin in uncoating. However, the suggested difference in molecular stoicohometry for the different sub-functions supports the notion that they are independent modulatory events (Eisenberg and Greene 2007).

7.3.7 *Different Routes to Recruiting Generic Chaperone Activity to Specific Synaptic Reactions?*

As there is no evidence for the selective targeting of Hsc70, auxilin is likely central to modulating this activity towards the endocytosis processes at the synapse. The steady-state levels and the neuronal expression of auxilin likely help underpin this synaptic selectivity. Auxilin does not appear to contain a defined synaptic targeting signal but it has distinct clathrin- and lipid-binding domains, which might provide bivalency to facilitate synaptic selectivity. Perhaps the combination of its neuron-specific expression and these molecular interactions ensure that auxilin targets synaptic vesicle retrieval with sufficient efficacy to not require more specific targeting mechanisms. This contrasts with CSP, which like auxillin, uses a DNA-J domain to selectively recruit Hsc70 to synaptic vesicles. Thus, both neuron-specific expression and specific synaptic vesicle targeting ensure that CSP's co-chaperone function is restricted to the synapse. This is particularly intriguing as the synaptic vesicle membrane that undergoes an auxilin/Hsc70-dependent retrieval will have an independent capacity to recruit Hsc70 through CSP. However, the potential cross talk and the reasons for distinct modes of Hsc70 recruitment have not been investigated.

7.3.8 *Are There More Synaptic Chaperones?*

The above discussion resolves clear exemplars of how chaperone modality is central to synaptic function. Chaperones re-distribute to synapses after stress (see Sect. 7.2.2) also underlining their crucial role for neuroprotection. There is additional evidence to suggest a more widespread role of such chaperone activity for synaptic function. Drugs that modulate Hsp70 and 90 have been shown to modify measures of synaptic transmission to imply a wider role for classic chaperones (Gerges et al. 2004). The relative autarky of the synapse requires that core cellular functions like protein biogenesis, mitochondrial integrity and protein degradation have a selective requirement at the synapse and hence for this reason alone chaperones associated with these processes and organelles will co-locate in the synapse. In contrast the particular demand of the synapse for generic activities may have forced the evolution of a specialized use of such pathways. Clues from this arise by the evidence that proteins with motifs suggestive of a chaperone modality appear to selectively impact on synaptic function (Granata et al. 2008). Overall, molecular investigations of the synapse provoke consideration of a growing list of synaptic mechanisms utilizing chaperone modality (Fujita et al. 2006; McEwen and Kaplan 2008).

7.4 Conclusions

CNS regions exhibit distinct profiles of molecular chaperone expression under basal, physiological conditions and during chronic neurodegeneration associated with the protein misfolding diseases. Additionally molecular chaperones localize to the synaptic compartment and we propose that synapses represent a "particular folding environment". Our discussion has outlined examples that highlight how expression and regulated protein interactions can be used to concentrate a generic chaperone modality to key aspects of synaptic function. Co-chaperones rather than direct chaperone targeting seem to be the most precise way of achieving specialization of chaperone function. The temporal and spatial demands that synaptic transmission places on cellular function makes a case for an evolved synaptic "chaperome". Indeed, the ancient basis of the synaptic specialization suggests that it is well placed to have co-evolved with the ancient "chaperome" in which such modality plays a pivotal role in regulating the highly interactive protein networks that support synaptic function (Grant and Ryan 2009). We speculate that additional synaptic chaperones exist and are used in the folding and re-folding dynamics that underpin these networks.

The consequence of an evolved and specialized "synaptic chaperome" would be that it provides a route through which perturbations (via genetic or extrinsic stressors) can selectively disrupt synaptic function to generate dysfunction. Indeed, it is in the area of disease and the emergence of protein misfolding as its core trigger that the concept of synaptic chaperones shows further relevance. It is already clear that such misfolding selectively targets the synaptic compartment and enough evidence exist to ascribe disruption through chaperone pathways as an important element in such disease processes. In addition, neurodegeneration is increasingly being shown to arise through the disruption of extrinsic chaperone or protein homeostatic pathways (Luo and Le 2010) or via the loss of specific proteins that harbour an intrinsic chaperone modality (Zhai et al. 2008). Therefore, further investigating the "synaptic chaperome" will lead to an increased understanding of not only synaptic function, but also how it is dysfunctional in disease.

References

Bender T, Leidhold C, Ruppert T, Franken S, Voos W (2010) The role of protein quality control in mitochondrial protein homeostasis under oxidative stress. Proteomics 10(7):1426–43.

Bechtold DA, Brown IR (2000) Heat shock proteins Hsp27 and Hsp32 localize to synaptic sites in the rat cerebellum following hyperthermia. Brain Res Mol Brain Res. 75(2):309–20.

Bechtold DA, Rush SJ, Brown IR (2000) Localization of the heat-shock protein Hsp70 to the synapse following hyperthermic stress in the brain. J Neurochem. 74(2):641–6.

Brown IR (2007) Heat shock proteins and protection of the nervous system. Ann N Y Acad Sci. 1113:147–58.

Buck TM, Kolb AR, Boyd CR, Kleyman TR, Brodsky JL (2010) The endoplasmic reticulum-associated degradation of the epithelial sodium channel requires a unique complement of molecular chaperones. Mol Biol Cell. 21(6):1047–58.

Chang WH, Cemal CK, Hsu YH, Kuo CL, Nukina N, Chang MH, Hu HT, Li C, Hsieh M (2005) Dynamic expression of Hsp27 in the presence of mutant ataxin-3. Biochem Biophys Res Commun. 336:258–67

Chen S, Brown IR (2007a) Translocation of constitutively expressed heat shock protein Hsc70 to synapse-enriched areas of the cerebral cortex after hyperthermic stress. J Neurosci Res. 85(2):402–9.

Chen S, Brown IR (2007b) Neuronal expression of constitutive heat shock proteins: implications for neurodegenerative diseases. Cell Stress Chaperones 12(1):51–8.

Clayton A, Turkes A, Navabi H, Mason MD, Tabi Z (2005) Induction of heat shock proteins in B-cell exosomes. J Cell Sci. 118:3631–8.

Collingridge GL, Isaac JT (2003) Functional roles of protein interactions with AMPA and kainate receptors. Neurosci Res. 47:3–15.

Craig EA, Huang P, Aron R, Andrew A (2006) The diverse roles of J-proteins, the obligate Hsp70 co-chaperone. Rev Physiol Biochem Pharmacol. 156:1–21.

D'Souza SM, Brown IR (1998) Constitutive expression of heat shock proteins Hsp90, Hsc70, Hsp70 and Hsp60 in neural and non-neural tissues of the rat during postnatal development. Cell Stress Chaperones 3(3):188–99.

Der Perng M, Quinlan R (2004) Neuronal diseases: small heat shock proteins calm your nerves. Curr Biol. 14:R625–6.

Di Domenico F, Sultana R, Tiu GF, Scheff NN, Perluigi M, Cini C, Butterfield AD (2010) Protein levels of heat shock proteins 27, 32, 60, 70, 90 and thioredoxin-1 in amnestic mild cognitive impairment: an investigation on the role of cellular stress response in the progression of Alzheimer disease. Brain Res. 1333:72–81.

Dittman J, Ryan TA (2009) Molecular circuitry of endocytosis at nerve terminals. Annu Rev Cell Dev Biol. 25:133–60.

Eisenberg E, Greene LE (2007) Multiple roles of auxilin and hsc70 in clathrin-mediated endocytosis. Traffic 8:640–6.

Foster JA, Brown IR (1996) Intracellular localization of heat shock mRNAs (hsc70 and hsp70) to neural cell bodies and processes in the control and hyperthermic rabbit brain. J Neurosci Res. 46(6):652–65.

Fujita M, Wei J, Nakai M, Masliah E, Hashimoto M (2006) Chaperone and anti-chaperone: two-faced synuclein as stimulator of synaptic evolution. Neuropathology 26:383–92.

Gerges NZ, Tran IC, Backos DS, Harrell JM, Chinkers M, Pratt WB, Esteban JA (2004) Independent functions of hsp90 in neurotransmitter release and in the continuous synaptic cycling of AMPA receptors. J Neurosci. 24:4758–66.

Granata A, Watson R, Collinson LM, Schiavo G, Warner TT (2008) The dystonia-associated protein torsinA modulates synaptic vesicle recycling. J Biol Chem. 283:7568–79.

Guo ZH, Mattson MP (2000) In vivo 2-deoxyglucose administration preserves glucose and glutamate transport and mitochondrial function in cortical synaptic terminals after exposure to amyloid beta-peptide and iron: evidence for a stress response. Exp Neurol. 166(1):173–9.

Hartl FU, Hayer-Hartl M (2009) Converging concepts of protein folding in vitro and in vivo. Nat Struct Mol Biol. 16(6):574–81.

Haslbeck M (2002) sHsps and their role in the chaperone network. Cell Mol Life Sci. 59:1649–57.

Hay DG, Sathasivam K, Tobaben S, Stahl B, Marber M, Mestril R, Mahal A, Smith DL, Woodman B, Bates GP (2004) Progressive decrease in chaperone protein levels in a mouse model of Huntington's disease and induction of stress proteins as a therapeutic approach. Hum Mol Genet 13:1389.

Hightower LE, Guidon PT (1989) Selective release from cultured mammalian cells of heat-shock (stress) proteins that resemble glia-axon transfer proteins. J Cell Physiol. 138:257–66.

Houenou LJ, Li L, Lei M, Kent CR, Tytell M (1996) Exogenous heat shock cognate protein Hsc 70 prevents axotomy-induced death of spinal sensory neurons. Cell Stress Chaperones 1(3):161–6.

Iwaki T, Wisniewski T, Iwaki A, Corbin E, Tomokane N, Tateishi J, Goldman JE (1992) Accumulation of alpha B-crystallin in central nervous system glia and neurons in pathologic conditions. Am J Pathol. 140:345–56.

Kabani M, Martineau CN (2008) Multiple hsp70 isoforms in the eukaryotic cytosol: mere redundancy or functional specificity? Curr Genomics. 9(5):338–48.

Kelty JD, Noseworthy PA, Feder ME, Robertson RM, Ramirez JM (2002) Thermal preconditioning and heat-shock protein 72 preserve synaptic transmission during thermal stress. J Neurosci. 22(1):RC193.

Kirkegaard T, Roth AG, Petersen NH, Mahalka AK, Olsen OD, Moilanen I, Zylicz A, Knudsen J, Sandhoff K, Arenz C, Kinnunen PK, Nylandsted J, Jäättelä M (2010) Hsp70 stabilizes lysosomes and reverts Niemann-Pick disease-associated lysosomal pathology. Nature 463(7280):549–53.

Klemenz R, Andres AC, Frohli E, Schafer R, Aoyama A (1993) Expression of the murine small heat shock proteins hsp 25 and alpha B crystallin in the absence of stress. J Cell Biol. 120:639–45.

Koren J III, Jinwal UK, Lee DC, Jones JR, Shults CL, Johnson AG, Anderson LJ, Dickey CA (2009) Chaperone signalling complexes in Alzheimer's disease. J Cell Mol Med. 13(4):619–30.

Lancaster GI, Febbraio MA (2005) Mechanisms of stress-induced cellular HSP72 release: implications for exercise-induced increases in extracellular HSP72. Exerc Immunol Rev. 11:46–52.

Lein ES et al. (2007) Genome-wide atlas of gene expression in the adult mouse brain. Nature 445:168–76.

Luo GR, Le WD (2010) Collective roles of molecular chaperones in protein degradation pathways associated with neurodegenerative diseases. Curr Pharm Biotechnol. 11:180–7.

Mambula SS, Calderwood SK (2006) Heat shock protein 70 is secreted from tumor cells by a nonclassical pathway involving lysosomal endosomes. J Immunol. 177:7849–57.

Martin J (1997) Molecular chaperones and mitochondrial protein folding. J Bioenerg Biomembr. 29(1):35–43.

Mayer MP, Bukau B (2005) Hsp70 chaperones: cellular functions and molecular mechanism. Cell Mol Life Sci. 62:670–84.

Manzerra P, Brown IR (1996) The neuronal stress response: nuclear translocation of heat shock proteins as an indicator of hyperthermic stress. Exp Cell Res. 229(1):35–47.

McEwen JM, Kaplan JM (2008) UNC-18 promotes both the anterograde trafficking and synaptic function of syntaxin. Mol Biol Cell. 19:3836–46.

Morgan A, Burgoyne RD (1995) Is NSF a fusion protein? Trends Cell Biol. 5:335–9.

Muchowski PJ, Wacker JL (2005) Modulation of neurodegeneration by molecular chaperones. Nat Rev Neurosci 6:11–22.

O'Connor V, Augustine GJ, Betz H (1994) Synaptic vesicle exocytosis: molecules and models. Cell 76:785–7.

Outeiro TF, Klucken J, Strathearn KE, Liu F, Nguyen P, Rochet J-C, Hyman BT, McLean PJ (2006) Small heat shock proteins protect against [alpha]-synuclein-induced toxicity and aggregation. Biochem Biophys Res Commun 351:631–8.

Ousman SS, Tomooka BH, van Noort JM, Wawrousek EF, O'Connor KC, Hafler DA, Sobel RA, Robinson WH, Steinman L (2007) Protective and therapeutic role for alphaB-crystallin in autoimmune demyelination. Nature 448:474–9.

Powers ET, Morimoto RI, Dillin A, Kelly JW, Balch WE (2009) Biological and chemical approaches to diseases of proteostasis deficiency. Annu Rev Biochem 78:959–91.

Püschel AW, O'Connor V, Betz H (1994) The N-ethylmaleimide-sensitive fusion protein (NSF) is preferentially expressed in the nervous system. FEBS Lett. 347:55–8.

Puthenveedu MA, Yudowski GA, von Zastrow M (2007) Endocytosis of neurotransmitter receptors: location matters. Cell 130:988–9.

Quraishe S, Asuni A, Boelens WC, O'Connor V, Wyttenbach A (2008) Expression of the small heat shock protein family in the mouse CNS: differential anatomical and biochemical compartmentalization. Neuroscience 153(2):483–91.

Quraishi H, Brown IR (1995) Expression of heat shock protein 90 (hsp90) in neural and nonneural tissues of the control and hyperthermic rabbit. Exp Cell Res. 219(2):358–63.

Renkawek K, De Jong WW, Merck KB, Frenken CW, Van Workum FP, Bosman GJ (1992) alpha B-crystallin is present in reactive glia in Creutzfeldt-Jakob disease. Acta Neuropathol. 83:324–7.

Renkawek K, Stege GJ, Bosman GJ (1999) Dementia, gliosis and expression of the small heat shock proteins hsp27 and alpha B-crystallin in Parkinson's disease. Neuroreport 10:2273–6.

Ryan TJ, Grant SG (2009) The origin and evolution of synapses. Nat Rev Neurosci. 10:701–12.

Rothman JE, Orci L (1992) Molecular dissection of the secretory pathway. Nature 355:409–15.

Rothman JE, Warren G (1994) Implications of the SNARE hypothesis for intracellular membrane topology and dynamics. Curr Biol. 4:220–33.

Sajjad MU, Samson B, Wyttenbach A (2010) Heat shock proteins: therapeutic drug targets for chronic neurodegeneration? Curr Pharm Biotechnol. 11(2):198–215.

Schwarz L, Vollmer G, Richter-Landsberg C (2010) The small heat shock protein HSP25/27 (HspB1) is abundant in cultured astrocytes and associated with astrocytic pathology in progressive supranuclear palsy and corticobasal degeneration. Int J Cell Biol 2010:717520.

Shinohara H, Inaguma Y, Goto S, Inagaki T, Kato K (1993) Alpha B crystallin and HSP28 are enhanced in the cerebral cortex of patients with Alzheimer's disease. J Neurol Sci. 119:203–8.

Slepnev VI, De Camilli P (2000) Accessory factors in clathrin-dependent synaptic vesicle endocytosis. Nat Rev Neurosci. 1:161–72.

Söllner T, Bennett MK, Whiteheart SW, Scheller RH, Rothman JE (1993) A protein assembly-disassembly pathway in vitro that may correspond to sequential steps of synaptic vesicle docking, activation, and fusion. Cell 75:409–18.

Stevens FJ, Argon Y (1999) Protein folding in the ER. Semin Cell Dev Biol. 10(5):443–54.

Südhof TC (2004) The synaptic vesicle cycle. Annu Rev Neurosci. 27 509–47.

Südhof TC, Rothman JE (2009) Membrane fusion: grappling with SNARE and SM proteins. Science 323:474–7.

Sutton RB, Fasshauer D, Jahn R, Brunger AT (1998) Crystal structure of a SNARE complex involved in synaptic exocytosis at 2.4 A resolution. Nature 395:347–53.

Stein A, Weber G, Wahl MC, Jahn R (2009) Helical extension of the neuronal SNARE complex into the membrane. Nature 460.525–8.

Takamori S, Holt M, Stenius K, Lemke EA, Grønborg M, Riedel D, Urlaub H, Schenck S, Brügger B, Ringler P, Müller SA, Rammner B, Gräter F, Hub JS, De Groot BL, Mieskes G, Moriyama Y, Klingauf J, Grubmüller H, Heuser J, Wieland F, Jahn R (2006) Molecular anatomy of a trafficking organelle. Cell 127:831–46.

Terada S, Kinjo M, Aihara M, Takei Y, Hirokawa N (2010) Kinesin-1/Hsc70-dependent mechanism of slow axonal transport and its relation to fast axonal transport. EMBO J. 29(4):843–54.

Tytell M (2005) Release of heat shock proteins (Hsps) and the effects of extracellular Hsps on neural cells and tissues. Int J Hyperthermia. 21(5):445–55.

van Noort JM (2008) Stress proteins in CNS inflammation. J Pathol. 214:267–75.

Vega VL, Rodríguez-Silva M, Frey T, Gehrmann M, Diaz JC, Steinem C, Multhoff G, Arispe N, De Maio A (2008) Hsp70 translocates into the plasma membrane after stress and is released into the extracellular environment in a membrane-associated form that activates macrophages. J Immunol. 180(6):4299–307.

Voos W, Röttgers K (2002) Molecular chaperones as essential mediators of mitochondrial biogenesis. Biochim Biophys Acta. 1592(1):51–62.

Whiteheart SW, Matveeva EA (2004) Multiple binding proteins suggest diverse functions for the N-ethylmaleimide sensitive factor. J Struct Biol. 146:32–43.

Wilhelmus MM, Otte-Höller I, Wesseling P, de Waal RM, Boelens WC, Verbeek MM (2006a) Specific association of small heat shock proteins with the pathological hallmarks of Alzheimer's disease brains. Neuropathol Appl Neurobiol. 32:119–30.

Wilhelmus MM, Boelens WC, Otte-Holler I, Kamps B, de Waal RM, Verbeek MM (2006b) Small heat shock proteins inhibit amyloid-beta protein aggregation and cerebrovascular amyloid-beta protein toxicity. Brain Res. 1089:67.

Winter U, Chen X, Fasshauer D (2009) A conserved membrane attachment site in alpha-SNAP facilitates N-ethylmaleimide-sensitive factor (NSF)-driven SNARE complex disassembly. J Biol Chem. 284:31817–26.

Wojcik SM, Brose N (2007) Regulation of membrane fusion in synaptic excitation-secretion coupling: speed and accuracy matter. Neuron 55:11–24.

Witt SN (2010) Hsp70 molecular chaperones and Parkinson's disease. Biopolymers 93(3):218–28.

Wyttenbach A, Sauvageot O, Carmichael J, Diaz-Latoud C, Arrigo AP, Rubinsztein DC (2002) Heat shock protein 27 prevents cellular polyglutamine toxicity and suppresses the increase of reactive oxygen species caused by huntingtin. Hum Mol Genet 11:1137–51.

Wyttenbach A (2004) Role of heat shock proteins during polyglutamine neurodegeneration: mechanisms and hypothesis. J Mol Neurosci. 23:69–96.

Wyttenbach A, Arrigo AP (2006) Role of heat shock proteins during neurodegeneration in Alzheimer's, Parkinson's and Huntington's disease. In: Heat Shock Proteins in Neural Cells. Edited by: C. Richter-Landsberg. Eurekah.com.

Wyttenbach A, Hands S, O'Connor V (2010) Small stress proteins in the central nervous system: from neurodegeneration to neuroprotection. In: Small stress proteins and human diseases. Edited by: AP Arrigo & S Simon. Nova Sciences, Hauppauge.

Xing Y, Böcking T, Wolf M, Grigorieff N, Kirchhausen T, Harrison SC (2010) Structure of clathrin coat with bound Hsc70 and auxilin: mechanism of Hsc70-facilitated disassembly. EMBO J. 29:655–65.

Yoo BC, Kim SH, Cairns N, Fountoulakis M, Lubec G (2001) Deranged expression of molecular chaperones in brains of patients with Alzheimer's disease. Biochem Biophys Res Commun. 280(1):249–58.

Zhai RG, Zhang F, Hiesinger PR, Cao Y, Haueter CM, Bellen HJ (2008) NAD synthase NMNAT acts as a chaperone to protect against neurodegeneration. Nature 452:887–91.

Chapter 8
Cysteine-String Protein's Role at Synapses

Konrad E. Zinsmaier and Mays Imad

Abstract Accumulating evidence illustrates the significance of chaperone systems for the regulation and maintenance of neuronal and synaptic function. The significance of synaptic chaperones is best illustrated by cysteine-string protein (CSP), a member of the DnaJ/Hsp40 family of Hsp70/Hsc70 co-chaperones. CSP recruits the ubiquitous chaperone Hsc70 to synaptic vesicles forming a chaperone complex that maintains synaptic function and prevents neurodegeneration. Here, we summarize studies that demonstrate CSP's neuroprotective role for synaptic function and discuss insights into its possible clientele (the proteins whose function it facilitates).

8.1 General Introduction

Synaptic transmission occurs at highly specialized cell–cell contact sites, termed synapses. In the presynaptic axon terminal, quantal packages of neurotransmitter are stored in synaptic vesicles. Regulated neurotransmitter release is initiated by a depolarization-dependent Ca^{2+} influx through voltage-gated Ca^{2+} channels that triggers the fusion of synaptic vesicles with the presynaptic membrane releasing neurotransmitter into the synaptic cleft. In turn, activation of specific neurotransmitter receptors in the postsynaptic cell membrane induces a postsynaptic response in less than 100 ms after the presynaptic trigger. After exocytosis, synaptic vesicle membranes and proteins are rapidly recaptured and locally recycled ensuring that synapses can faithfully sustain transmitter release during prolonged periods of high nerve activity.

Presynaptic terminals resemble specialized secretory machines that exhibit an unparalleled autonomy and durability. Both Ca^{2+} signaling and synaptic vesicle fusion are driven by a series of protein–protein and protein–lipid interactions that involve a complex interplay among a specialized set of synaptic proteins and the

K.E. Zinsmaier (✉)
Department of Neuroscience, University of Arizona Gould-Simpson Building 627, P.O. Box 210077, 1040 E. 4th Street, Tucson, AZ 85721–0077, USA
e-mail: kez4@email.arizona.edu

A. Wyttenbach and V. O'Connor (eds.), *Folding for the Synapse*,
DOI 10.1007/978-1-4419-7061-9_8,

cytoskeletal framework (reviewed in Slepnev and De Camilli 2000; Galli and Haucke 2001; Jahn et al. 2003; Galli and Haucke 2004; Rossi et al. 2004; Sudhof 2004; Rohrbough and Broadie 2005; Rizo and Rosenmund 2008). Not surprisingly, molecular chaperones are vital components of this crowded environment. Chaperones facilitate and maintain presynaptic function by controlling protein folding as well as specific protein interactions that mediate or regulate presynaptic signaling pathways (Chamberlain and Burgoyne 2000; Zinsmaier and Bronk 2001; Evans et al. 2003; Zhao et al. 2008). Among these interactions is the controlled formation and disassembly of the core complex of synaptic vesicle fusion, which is dependent on the chaperone activity of NSF (also see Chap. 7 by A. Wyttenbach, S. Quraishe, and V. O'Connor). Loss of chaperone systems typically causes an accumulation of abnormally folded and dysfunctional proteins that often jeopardize neuronal and/or synaptic function and cause neurodegenerative diseases (reviewed in Zoghbi and Orr 2000; Goedert 2001; Kobayashi and Sobue 2001; Selkoe 2001; Muchowski 2002; Bonini and Fortini 2003; Goldstein 2003; Levine et al. 2004; Marsh and Thompson 2004; Muchowski and Wacker 2005), many of which are discussed in this book.

8.2 Synaptic Specializations of the Hsp70 Chaperone System

At synapses, the critical and versatile role of chaperones is best exemplified by the pleiotropic use of an elaborate chaperone network that uses 70 kD heat shock protein (Hsp70) in several distinct ways. Hsp70 proteins are highly conserved and critical for a broad range of processes (see Chaps. 1–3). The multiple genes encoding Hsp70s fall into two classes: genes whose expression is rapidly induced in response to a conditioning heat shock (Hsp70) and genes that are constitutively expressed (heat shock cognate protein, Hsc70). The chaperone activity of Hsp70/Hsc70 proteins is based on their ATPase domain and substrate-binding domain, the latter recognizing short hydrophobic polypeptide stretches of "client proteins." Substrate binding is regulated by the nucleotide-bound state such that ATP binding increases substrate association and dissociation rates, leading to a rapid exchange of polypeptide substrates due to a weaker affinity. In the ADP-bound state, tightening of the substrate-binding pocket reduces substrate dissociation, enabling stable binding due to higher affinity (reviewed in Bukau and Horwich 1998; Hartl and Hayer-Hartl 2002; Young et al. 2003; Mayer and Bukau 2005; Brodsky and Chiosis 2006; Daugaard et al. 2007; Morano 2007). The functional relevance of this generic nucleotide loading is discussed elsewhere in this book (see Chap. 3 and Fig. 3.1 by J. Höhfeld and colleagues).

The functional versatility of Hsp70 proteins is specified and controlled by numerous co-factors, which are localized to various intracellular membranes or elements of the cytoskeleton enabling conformational work in diverse functional contexts. The numerous co-chaperones fall into three different classes that are defined by three specific protein domains (Young et al. 2003): J domains, BAG domains

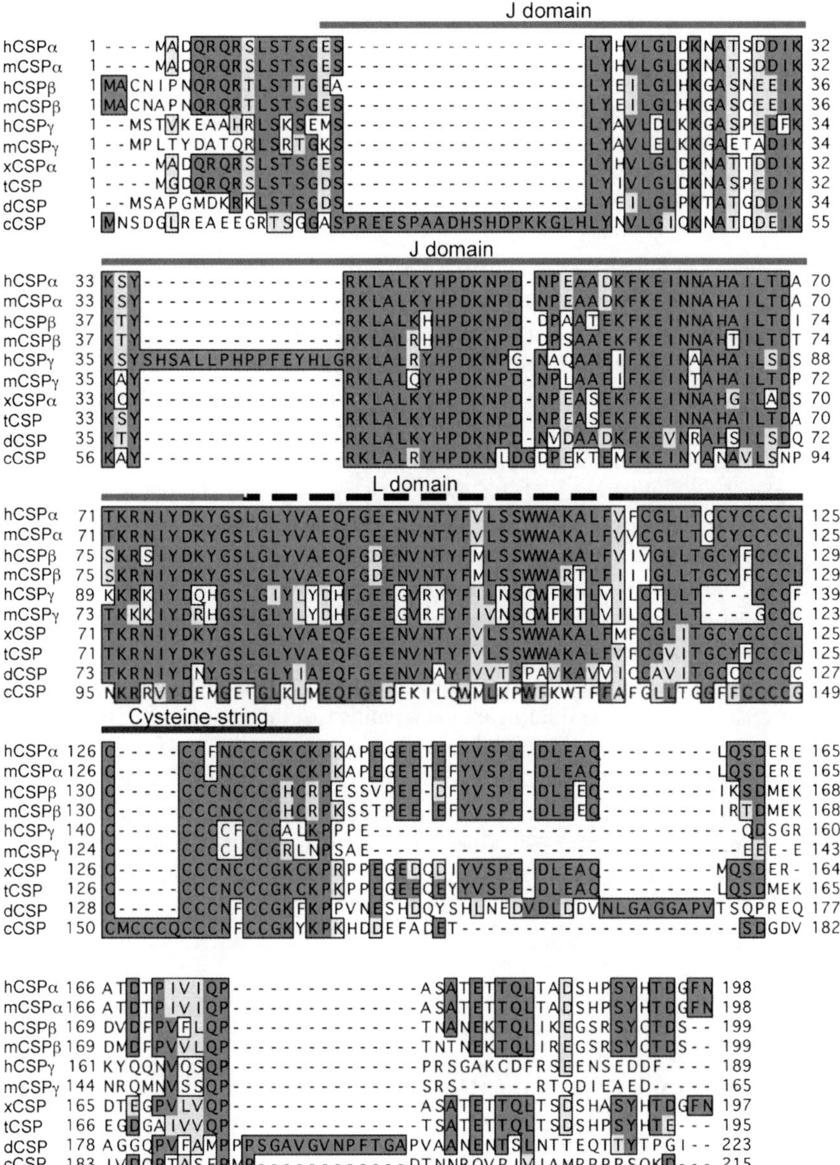

Fig. 8.1 Cysteine-String Proteins (CSPs) are Evolutionary Conserved. Alignment of CSPs from human (hCSPα, hCSPβ and hCSPγ), mouse (mCSPα, mCSPβ and mCSPγ), *Xenopus* (xCSP), *Torpedo* (tCSP), *Drosophila melanogaster* (dCSP), and *Caenorhabditis. elegans* (cCSP). Identical residues are shaded in *dark grey* and similar residues are shaded in *light grey*. The locations of J domain (*light grey line*), L domain, (*dashed line*) and cysteine-string domain (*black line*) are indicated

(see Chap. 3 by J. Höhfeld and colleagues), and tetratricopeptide repeat (TPR) clamp domains. J-domain-containing DnaJ/Hsp40 proteins were the first recognized co-chaperones of the Hsp70 family (Caplan et al. 1993). J domains are characterized by a functionally critical and conserved *His-Pro-Asp* tripeptide motif and typically stimulate ATP hydrolysis of Hsp70/Hsc70 proteins stabilizing substrate binding (Bukau and Horwich 1998; Hartl and Hayer-Hartl 2002). In contrast, BAG domains favor substrate release by triggering nucleotide exchange of ADP for ATP (Höhfeld and Jentsch 1997; Sondermann et al. 2001; Takayama and Reed 2001). TPR clamp domains recognize the C-terminal of Hsp90, Hsc70, or both and provide links for Hsc70 to "client proteins" and related pathways, like the proteasomal pathway (Scheufler et al. 2000; Sondermann et al. 2001; Young et al. 2003).

J-domain-containing proteins are typically localized to specific cellular compartments providing spatial and temporal specificity to Hsp70 chaperone systems. To date, eight J-domain-containing proteins have been identified in neurons: auxilin, GAK, CSP, Rme-8, Hsp40, Hsj1, Mrj, and Rdj2 (Cyr et al. 1994; Cheetham and Caplan 1998; Ohtsuka and Hata 2000; Walsh et al. 2004; Hennessy et al. 2005; Qiu et al. 2006; Zhao et al. 2008). Of these, at least three are critical for maintaining synaptic function: auxilin, GAK, and CSP.

Auxilin and GAK mediate steps of synaptic vesicle recycling, while CSP mediates steps of synaptic vesicle exocytosis. During synaptic vesicle endocytosis, auxilin binds to the assembled clathrin coat, which provides a critical scaffold for vesicular membrane internalization. Auxilin then recruits Hsc70 via its J domain to clathrin-coated vesicles, facilitating their uncoating together with GAK (Ungewickell et al. 1995; Wilbur et al. 2005; Eisenberg and Greene 2007; Ungewickell and Hinrichsen 2007). CSP is associated with synaptic vesicles and recruits Hsc70 via its J domain to facilitate aspects of Ca^{2+}-triggered neurotransmitter exocytosis (reviewed in Umbach et al. 1995; Buchner and Gundersen 1997; Zinsmaier 1997; Chamberlain and Burgoyne 2000; Burgoyne et al. 2001; Zinsmaier and Bronk 2001; Evans and Morgan 2003; Evans et al. 2003; Zhao et al. 2008). Here, we review the biochemical and genetic evidence that illustrates CSP's critical role for synaptic function.

8.3 Evolutionary Conservation of CSP

CSP was first discovered in *Drosophila* as a synapse-associated antigen (Zinsmaier et al. 1990) and has now been identified in many species, ranging from worms to humans (Gundersen and Umbach 1992; Mastrogiacomo and Gundersen 1995; Coppola and Gundersen 1996; Chamberlain and Burgoyne 1996b; Chin et al. 1997; Mastrogiacomo et al. 1998b; Eybalin et al. 2002). Mammalian CSP is encoded by three different genes (Fig. 8.1): CSPα, CSPβ, and CSPγ. Mammalian CSPα expresses two protein isoforms: CSPα-1 and 2. CSPα-2 constitutes a C-terminally truncated protein variant that is derived from a differentially spliced RNA transcript (Chamberlain and Burgoyne 1996b). In contrast, the *Drosophila* genome contains only a single gene that expresses four different protein isoforms, which are derived from at least three alternatively spliced RNAs (Zinsmaier et al. 1994; Eberle et al. 1998).

Phylogenetic analysis suggests that mammalian CSPα and β are closely related to each other while CSPγ is a more distant relative. Overall, CSPγ shares less homology with mammalian CSPα than with any of the *Drosophila* CSP (dCSP) isoforms (Evans et al. 2003). Mammalian CSPα and β are almost equally related to *Drosophila* dCSP (~50 and 45% identity, respectively) and therefore likely paralogs of dCSP.

CSPs exhibit three signature domains, including the centrally located and name-giving cysteine-string domain, an N-terminal J domain, and a linker domain that is sandwiched between both the J and cysteine-string domain (Fig. 8.1). In addition to these key domains, mammalian CSP isoforms contain a conserved phosphorylation site at the extreme N-terminus (Evans et al. 2001). However, shortly after the cysteine-string domain, the entire C-terminal region of CSP exhibits astonishingly little homology, even between mammalian isoforms (Fig. 8.1).

8.4 CSP Expression in Neuronal and Nonneuronal Cells

CSPα and dCSP are associated with synaptic vesicles (Mastrogiacomo et al. 1994; Zinsmaier et al. 1994; van deGoor et al. 1995; van deGoor and Kelly 1996; Eybalin et al. 2002). Indirect evidence suggests that the same is the case for mammalian CSPβ (Eybalin et al. 2002; Schmitz et al. 2006) but not for CSPγ, whose RNA is mostly expressed in testis (Fernandez-Chacon et al. 2004). CSPα accounts for ~1% of total synaptic vesicle protein with an estimated 8 monomers per vesicle (Mastrogiacomo et al. 1994). In addition, CSPα is associated with numerous secretory vesicles in endocrine, neuroendocrine, and exocrine cells (Braun and Scheller 1995; Chamberlain et al. 1996a; Pupier et al. 1997; Zhao et al. 1997; Brown et al. 1998; Redecker et al. 1998; Zhang et al. 1998).

Mammalian CSPα, like its *Drosophila* homologs, is widely expressed in the nervous system but also found in a number of nonneuronal tissues, including kidney, liver, pancreas, spleen, lung, and adrenal gland (Kohan et al. 1995; Coppola and Gundersen 1996; Chamberlain and Burgoyne 1996b; Eberle et al. 1998). In the mammalian brain, CSPα protein is particularly abundant in synapse-rich neuropil regions of the CNS. White matter tracts typically do not show significant expression of CSPα (Kohan et al. 1995). CSPα is expressed at axodendritic synapses such as mossy fibers contacting pyramidal cells in the hippocampus, axosomatic synapses in the cochlear nuclei, and at neuromuscular junctions (NMJs) (Kohan et al. 1995).

High levels of CSPα are found in the accessory olfactory bulb, cerebellum, cochlear nuclei, globus pallidus, hippocampus, interpeduncular nucleus, main olfactory bulb, median eminence, olivary pretectal nucleus, retina, the dorsal horn of the spinal cord, substantia innominata, the pars reticulata of the substantia nigra, and the superior colliculus (Kohan et al. 1995). In many of these brain regions, the high levels of CSPα immunoreactivity are likely due to the high density of synapses, like for the mossy fiber boutons in the hippocampus and in layers of the retina where CSP is predominantly found in the inner and outer plexiform layers (Kohan et al. 1995; Schmitz et al. 2006).

CSPα is also expressed in primary auditory neurons, sensory inner hair cells, and presynaptic differentiations of lateral and medial olivocochlear terminals whose cell bodies lie outside the auditory brainstem (Eybalin et al. 2002; Schmitz et al. 2006). CSPα is not expressed by sensory outer hair cells (Eybalin et al. 2002). Interestingly, single-cell RT-PCR showed that CSPβ but not CSPα is the major CSP isoform that is expressed in hair cells. In contrast, the retina expresses only CSPα (Schmitz et al. 2006).

In addition to its wide expression, CSP is also differentially expressed during development and appears to be subjected to functionally significant phosphorylation. Interestingly, the amount of phosphorylated CSP differs between distinct brain regions during and after development (Evans and Morgan 2005). Similar to that of other synaptic vesicle proteins, CSPα expression levels in the forebrain and cerebellum increase 10–20-fold from E18 to adulthood. However, phosphorylated CSP increases only four- to fivefold, implying that CSP is phosphorylated to a greater extent early in postnatal development than in adulthood (Evans and Morgan 2005). These expression patterns support a possible role for CSP prior to or in early synapse formation (Evans and Morgan 2005). In contrast to mammalian CSPα, little is known about the protein expression pattern of CSPβ and γ, whose mRNAs are abundantly expressed in testis but hardly detectable in other tissues with one exception CSPβ's strong expression in auditory hair cells (Fernandez-Chacon et al. 2004; Schmitz et al. 2006).

8.5 Significance of CSP for Synaptic Function

8.5.1 Neurological and Neurodegenerative Defects in Mice and Flies Lacking CSP

Major insights into the neuroprotective role of CSP came from genetic studies in flies and mice (Umbach et al. 1994; Zinsmaier et al. 1994; Heckmann et al. 1997; Umbach and Gundersen 1997; Eberle et al. 1998; Ranjan et al. 1998; Umbach et al. 1998; Dawson-Scully et al. 2000; Bronk et al. 2001; Fernandez-Chacon et al. 2004; Bronk et al. 2005; Chandra et al. 2005; Schmitz et al. 2006; Dawson-Scully et al. 2007). In *Drosophila*, deletion of the entire dCSP gene causes temperature-sensitive paralysis and semi-lethality. At room temperature, only 4% of the expected flies develop to adulthood. Adult survivors progressively exhibit sluggishness, spasmic jumping, shaking, and uncoordinated locomotion ending in paralysis and premature death (Zinsmaier et al. 1994; Eberle et al. 1998). The progressively developing neurological deficits of adult dCSP deletion mutants are likely due to neurodegeneration (Zinsmaier et al. 1994). Overexpression of dCSP also causes temperature-sensitive lethality. Only below 25°C can some animals survive to adulthood but then exhibit a severely reduced lifespan and numerous developmental defects (Nie et al. 1999).

In mice, deletion of CSPα does not affect embryos and newborns in the first 2 postnatal weeks. Thereafter, CSPα KO mice stop gaining weight and progressively develop severe motor and sensory impairments that lead to paralysis, blindness, and premature death by postnatal day (P) 40–60 (Fernandez-Chacon et al. 2004; Chandra et al. 2005; Schmitz et al. 2006). Starting at P15, CSPα KO mice show progressive loss of spontaneous activity and become ataxic. At the same time, mutant mice develop muscle weakness as evident by their tendency to clasp their hind limbs when suspended by the tail and a lack of gripping strength on fore- and hind limbs. Concomitantly, mutant mice exhibit a progressively poor posture, are unable to stay on a turning rod, and have difficulties getting up. At ~P30, CSPα KO mice enter into a lethargic state and die within the second postnatal month, never surviving beyond 3 months (Fernandez-Chacon et al. 2004; Chandra et al. 2005).

The time course of the progressively developing behavioral deficits of CSPα KO mice indicates a neurodegenerative process that becomes apparent after a lag period. Consistently, NMJs of CSPα KO mice exhibit age-dependent morphological changes that are symptomatic of degenerative processes including an abnormally large amount of multilamellar bodies and protrusions by Schwann cells (Fig. 8.2). These defects are evident at all ages but more pronounced at later stages. Similar signs of degeneration are found at CNS synapses (Fernandez-Chacon et al. 2004).

8.5.2 Synaptic Defects in Mice and Flies Lacking CSP

The neurodegeneration in mice and flies lacking CSP highlights the neuroprotective role of CSP. However, CSP also contributes to various aspects of synaptic function before neurodegeneration is apparent. At larval NMJs of *Drosophila* dCSP deletion mutants, evoked neurotransmitter release is reduced by ~50% at room temperature while it is fully abolished at elevated temperatures (Fig. 8.3a). At room temperature, the Ca^{2+} sensitivity but not Ca^{2+} cooperativity of evoked release is reduced (Fig. 8.3b), and evoked release is desynchronized (Umbach et al. 1994; Zinsmaier et al. 1994; Heckmann et al. 1997; Eberle et al. 1998; Dawson-Scully et al. 2000; Bronk et al. 2005). The defect in evoked release is restored by transgenic expression of CSP (Umbach et al. 1994; Nie et al. 1999).

The loss of transmitter release at dCSP mutant larval NMJs is not due to a block in synaptic vesicle recycling and a subsequent depletion of releasable vesicles (Umbach et al. 1994; Ranjan et al. 1998; Dawson-Scully et al. 2000; Dawson-Scully et al. 2007), since it can be overcome by elevated extracellular Ca^{2+} levels or by accumulating intracellular Ca^{2+} during high-frequency stimulation (Fig. 8.3c). In addition, a lack of presynaptic Ca^{2+} influx through voltage-gated Ca^{2+} channels cannot explain the loss of evoked neurotransmitter release. With increasing temperature, tetanically evoked presynaptic Ca^{2+} signals are increasingly larger at dCSP mutant terminals than in control (Dawson-Scully et al. 2000). The elevated

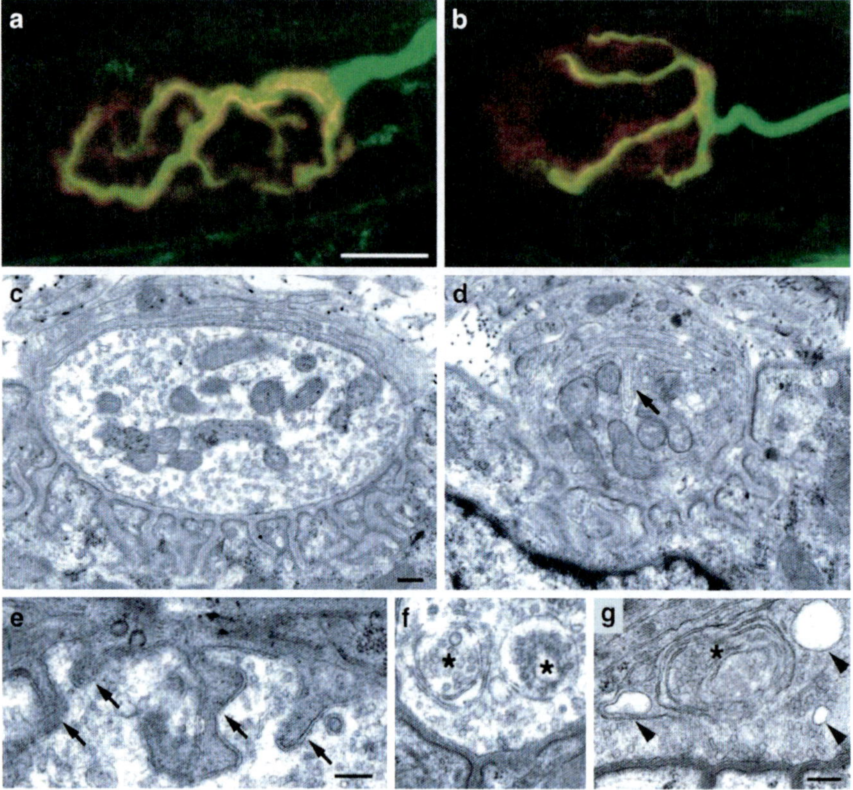

Fig. 8.2 Degeneration of NMJs in CSPα KO Mice. (**a–b**) Confocal images of NMJs from wild-type (**a**) and CSPα KO mice (**b**) double labeled with rhodamine-tagged α-bungarotoxin (*red*) and antibodies to neurofilaments and synaptic vesicles (*green*). (**c–g**) EM images of NMJs from wild-type (**c**) and CSPα KO mice (**d**–**g**). Arrows in panels (**d**) and (**e**) identify Schwann cell intrusions, asterisks in panels (**f**) and (**g**) indicate multilamellar bodies, and arrowheads in panel (**g**) indicate vacuoles. Images **a**–**f** are from P23 animals; image **g** is from a P17 animal (calibration bars: image **a**, 10 μm; images **c**, **e**, and **g**, 0.25 μm). Reprinted from Fernandez-Chacon et al. (2004) with permission from Elsevier

Ca^{2+} levels are likely due to a defect in presynaptic Ca^{2+} clearance, since tetanically-induced Ca^{2+} signals exhibit a prolonged decay at higher temperatures, and because single action-potential induced Ca^{2+} signals are normal at any examined temperature (Bronk et al. 2005; Dawson-Scully et al. 2007). Consistently, presynaptic Ca^{2+} levels at rest are normal at 24°C but elevated at 34°C (Dawson-Scully et al. 2000). Hence, loss of evoked neurotransmitter release at larval NMJ of dCSP deletion mutants is likely caused by an impaired step of Ca^{2+}-triggered exocytosis (Dawson-Scully et al. 2000; Bronk et al. 2005; Dawson-Scully et al. 2007).

Larval *Drosophila* NMJs of dCSP deletion mutants exhibit a reduced number of synaptic boutons while presynaptic overexpression of dCSP increases the number of boutons (Dawson-Scully 2003; Bronk et al. 2005; Dawson-Scully et al. 2007), which suggests an additional role of CSP in synaptic development. Although the

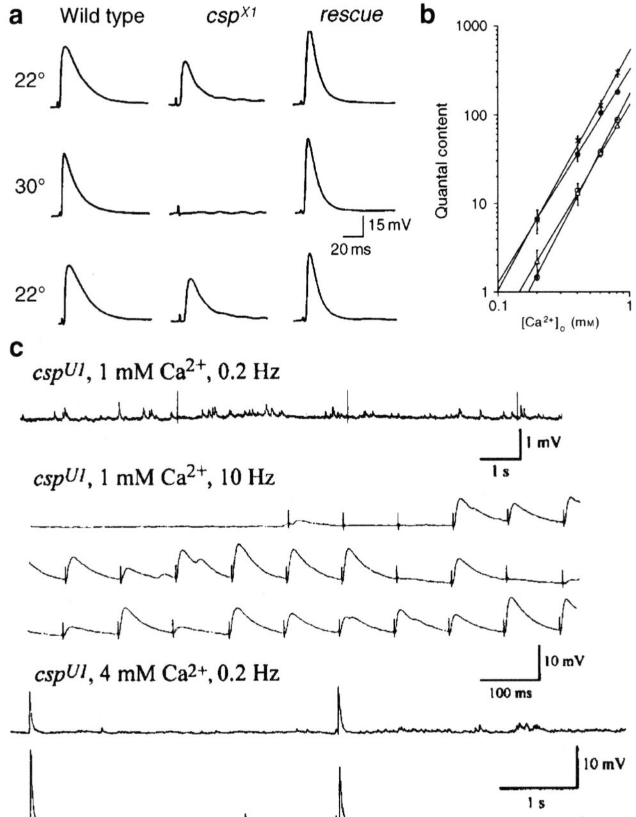

Fig. 8.3 Loss of Evoked Neurotransmitter Release at NMJs of *Drosophila* dCSP Mutants. (**a**) Intracellular recordings of nerve-evoked excitatory junction potentials (EJPs) from larval body wall muscles. Recordings are from control, a dCSP gene deletion mutant (csp^{X1}), and from the csp^{X1} mutant expressing normal dCSP protein (rescue). At high temperatures (30°C) EJPs are abolished in the mutant (small spike represents stimulus artefact). After returning to room temperature, mutant EJPs recover. (**b**) Quantal content (number of released vesicles) as a function of extracellular $[Ca^{2+}]$ for control (crosses), two dCSP mutant alleles (csp^{X1}, *open circles*; csp^{R1}, *open triangles*), and the rescue (*closed circles*). (**c**) Rescue of the temperature-sensitive block of evoked release at 30°C in dCSP mutants. When single stimuli (0.2 Hz) failed to evoke neurotransmission in dCSP gene deletion mutants (csp^{U1}) at 30°C, it is possible to partially rescue the temperature-sensitive defect with high-frequency stimulation or high extracellular Ca^{2+}. Evoked EJPs are blocked in csp^{U1} null mutants with 0.2 Hz stimulation and 1 mM $[Ca^{2+}]_e$. High-frequency stimulation partially rescues 10 Hz the temperature-sensitive block of release observed at 0.2 Hz stimulation while increasing the extracellular Ca^{2+} to 4 mM fully restores faithful and stable synaptic transmission at 0.2 Hz stimulation. Reprinted from Umbach et al. (1994) and Dawson-Scully et al. (2000) with permission from Elsevier and the Society for Neuroscience, respectively

growth defect of dCSP deletion mutants contributes to the intracellularly recorded 50% reduction of evoked release, it is not the major factor. Focal recordings from individual synaptic boutons of dCSP mutant NMJs showed that, on the level of a single bouton, the loss of evoked release is even more pronounced, as it is reduced by ~70% (Dawson-Scully et al. 2007). Much in contrast to photoreceptor terminals of adult dCSP mutant animals, larval NMJs show no signs of neurodegeneration and no loss of synaptic vesicles (Zinsmaier et al. 1994; Dawson-Scully et al. 2007). In conclusion, the genetic analysis of *Drosophila* deletion mutants suggests that CSP is required for several functionally independent synaptic mechanisms facilitating synaptic growth, Ca^{2+}-triggered synaptic vesicle fusion, thermal protection of the evoked release machinery, and presynaptic Ca^{2+} homeostasis at rest and during high-frequency stimulation.

Synapses of CSPα KO mice exhibit an age-dependent deterioration of synaptic function that correlates well with a decline in synaptic structure and progressively developing neurological symptoms. At the Calyx of Held of young CSPα KO mice (P11-P13), evoked neurotransmitter release, presynaptic Ca^{2+} entry through P/Q-type or N type Ca^{2+} channels, and G protein-mediated coupling of $GABA_B$ receptors to P/Q-type Ca^{2+} channels are normal. However, thereafter synaptic function deteriorates progressively with age (Fernandez-Chacon et al. 2004). By P20-P23, evoked release is decreased ~twofold in KO mice, asynchronous, and exhibits abnormal release kinetics. Interestingly, the average amplitude of spontaneous release events is larger at the Calyx of Held. This effect may be due to a presynaptic increase in glutamate loading, larger synaptic vesicles, or an enhanced density of postsynaptic glutamate receptors, and might reflect a homeostatic mechanism by which the synapse partially compensates for the decrease in transmitter release (Fernandez-Chacon et al. 2004).

Nerve-induced compound muscle action potentials (CMAPs) are normal in CSPα KO mice by P23 but drop significantly at P43-P47, indicating that the initially robust function of the NMJ deteriorates with age (Fernandez-Chacon et al. 2004). Consistently, intracellular recordings from CSPα mutant NMJs revealed that evoked neurotransmitter release is normal during the first 2 weeks of life with the exception of sporadically appearing bursts of spontaneous release events. However, by P18, evoked release is significantly reduced and failures of transmission become apparent (Fig. 8.4). By P30, the reduction of evoked release is quantitatively similar to that seen at the CSPα mutant Calyx of Held (see below), dCSP mutant NMJs, and frog NMJs upon injection of CSP antibodies (Umbach et al. 1994; Zinsmaier et al. 1994; Poage et al. 1999; Dawson-Scully et al. 2000; Fernandez-Chacon et al. 2004; Dawson-Scully et al. 2007).

As seen at fly dCSP mutant NMJs (Dawson-Scully et al. 2000), the defect in evoked release at NMJs of CSPα KO mice can be restored by high extracellular Ca^{2+} levels (>2 mM), by repetitive high-frequency stimulation accumulating intracellular Ca^{2+} (Fig. 8.3b–f), and by application of phorbol ester (Ruiz et al. 2008), the latter potentiating transmitter release by increasing the size of the readily releasable pool and the Ca^{2+} sensitivity of the fusion machinery through protein kinase C or Munc-13 activation (Shapira et al. 1987; Betz et al. 1998; Stevens and Sullivan 1998; Yawo 1999; Waters and Smith 2000; Wu and Wu 2001; Rhee et al. 2002; Lou et al. 2005; Virmani et al. 2005). Since the dependence of evoked release on the activation of P/Q-type voltage-gated Ca^{2+} channels is not altered at CSPα KO NMJs, it seems that a defect of Ca^{2+} entry does not reduce evoked release.

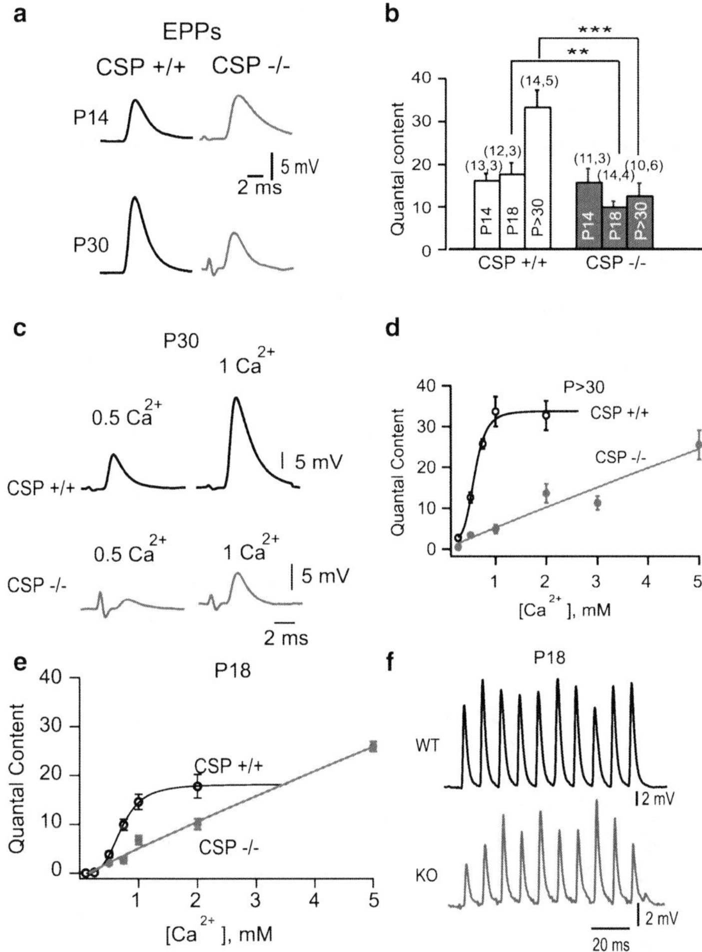

Fig. 8.4 Loss of Evoked Neurotransmitter at NMJs of CSPα KO Mice. (**a**) Representative excitatory potsynaptic potential (EPP) traces at P14 and P30 in CSPα KO and WT mice. (**b**) Lower mean quantal content (QC) in WT and CSPα KO at P14, P18 and $P > 30$ in 2 mm $[Ca^{2+}]_e$. (**c–e**) Rescue of evoked neurotransmitter release at CSPα KO terminals by elevated extracellular Ca^{2+}. Representative WT and CSPα KO EPP traces at different $[Ca^{2+}]_e$ from P30 mice (**c**). Ca^{2+} dependency of quantal content QC at various $[Ca^{2+}]_e$ from $P > 30$ (**d**) and P18 (**e**) mice. Data were fit with a Hill function in controls and with a power function in mutants. (**f**) Repetitive stimulation at 100 Hz partially restores neurotransmitter release in CSPα KO mice at P18. Reprinted from Ruiz et al. (2008) with permission from Wiley-Blackwell

Consequently, the loss of evoked release at CSP mutant NMJs of both flies and mice is likely due to a defect in Ca^{2+} triggering exocytosis at a step close to vesicle fusion. This defect may include a number of known protein interactions of CSP, including synaptotagmin, syntaxin or synaptobrevin (Nie et al. 1999; Wu et al. 1999; Evans and Morgan 2002; Chandra et al. 2005). Alternatively, it may be due to an unidentified vesicle pool that remains intact after partial neurodegeneration in CSPα-deficient terminals but is less sensitive to Ca^{2+} (Ruiz et al. 2008).

8.6 Protein Domains and Biochemical Interactions of CSP

8.6.1 The J domain of CSP

The J domain is the signature domain of pro- and eukaryotic DnaJ/Hsp40 proteins through which they associate and activate Hsp70 chaperones (Szyperski et al. 1994; Tsai and Douglas 1996; Corsi and Schekman 1997). DnaJ/Hsp40 proteins have been categorized into three groups, termed type I–III (Cheetham and Caplan 1998). Type I and II proteins likely function similarly and bind to nonnative substrates (Kelley 1998). In contrast, type III proteins may not bind to nonnative polypeptides and thus may not function as molecular chaperones on their own (Qiu et al. 2006). CSPα, β, and γ proteins have been classified as type III DnaJ/Hsp40 proteins, although the J domain of CSPγ (DnaJC5G, Fig. 8.1) exhibits a unique 16-residue insertion in helix II (Cheetham and Caplan 1998; Evans et al. 2003; Qiu et al. 2006). Of the 41 human J-domain-containing proteins, all but CSPγ and DnaJC8 have been suggested to interact with Hsp70/Hsc70 proteins (Qiu et al. 2006).

The structure of the J domain has been determined in a number of DnaJ/Hsp40 proteins (Hennessy et al. 2005), including that of human auxilin (Jiang et al. 2003; Gruschus et al. 2004b). With few exceptions, the 70 amino acid long J domain is usually found in the N-terminal region and superficially resembles a finger fitting into a binding pocket of Hsp70 proteins (Kelley 1998, 1999). J domains typically consist of four helices and a loop region between helices II and III that contains a highly conserved HPD motif (Qian et al. 1996). In auxilin, the J domain contains an extra loop region between helices I and II (Jiang et al. 2003). The flexible structure of the J domain facilitates binding to Hsp70 proteins by an induced-fit mechanism, with the HPD motif aiding in the alteration of the orientation of the charged residues in helix II, such that helix II can interact correctly with the ATPase domain of Hsp70 (Huang et al. 1999; Berjanskii et al. 2002; Landry 2003). The role of the HPD motif has been precisely modeled for the J-domain-mediated interaction between auxilin (DnaJC6) and Hsc70 (Gruschus et al. 2004a). In this model, H874 of the HPD loop of the auxilin J domain forms transient hydrogen bonds with A148, Y149, and N174 of the Hsc70 ATPase domain, whereas D876 of the HPD loop forms a salt bridge with R171 of the Hsc70 ATPase domain.

Not surprisingly, the J domain of CSPα facilitates binding to Hsc70 and Hsp70 proteins in vitro and in vivo, and stimulates in vitro their ATPase activities 12–14-fold (Braun et al. 1996; Chamberlain and Burgoyne 1997b). Mutational analysis further suggests that the J domain is essential and sufficient for CSP-triggered ATPase stimulation of Hsc70 (Braun et al. 1996; Chamberlain and Burgoyne 1997b; Zhang et al. 1999). Point mutations in the HPD motif (H43Q and D45N) of the J-domain abolished activation of Hsc70 (Zhang et al. 1999). Together, these findings suggest that CSP assembles an enzymatically active Hsp70/Hsc70 chaperone complex. Since Hsc70 is strongly expressed at synaptic terminals (Ungewickell et al. 1995), it is assumed that Hsc70 but not Hsp70 is the major binding partner of CSP, which is supported by co-immunoprecipitations of CSP and Hsc70 from isolated synaptic vesicle fractions (Tobaben et al. 2001). Consistently, partial loss of dHsc70 function at synaptic terminals of *Drosophila* causes similar phenotypes as loss of dCSP (Bronk et al. 2001). In addition, heterozygous mutations in dCSP and dHsc70 cause "synthetic phenotypes" at synaptic terminals, suggesting that both proteins act in a common pathway (Bronk et al. 2001).

The function of the J domain in CSPβ is likely conserved since it is structurally similar to that of CSPα (~71%, Fig. 8.1), although this needs experimental confirmation. In contrast, the significant disruption of the J domain in CSPγ by the 16-residue insertion (Fig. 8.1) makes it questionable whether it interacts with Hsp70/Hsc70 proteins.

In addition to binding Hsp70 proteins, CSPα's J domain also binds the G protein subunit $G\alpha_s$ but not Gβγ subunits (Magga et al. 2000; Natochin et al. 2005). Specifically, CSPα targets the inactive GDP-bound form of $G\alpha_s$ promoting GDP/GTP exchange. This guanine nucleotide exchange activity of CSPα is regulated by Hsc70/SGT (Natochin et al. 2005). This interaction is significant for CSPα's modulation of Gβγ-mediated inhibition of N-type Ca^{2+} channels (Magga et al. 2000; Miller et al. 2003a; Miller et al. 2003b), but the downstream effect of the CSPα-induced increase in cAMP levels remains to be established.

8.6.2 The Cysteine-String Domain of CSP

The signature domain of CSP proteins is a unique cysteine-string domain that contains 14 cysteines over a span of 24 amino acids in vertebrate CSP (Fig. 8.1). This domain also matches a "C-x(2)-Cx(2)-C-x(3)-C" signature suggestive of a putative ferredoxin-like iron-sulfur (4Fe-4S) binding region. However, because of the extensive lipidation of the cysteine-string domain, it is unclear whether the ferredoxin-like domain is functional in vivo (Swayne et al. 2003).

The cysteine-string domain (residues 113–136 in CSPα) is extensively palmitoylated on up to 14 cysteine residues (Gundersen et al. 1994). Although CSP lacks transmembrane-spanning sequences, chemical depalmitoylation does not displace it from membranes (van deGoor and Kelly 1996; Mastrogiacomo et al. 1998a; Chamberlain and Burgoyne 1998b). However, genetic modifications of cysteine-residues in the string domain prevent proper localization of CSP to vesicles, suggesting that palmitoylation of the cysteine-string domain is required for proper protein sorting and/or targeting (Chamberlain and Burgoyne 1998b; Arnold et al. 2004). In the absence of palmitoylation, the minimal membrane-targeting sequence of CSPα comprises amino acids 106–136 (Greaves and Chamberlain 2006), which is consistent with an in silico analysis suggesting that residues 108–130 have a tendency to move from the aqueous environment to the membrane interface without traversing the bilayer (Boal et al. 2007). Unpalmitoylated CSP is not properly trafficked in PC12 cells and localizes to ER membranes instead of granule membranes. Hence, the cysteine-string domain likely acts as both a membrane-targeting sequence and a palmitoylation domain (Greaves and Chamberlain 2006).

Drosophila CSP is palmitoylated in the nervous system by the palmitoyl transferase Huntingtin-interacting protein 14 (Hip14), also known as DHHC17 in mammals (Ohyama et al. 2007; Stowers and Isacoff 2007). At Hip14 mutant synapses, both CSP and SNAP25 are mislocalized (Ohyama et al. 2007; Stowers and Isacoff 2007). Like *Drosophila* dCSP deletion mutants, Hip14 mutants show defects in evoked neurotransmitter release at room temperature and a nearly complete loss of evoked release at elevated temperatures. This defect could be partially rescued when CSP was artificially targeted to synaptic vesicles by expressing a chimaeric CSP protein that contained the membrane domain of synaptobrevin. Hence, palmitoylation-mediated synaptic vesicle association is critical for normal CSP function (Ohyama et al. 2007).

In mammals, a total of four palmitoyl transferases have been identified that palmitoylate CSP and promote stable membrane attachment, including DHHC3, DHHC7, DHHC15, and DHHC17 (Greaves et al. 2008). Mechanistically, it has been proposed that CSP may utilize the weak membrane affinity of the cysteine-string domain to associate with membranes transiently to "sample" them for palmitoyl transferases. Upon association with Golgi membranes, CSP is then recognized by DHHC3/7/15/17, palmitoylated and anchored in the membrane, which in turn, facilitates forward protein trafficking of CSP (Greaves et al. 2008).

Beyond vesicle targeting, the cysteine-string domain of CSP also mediates protein interactions, namely binding to SGT (see below). In addition, it has been suggested that the lipidated cysteine-string domain may act as a "fusion promoting agent." The "acyl-flip" and "bilayer collapse" models describe a mechanism for membrane fusion by switching the association of lipidated cysteines from the vesicular to the plasma membrane (Gundersen et al. 1995). Although this idea is intriguing, it is unlikely to be the case since there is no evidence for repeated cycles of CSP lipidation.

8.6.3 The Linker Domain of CSP

The third conserved domain of CSP, the unique linker (L) domain, is located between the cysteine-string and the J domain at the N-terminal (Fig. 8.1). Biochemically, the L domain of CSPα (residues 83–112) binds to the G protein subunit Gβ and/or Gαβγ (Miller et al. 2003b), which is functionally relevant for CSPα's role as GDP/GTP exchange factor for $G\alpha_s$ and the inhibition of N-type Ca^{2+} channels (Magga et al. 2000; Miller et al. 2003a; Miller et al. 2003b; Natochin et al. 2005). In addition, the L domain together with the cysteine-string domain (residues 83–136, Fig. 8.1) are critical for forming CSP oligomers (Swayne et al. 2003).

Overexpression of CSPα-1 containing a point mutation in the linker domain (E93V) does not interfere with insulin secretion, or the activation of Hsc70 ATPase (Zhang et al. 1999). Studies with *Drosophila* dCSP suggest that the L domain is important, although not essential, for regulating presynaptic Ca^{2+} levels (Bronk et al. 2005).

8.7 The Trimeric CSP/Hsc70/SGT Complex

The enzymatically active CSP/Hsc70 complex likely requires association of small glutamine-rich tetratricopeptide (TPR)-containing protein (SGT) (Tobaben et al. 2001; Tobaben et al. 2003). SGT was originally identified by its interaction with HIV virus type 1 and parvovirus H1 (Callahan et al. 1998; Cziepluch et al. 1998) and is expressed in mammals via two isoforms: αSGT is widely expressed in neuronal and nonneuronal tissues while βSGT is preferentially expressed in neurons (Tobaben et al. 2001; Tobaben et al. 2003).

CSPα, SGT, and Hsc70 mRNA expression in rat brain show a high degree of overlap. All three mRNAs are widely expressed, with highest levels in the cerebellum and the hippocampus and lower levels in the cortex, olfactory bulb, thalamus, and striatum. Furthermore, CSP, SGT, and Hsc70 co-immunoprecipitate together from purified synaptic vesicles, suggesting that they form a complex on the surface of synaptic vesicles (Tobaben et al. 2001). SGT proteins contain an N-terminal oligomerization domain, a centrally located TPR repeat domain, and a C-terminally located polyglutamine (polyQ)-rich domain. A TPR repeat consists of a degenerate 34-residue motif that mediates a variety of protein–protein interactions, including binding to Hsc70 (Scheufler et al. 2000). TPR-containing co-chaperones can bind either to internal sequences or to the very C terminus of Hsc70 and may at least partially determine the fate of a client protein (Frydman and Hohfeld 1997; Irmer and Hohfeld 1997; Demand et al. 1998; Liu et al. 1999; Jiang et al. 2001).

The TPR repeat domain of SGT contains three TPRs, like the Hsp90-organizing protein (Hop), which links Hsc70 and Hsp90 (Frydman and Hohfeld 1997). Binding of Hop to Hsc70 is mediated by a TPR carboxylate clamp mechanism in which the two carboxylate groups of an aspartate in the C terminus of Hsc70 serve as an anchor for electrostatic interactions with basic and polar residues of the surrounding TPRs (Scheufler et al. 2000). Since mutations in the evolutionary conserved basic and polar residues of the three TPRs in αSGT and βSGT perturb the binding to Hsc70, it seems likely that a clamp mechanism similar to that described for the Hop-Hsc70 interaction may at least partially mediate the association of SGT with Hsc70 (Tobaben et al. 2003).

The TPR domains of both SGTs are necessary for binding to CSP (Tobaben et al. 2001; Tobaben et al. 2003). Interestingly, SGTs bind to both CSPα-1 and the C-terminally truncated isoform CSPα-2, although neither contains an acidic residue at its C terminal that could mediate a two-carboxylate clamp mediated association (Tobaben et al. 2003). Consistently, SGT mutations that interfere with the two-carboxylate clamp mediated association of Hsc70 have no effect on CSP binding, suggesting that the TPR domain can operate in two different binding modes. The association of SGT with CSP depends on interactions of the TPR domain with the hydrophobic cysteine-string of CSP but the precise mechanism remains to be determined (Tobaben et al. 2003).

Several biochemical features listed below highlight the significance of the CSP/SGT/Hsc70 complex as a valid chaperone system (Fig. 8.5).

1. The complex undergoes an association–dissociation cycle that is ATP-dependent and driven by ATP hydrolysis. While SGT alone binds CSP in a nucleotide-independent manner, Hsc70 binds CSP preferentially in the presence of ATP

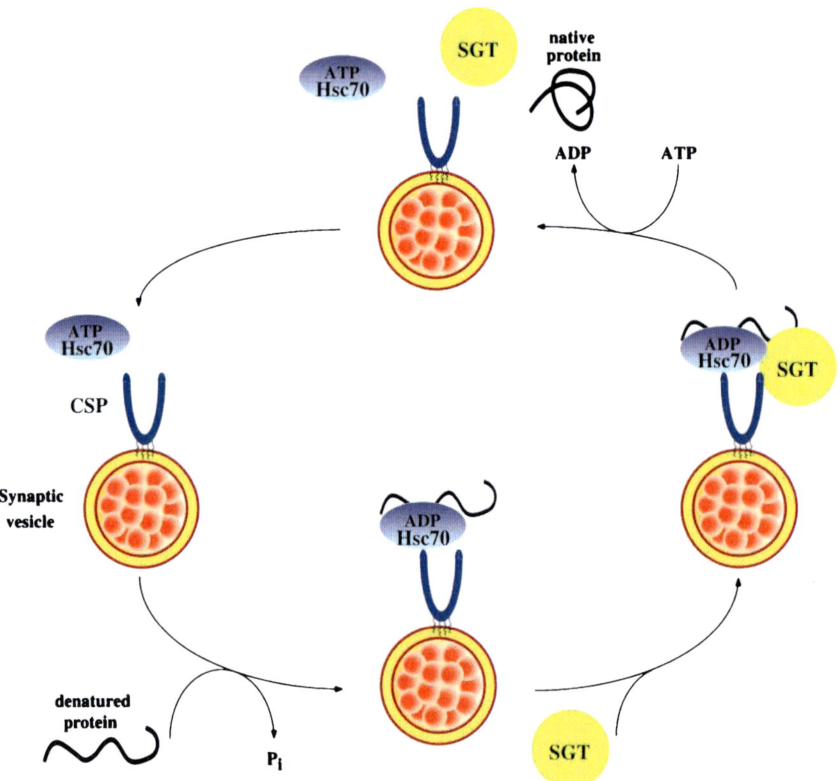

Fig. 8.5 Model of the CSPα/Hsc70/SGT Chaperone Complex. CSPα is a synaptic vesicle protein inserted into the vesicle membrane by its hydrophobic palmitoyl side chains attached to cysteine residues. CSPα recruits Hsc70 and SGT and forms a trimeric protein complex at the synaptic vesicle. In the presence of ATP and available substrate, this protein complex dissociates. As a consequence of ATP hydrolysis and chaperone catalysis, protein substrates in the vicinity of the synaptic vesicle surface are likely refolded and reactivated. The physiological substrate for this chaperone machine remains unknown. However, presynaptic Ca^{2+} channels or SNARE proteins could be physiological substrates. Reprinted from Tobaben et al. (2001) with permission from Elsevier

while binding between SGT and Hsc70 depends on ADP. All three proteins form a complex in the presence of ADP, while ATP dissociates the complex (Tobaben et al. 2001).

2. Simultaneous binding of CSP and SGT to Hsc70 maximally stimulates the ATPase activity of Hsc70. CSP alone stimulates the Hsc70 ATPase by a factor of ~12 while the stimulation by SGT is only ~threefold. However, together CSP and SGT stimulate the ATPase activity ~19-fold (Tobaben et al. 2001).
3. The trimeric complex functions as a chaperone and exhibits a strong, ATP-dependent protein refolding activity while SGT alone exhibits only a weak and ATP-independent chaperone activity (Tobaben et al. 2001). Hence, these findings suggest that CSPα recruits Hsc70 and SGT to synaptic vesicles. ATP hydrolysis and chaperone catalysis then occurs in the presence of ATP and available client protein, re-activating associated client proteins and dissociating the chaperone complex (Tobaben et al. 2001; Tobaben et al. 2003).

Despite this convincing in vitro data, the physiological significance and role of SGT at synapses remains to be determined. Overexpression of αSGT in cultured hippocampal cells impairs evoked neurotransmitter release by reducing the probability of release and the size of the readily releasable vesicle pool (Tobaben et al. 2001). However, these dominant effects of SGT overexpression could be caused by titration of CSP, Hsc70, or other unknown binding proteins. Deletion of SGT in *Drosophila* does not significantly impair synaptic transmission nor does it cause neurodegeneration (Zinsmaier, personal communication).

The physiological function of SGT binding to Hsc70 also remains controversial. While αSGT potentiates CSPα-stimulated ATPase and refolding activities of Hsc70 (Tobaben et al. 2001), it inhibits these activities of Hsc70 when stimulated by other DnaJ/Hsp40 proteins (Wu et al. 2001; Angeletti et al. 2002). These opposing effects may indicate that SGT's net effect on Hsc70 may depend on the corresponding molecular context. Alternatively, it is possible that the binding of SGT to CSP represents a "client protein" interaction since the cysteine-string domain of CSP exhibits structural similarities with domains of peptides from cell surface proteins that presumably interact with SGT due to their misfolded nature (Tobaben et al. 2003). However, it is not clear whether the cysteine-rich and hydrophobic structure of CSP's cysteine-string domain indeed requires constant chaperoning.

8.8 CSP's Neuroprotective Role at Synapses: A General or Substrate-Specific Synaptic Chaperone System?

The biochemical and genetic analyses of CSP function in various systems (Sects. 8.5 and 8.6) provide compelling evidence that CSP alone, in cooperation with Hsc70, or in cooperation with Hsc70 and SGT, provides an essential chaperone activity for synaptic structure and function. While there is little to no doubt that CSP resembles a synaptic chaperone system, it is not resolved which molecules and/or synaptic mechanisms are targeted by CSP. Furthermore, it is not known how CSP prevents neurodegeneration, which could be caused by loss of a particular synaptic function, by a general accumulation of misfolded proteins, or by an accumulation of a few specific proteins or compounds that have a high neurotoxic potential. Hence, the question is whether CSP controls specific proteins, protein complexes, and protein interactions or whether it is a more universal chaperone system that is part of a salvage process recycling damaged components at synaptic terminals.

Initially, studies in *Drosophila* showed that dCSP is required for facilitating evoked release and presynaptic Ca^{2+} homeostasis before signs of neurodegeneration are apparent (Umbach et al. 1994; Zinsmaier et al. 1994; Ranjan et al. 1998; Dawson-Scully et al. 2000; Bronk et al. 2005; Dawson-Scully et al. 2007). This temporal separation of synaptic deficits from neurodegeneration implied that dCSP facilitates specific steps of transmitter release that when defective trigger neurodegeneration over time. However, the phenotype of CSPα KO mice, at least initially, argued against this idea, as newborn KO mice show no behavioral and synaptic deficits during the first 2 weeks after birth and no temporal separation between synaptic deficits and neurodegeneration (Fernandez-Chacon et al. 2004; Chandra et al. 2005; Ruiz et al. 2008). Furthermore, at NMJs of CSPα KO mice structural abnormalities consistent with degeneration are apparent before synaptic function becomes impaired while first signs of synaptic dysfunction and degeneration coincide at the Calyx of Held (Fernandez-Chacon et al. 2004; Ruiz et al. 2008).

Although the phenotype of CSPα KO mice fits a general, use-dependent chaperone function (Fernandez-Chacon et al. 2004), it does not rule out the alternative possibility, that CSPα controls the activity of specific proteins like Ca^{2+} channels, synaptotagmin, or syntaxin 1A. Considering that CSPα's chaperone function is redundant with another chaperone, then high levels of expression of this unknown chaperone could "mask" synaptic phenotypes in CSPα KO mice during the first two postanatal weeks. This possibility seemed initially unlikely because no protein closely related to CSPα function was known to be expressed in mammalian brain at significant levels (Fernandez-Chacon et al. 2004). However, this notion changed after the surprising discovery that overexpression of mouse or human α-synuclein in CSPα KO mice restores the lethality of CSPα mutant animals (Chandra et al. 2005). It also reverses the weight loss, motor impairment, and neurodegenerative phenotypes caused by the lack of CSPα in a cell-autonomous manner while deletion of endogenous synucleins accelerates the lethality and neurodegeneration of CSPα-deficient mice. Finally, overexpression of α-synuclein also corrects the decreased protein levels of Hsc70 and Hsp70 and partly suppresses the defect in SNARE complex assembly of CSPα KO mice, although it did not reverse the decline in SNAP-25 levels (Chandra et al. 2005).

Further support for target-specific functions of CSP comes from *Drosophila*. If one assumes that the CSP system acts exclusively as a general, use-dependent chaperone, then one would expect that all synaptic defects of dCSP mutant animals are mechanistically linked through J-domain-mediated interactions with Hsc70. However, this is not the case (Bronk et al. 2005). Expression of mutant dCSP lacking the J or L domain in dCSP deletion mutant motor neurons differentially restores the various synaptic defects, suggesting that CSP mediates several biochemically independent mechanisms that control synaptic growth, Ca^{2+}-triggered neurotransmitter exocytosis, thermal protection of evoked release, and presynaptic Ca^{2+} levels. Importantly, this analysis showed that the J domain and thereby interactions of CSP with Hsc70 are not essential for all functions of CSP (Bronk et al. 2005).

The temperature-dependent deterioration of evoked release at *Drosophila* NMJs of dCSP null mutants resembles a classical chaperone defect and has been widely regarded as evidence for a role of the CSP-Hsc70 interaction in stabilizing proteins against thermal stress (for review, see Chamberlain and Burgoyne 2000; Zinsmaier and Bronk 2001). Hence, it has been surprising that expression of J domain mutant dCSP proteins (deletion and a mutation in the HPD motif) fully restores a normal thermo-tolerance of evoked release in dCSP deletion mutants, suggesting that dCSP can act as a general Hsc70-independent chaperone (Bronk et al. 2005). Consistently, an Hsc70-independent chaperone function has also been indicated by in vitro studies using mammalian CSPα, which can suppress thermally induced protein aggregation independent of mutations in the J domain (Chamberlain and Burgoyne 1997a).

Taken together, it seems likely that the CSP chaperone system controls specific protein activities but also contributes to "classical" chaperone activities maintaining proper folding of a number of synaptic proteins.

8.9 Individual Synaptic Functions of CSP

In view of the evidence that CSP may target specific proteins and facilitate distinct mechanisms of synaptic function, we will discuss the nature of molecular interactions that support these functions (Fig. 8.6).

A number of specific synaptic functions have been suggested for CSP (Fig. 8.6). (1) CSP may facilitate Ca^{2+}-triggered vesicle fusion through interactions with SNARE proteins and synaptotagmin, which is likely controlled by protein kinase A- and B/Akt-dependent phosporylation of CSP (Brown et al. 1998; Zhang et al. 1998; Chamberlain and Burgoyne 1998a; Nie et al. 1999; Wu et al. 1999; Zhang et al. 1999; Dawson-Scully et al. 2000; Graham and Burgoyne 2000; Evans et al. 2001; Evans and Morgan 2002; Chandra et al. 2005; Evans et al. 2006; Swayne et al. 2006). (2) CSP may directly or indirectly modulate presynaptic Ca^{2+} channels (Gundersen and Umbach 1992; Umbach and Gundersen 1997; Brown et al. 1998; Leveque et al. 1998; Umbach et al. 1998; Zhang et al. 1998; Chamberlain and Burgoyne 1998a; Zhang et al. 1999; Burgoyne et al. 2000; Graham and Burgoyne 2000; Magga et al. 2000; Chen et al. 2002). (3) CSP together with SGT and Hsc70 may act as a guanine nucleotide exchange factor (GEF) for the G protein subunit $G\alpha_s$, which likely mediates a tonic G protein-mediated inhibition of Ca^{2+} channels (Magga et al. 2000; Miller et al. 2003a; Miller et al. 2003b; Natochin et al. 2005) but could also mediate synaptic growth (Bronk et al. 2005). (4) In vitro studies implicate CSP in the regulation of vesicular γ-aminobutyric acid (GABA) uptake through an indirect association with glutamate decarboxylase (GAD) and the

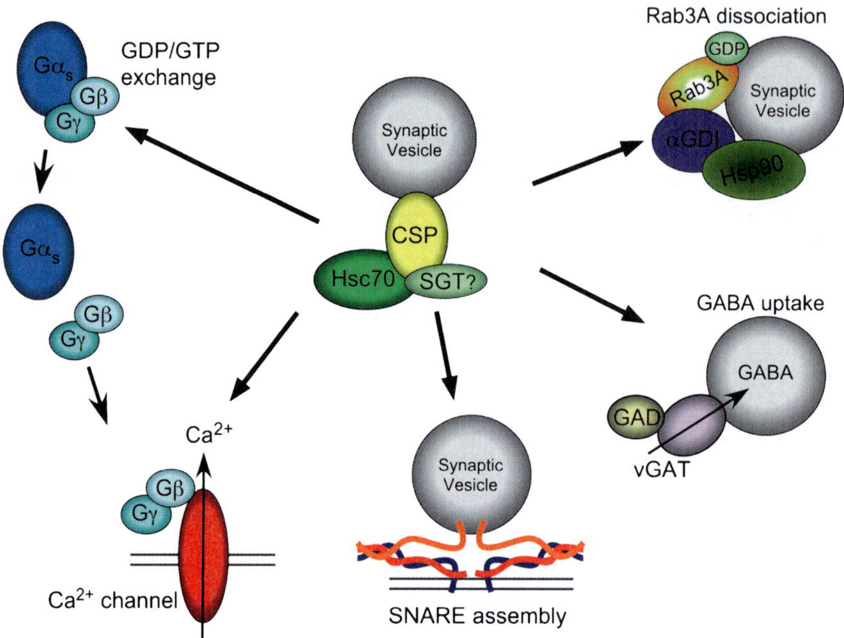

Fig. 8.6 Protein Interactions of CSP at Synaptic Terminals. CSP alone or in association with Hsc70 and SGT has been suggested to promote GDP/GTP exchange for the G protein Gα_s, to promote G$\beta\gamma$-mediated inhibition of presynaptic Ca^{2+} channels, to directly modulate presynaptic Ca^{2+} channels, to directly modulate presynaptic Ca^{2+} channel activity through interactions with the synprint site, to facilitate synaptic vesicle fusion by modulating and/or chaperoning SNARE complex assembly, to complex GAD with vGAT on SV membranes for GABA synthesis and vesicular GABA uptake, and to remove Rab3A-GDP from the SV membrane through interactions with Hsp90/Hsc70/αGDI complex

vesicular GABA transporter (Hsu et al. 2000; Jin et al. 2003). (5) CSP may facilitate vesicle membrane dissociation of RAB3A-GDP by recruiting α-GDP-dissociation factor to the vesicle membrane (Sakisaka et al. 2002).

8.9.1 CPS's Role for Synaptic Vesicle Exocytosis

First evidence for a role of CSP close to a step of vesicle fusion came from overexpression studies. Overexpression of CSPα in PC12 cells enhanced dopamine exocytosis (Chamberlain and Burgoyne 1998a) while overexpression of CSPα or antisense CSP inhibited exocytosis in insulin-secreting cells (Brown et al. 1998; Zhang et al. 1998; Zhang et al. 1999). Furthermore, overexpression of CSPα in bovine adrenal chromaffin cells affected catecholamine exocytosis by reducing the number of release events and slowing their release kinetics, suggesting that CSP modulates proteins that are involved at a step close to membrane fusion (Graham and Burgoyne 2000).

Evidence for a direct protein interaction of dCSP with syntaxin, a critical member of the SNARE complex of vesicle fusion, came from immunoprecipitations and pull-down assays of recombinant proteins from *Drosophila* (Nie et al. 1999; Wu et al. 1999) and mammalian systems (Magga et al. 2000; Chamberlain et al. 2001; Evans et al. 2001; Evans and Morgan 2002; Swayne et al. 2006). Similarly, CSPα also interacts with the SNARE synaptobrevin (Leveque et al. 1998; Boal et al. 2004) and the Ca^{2+} sensor synaptotagmin (Evans and Morgan 2002). Importantly, not only CSPα but also Hsc70 associates with syntaxin 1A and the SNARE complex, implicating both as a target of chaperone activity (Swayne et al. 2006). In addition, overexpression of dCSP, which by itself has no apparent effect on evoked release at larval *Drosophila* NMJs, suppresses the decrease of evoked release induced by the overexpression of syntaxin 1A, suggesting that dCSP modulates protein–protein interactions of syntaxin (Nie et al. 1999).

Molecular analysis of CSPα KO mice further corroborated a role for CSP in modulating syntaxin protein interactions. Quantitative immunoblotting of more than 30 proteins from CSPα KO mice revealed reduced levels of the synaptic SNARE proteins SNAP-25 and SNAP23, α-synuclein, Hsp70, and Hsc70 (Chandra et al. 2005). In addition, lack of CSPα reduces SNARE complex assembly. This reduction in SNARE complex assembly is functionally relevant because transgenic overexpression of normal α-synuclein in CSPα KO mice corrects not only the neurological and neurodegenerative defects but also the decreased levels of Hsc70 and Hsp70 and partially suppresses the defect in SNARE complex assembly (Chandra et al. 2005). Hence, these findings suggest that the reduced levels of SNARE complex assembly may cause at least some of the synaptic deficits of CSPα mutant animals and that an accumulation of misfolded SNARE proteins may trigger the observed neurodegeneration.

Interestingly, the interaction between CSPα and syntaxin exhibits a phosphorylation dependency. Mammalian CSPα is phosphorylated by cAMP-dependent protein kinase A (PKA) and protein kinase B/Akt at Ser10. Phosphorylation of CSPα reduces binding to syntaxin tenfold (Evans et al. 2001) and abolishes the slowing of release kinetics upon overexpression of CSPα in chromaffin cells but not the gross inhibition of catecholamine exocytosis (Evans et al. 2001; Evans et al. 2006). In addition, the interaction of CSPα with synaptotagmin exhibits a similar phospho dependency (Evans and Morgan 2002). Since changes in neurotransmitter release kinetics are likely to underlie forms of synaptic plasticity, such as long-term potentiation and long-term depression, it is possible that the regulation of presynaptic Akt or PKA activity may act via CSP phosphorylation to fine-tune the kinetics of neurotransmitter release (Evans et al. 2006).

8.9.2 CSP's Role for Presynaptic Ca^{2+} Channels and Ca^{2+} Homeostasis

Early on, CSP has been suggested to modulate the activity of presynaptic Ca^{2+} channels but the significance of this activity for synaptic function remains a matter of debate. Neither CSPα KO mice nor dCSP null mutant flies provide so far direct evidence for Ca^{2+} channel regulation (Dawson-Scully et al. 2000;

Fernandez-Chacon et al. 2004; Bronk et al. 2005; Ruiz et al. 2008). However, dCSP mutant flies exhibit increased presynaptic Ca^{2+} levels, which could be caused by a de-inhibition of Ca^{2+} channels or defects in Ca^{2+} clearance (Dawson-Scully et al. 2000). NMJs of CSPα KO mice may also exhibit increased Ca^{2+} levels since the spontaneous bursts of quantal neurotransmitter release could be due to fluctuating presynaptic Ca^{2+} levels (Ruiz et al. 2008), similar to the spontaneous Ca^{2+} "flashes" that occasionally occur in *csp* null NMJs in *Drosophila* (Dawson-Scully et al. 2000).

Initially, CSP has been hypothesized to promote transmitter release by increasing the activity of N-type Ca^{2+} channels since co-injection of Torpedo CSP RNA in frog oocytes increased Ca^{2+} currents (Gundersen and Umbach 1992). Consistently, recombinant *Drosophila* dCSP increases Ca^{2+} currents at chick ciliary calyces, presumably by recruitment of dormant channels (Chen et al. 2002). In contrast, transient co-expression of CSP with N-type Ca^{2+} channels produces a tonic, Gβγ-mediated inhibition of N-type Ca^{2+} channel currents in tsa-201 cells, reducing current densities by ~twofold and slowing current kinetics (Magga et al. 2000; Miller et al. 2003b). Hence, the mode of the proposed regulation of Ca^{2+} channels may depend on the type of synapse or the synaptic protein context.

CSPα regulates heterotrimeric G protein function in association with Hsc70 and SGT by preferentially targeting the inactive GDP-bound form of $G\alpha_s$, promoting GDP/GTP exchange and an increase in cytosolic cAMP levels (Natochin et al. 2005). The downstream effects of the CSPα-induced increase in cAMP remain to be established but exchange of GDP for GTP on Gα releases Gβγ to interact with, for example, N-type Ca^{2+} channels. CSPα binds to $G\alpha_s$- and βγ-subunits through separate binding sites in vitro and in vivo. The first site, located in the J domain, binds $G\alpha_s$ but not Gβγ subunits. The second site, located in the linker domain, binds free Gβγ subunits and $G\alpha_s\beta\gamma$ complexes (Magga et al. 2000; Miller et al. 2003b; Natochin et al. 2005). Interaction with either binding site elicits a tonic inhibition of N-type Ca^{2+} channels, although through distinct mechanisms. Ca^{2+} channel inhibition by the linker but not the J domain is blocked by co-expression of the synaptic protein interaction site (synprint) of the N-type Ca^{2+} channel. Hence, distinct binding sites of CSP may play a role in modulating G protein function and G protein-mediated inhibition of Ca^{2+} channels (Miller et al. 2003b). Notably, syntaxin 1A and CSPα bind to similar regions of the synprint site and assembly of the CSPα/Hsc70/ SGT complex does not alter binding of CSPα to the synprint site (Swayne et al. 2006). A physical interaction of CSP with the synprint site of P/Q-type Ca^{2+} channels has also been observed (Leveque et al. 1998), although no functional data are available.

Interestingly, CSP function can be compromised by pathological conditions found in a number of neurodegenerative diseases that are characterized by expanded polyglutamine (polyQ) tracts. Co-expression of CSP with huntingtin exon 1 containing expanded polyQ tracts ($\text{huntingtin}^{\text{exon1/exp}}$) and N-type Ca^{2+} channels eliminates the tonic inhibition of Ca^{2+} channels by sequestering CSP and blocking its association with G proteins. In contrast, expression of huntingtin exon 1 without an expanded polyQ repeat has no effect. Hence, these findings indicate a potential dysfunction of CSP in Huntington's disease (Miller et al. 2003a).

8.10 Concluding Remarks

Alzheimer's disease (AD), Parkinson's disease (PD), and polyQ diseases, as exemplified by Huntington's disease, comprise the majority of protein conformation diseases that are associated with neurodegeneration. A common feature of these otherwise diverse disorders is that they are caused by mutations or cellular events that lead to an intra- and/or extracellular aggregation of misfolded proteins that share common biochemical characteristics. Historically, these protein deposits were thought to directly trigger neurodegeneration. However, rapidly increasing evidence suggests that diffusible assemblies, instead of mature deposits, may confer toxicity and trigger neurodegeneration (reviewed in Zoghbi and Orr 2000; Goedert 2001; Selkoe 2001; Muchowski 2002; Walsh and Selkoe 2004; Muchowski and Wacker 2005).

A remarkable feature of the aggregating proteins in AD, PD, and polyQ diseases is that, although they are unrelated, they typically co-localize with components of the protein quality control machinery, including various chaperones mostly of the Hsp family and components of protein degradation systems (Sherman and Goldberg 2001; Muchowski and Wacker 2005). This co-localization might indicate a failed attempt to refold or degrade the aggregated proteins that ended up in an irreversible sequestration and subsequent loss of chaperone activity. Alternatively, the co-localization might indicate an action promoting a spatial sequestration of potentially toxic protein forms into inclusion bodies. Consistent with these ideas, there is now numerous evidence supporting a pivotal role of Hsp70/Hsc70 proteins in regulating the deposition and toxicity of various disease-causing proteins (Kobayashi and Sobue 2001; Bonini and Fortini 2003; Levine et al. 2004; Marsh and Thompson 2004; Muchowski and Wacker 2005). For example, overexpression of Hsp70/Hsc70 suppresses the toxicity of pathogenic Aβ, α-synuclein, and proteins containing expanded polyQ repeats in cell culture and animal models of AD, PD, and polyQ diseases (Warrick et al. 1998; Chai et al. 1999; Warrick et al. 1999; Jana et al. 2000; Krobitsch and Lindquist 2000; Muchowski et al. 2000; Auluck and Bonini 2002; Auluck et al. 2002; McLean et al. 2002; Dou et al. 2003; Gunawardena et al. 2003; Klucken et al. 2004; Magrane et al. 2004).

A number of findings indicate that CSP may play a pivotal role in some neurodegenerative diseases of humans. As indicated by the massive neurodegeneration in CSP mutant flies and mice (Zinsmaier et al. 1994; Fernandez-Chacon et al. 2004), interference with CSP function is potentially a mechanism of causing neurodegenerative diseases. Notably, CSPα KO mice exhibit a neurodegenerative phenotype that is clinically, although not pathologically, reminiscent of amyotrophic lateral sclerosis (Fernandez-Chacon et al. 2004). While the precise sequence of pathogenic events in CSPα mutants remains to be identified, it appears that components of the release machinery, like SNARE proteins, become secondarily disabled and neurotoxic (Chandra et al. 2005). It is plausible that accumulation of disabled SNAREs may lead to similar neurotoxic products as botulinum and tetanus toxin-induced cleavage of SNARE proteins, which cause irreversible degeneration of synaptic terminals (Schiavo et al. 2000). Alternatively, CSP's control over presynaptic Ca^{2+} homeostasis (Dawson-Scully et al. 2000; Bronk et al. 2005) could lead to mitochondrial Ca^{2+} overload and degeneration (Demaurex and Distelhorst 2003).

Genetic or environmental factors that can sequester CSPα or induce changes in CSPα activity could contribute to common neurodegenerative diseases. For example, the sequestration of CSP by expanded polyQ stretches (Miller et al. 2003a) could be a factor that contributes to the neurodegeneration of polyQ-induced diseases. In addition, SGT mutations suppress the toxicity of human Aβ peptide in *C. elegans* (Fonte et al. 2002). Finally, expression of CSPα is significantly altered by antidepressants (Cordeiro et al. 2000a, b; Yamada et al. 2001; Cordeiro et al. 2003, 2004), amphetamines (Bowyer and Davies 1999), and diabetes (Zhang et al. 2002). Together, these studies indicate that changes in CSPα activity may have significant consequences for the nervous system. Future studies will have to explore whether CSPα is indeed a critical factor in polyQ and other neurodegenerative diseases.

Acknowledgements This work has been supported by grants to K.E.Z. from NINDS (R03 NS057215–01, R21 NS055202–02) and a PERT fellowship to M.I. (K12 GM000708).

References

Angeletti PC, Walker D, Panganiban AT (2002) Small glutamine-rich protein/viral protein U-binding protein is a novel cochaperone that affects heat shock protein 70 activity. Cell Stress Chaperones 7:258–268

Arnold C, Reisch N, Leibold C, Becker S, Prufert K, Sautter K, Palm D, Jatzke S, Buchner S, Buchner E (2004) Structure-function analysis of the cysteine string protein in Drosophila: cysteine string, linker and C terminus. J Exp Biol 207:1323–1334

Auluck PK, Bonini NM (2002) Pharmacological prevention of Parkinson disease in Drosophila. Nat Med 8:1185–1186

Auluck PK, Chan HY, Trojanowski JQ, Lee VM, Bonini NM (2002) Chaperone suppression of alpha-synuclein toxicity in a Drosophila model for Parkinson's disease. Science 295:865–868

Berjanskii M, Riley M, Van Doren SR (2002) Hsc70-interacting HPD loop of the J domain of polyomavirus T antigens fluctuates in ps to ns and micros to ms. J Mol Biol 321:503–516

Betz A, Ashery U, Rickmann M, Augustin I, Neher E, Sudhof TC, Rettig J, Brose N (1998) Munc13–1 is a presynaptic phorbol ester receptor that enhances neurotransmitter release. Neuron 21:123–136

Boal F, Le Pevelen S, Cziepluch C, Scotti P, Lang J (2007) Cysteine-string protein isoform beta (Cspbeta) is targeted to the trans-Golgi network as a non-palmitoylated CSP in clonal beta-cells. Biochim Biophys Acta 1773:109–119

Boal F, Zhang H, Tessier C, Scotti P, Lang J (2004) The variable C-terminus of cysteine string proteins modulates exocytosis and protein–protein interactions. Biochemistry (Mosc) 43:16212–16223

Bonini NM, Fortini ME (2003) Human neurodegenerative disease modeling using Drosophila. Annu Rev Neurosci 26:627–656

Bowyer JF, Davies DL (1999) Changes in mRNA levels for heat-shock/stress proteins (Hsp) and a secretory vesicle associated cysteine-string protein (Csp1) after amphetamine (AMPH) exposure. Ann N Y Acad Sci 890:314–329

Braun J, Wilbanks SM, Scheller RH (1996) The cysteine string secretory vesicle protein activates Hsc70 ATPase. J Biol Chem 271:25989–25993

Braun JEA, Scheller RA (1995) Cysteine string protein, a DnaJ family member, is present on diverse secretory vesicles. Neuropharmacology 34:1361–1369

Brodsky JL, Chiosis G (2006) Hsp70 molecular chaperones: emerging roles in human disease and identification of small molecule modulators. Curr Top Med Chem 6:1215–1225

Bronk P, Nie Z, Klose MK, Dawson-Scully K, Zhang J, Robertson RM, Atwood HL, Zinsmaier KE (2005) The multiple functions of cysteine-string protein analyzed at Drosophila nerve terminals. J Neurosci 25:2204–2214

Bronk P, Wenniger JJ, Dawson-Scully K, Guo X, Hong S, Atwood HL, Zinsmaier KE (2001) Drosophila Hsc70–4 is critical for neurotransmitter exocytosis in vivo. Neuron. 30:475–488

Brown H, Larsson O, Branstrom R, Yang SN, Leibiger B, Leibiger I, Fried G, Moede T, Deeney JT, Brown GR, Jacobsson G, Rhodes CJ, Braun JE, Scheller RH, Corkey BE, Berggren PO, Meister B (1998) Cysteine string protein (CSP) is an insulin secretory granule-associated protein regulating beta-cell exocytosis. EMBO J 17:5048–5058

Buchner E, Gundersen CB (1997) The DnaJ-like cysteine string protein and exocytotic neurotransmitter release. TINS 20:223–227

Bukau B, Horwich AL (1998) The Hsp70 and Hsp60 chaperone machines. Cell 92:351–366

Burgoyne RD, Fisher RJ, Graham ME, Haynes LP, Morgan A (2001) Control of membrane fusion dynamics during regulated exocytosis. Biochem Soc Trans 29:467–472

Burgoyne RD, Graham ME, Fisher RJ (2000) Cysteine string protein controls late steps in exocytosis. Methods Find Exp Clin Pharmacol 22:355–355

Callahan MA, Handley MA, Lee YH, Talbot KJ, Harper JW, Panganiban AT (1998) Functional interaction of human immunodeficiency virus type 1 Vpu and Gag with a novel member of the tetratricopeptide repeat protein family. J Virol 72:5189–5197

Caplan AJ, Cyr DM, Douglas MG (1993) Eukaryotic homologues of Escherichia coli dnaJ: a diverse protein family that functions with hsp70 stress proteins. Mol Biol Cell 4:555–563

Chai Y, Koppenhafer SL, Bonini NM, Paulson HL (1999) Analysis of the role of heat shock protein (Hsp) molecular chaperones in polyglutamine disease. J Neurosci 19:10338–10347

Chamberlain LH, Burgoyne RD (1996b) Identification of a novel cysteine string protein variant and expression of cysteine string proteins in non-neuronal cells. J Biol Chem 271:7320–7323

Chamberlain LH, Burgoyne RD (1997a) The molecular chaperone function of the secretory vesicle cysteine string proteins. J Biol Chem 272:31420–31426

Chamberlain LH, Burgoyne RD (1997b) Activation of the ATPase activity of heat-shock proteins Hsc70/Hsp70 by cysteine-string protein. Biochem J 322:853–858

Chamberlain LH, Burgoyne RD (1998a) Cysteine string protein functions directly in regulated exocytosis. Mol Biol Cell 9:2259–2267

Chamberlain LH, Burgoyne RD (1998b) The cysteine-string domain of the secretory vesicle cysteine-string protein is required for membrane targeting. Biochem J 335:205–209

Chamberlain LH, Burgoyne RD (2000) Cysteine string protein: the chaperone at the synapse. J Neurochem 74:1781–1789

Chamberlain LH, Graham ME, Kane S, Jackson JL, Maier VH, Burgoyne RD (2001) The synaptic vesicle protein, cysteine-string protein, is associated with the plasma membrane in 3T3-L1 adipocytes and interacts with syntaxin 4. J Cell Sci 114:445–455

Chamberlain LH, Henry J, Burgoyne RD (1996a) Cysteine string proteins are associated with chromaffin granules. J Biol Chem 271:19514–19517

Chandra S, Gallardo G, Fernandez-Chacon R, Schluter OM, Sudhof TC (2005) Alpha-synuclein cooperates with CSPalpha in preventing neurodegeneration. Cell 123:383–396

Cheetham ME, Caplan AJ (1998) Structure, function and evolution of DnaJ: conservation and adaptation of chaperone function. Cell Stress Chaperones 3:28–36

Chen S, Zheng X, Schulze KL, Morris T, Bellen H, Stanley EF (2002) Enhancement of presynaptic calcium current by cysteine string protein. J Physiol 538:383–389

Chin H, Lyu MS, Kwon OJ, Kozak CA (1997) Genetic mapping of the gene encoding cysteine string protein. Mamm Genome 8:456–457

Coppola T, Gundersen C (1996) Widespread expression of human cysteine string proteins. FEBS Lett 391:269–272

Cordeiro ML, Gundersen CB, Umbach JA (2003) Dietary lithium induces regional increases of mRNA encoding cysteine string protein in rat brain. J Neurosci Res 73:865–869

Cordeiro ML, Gundersen CB, Umbach JA (2004) Convergent effects of lithium and valproate on the expression of proteins associated with large dense core vesicles in NGF-differentiated PC12 cells. Neuropsychopharm 29:39–44

Cordeiro ML, Umbach JA, Gundersen CB (2000a) Lithium ions enhance cysteine string protein gene expression in vivo and in vitro. J Neurochem 74:2365–2372

Cordeiro ML, Umbach JA, Gundersen CB (2000b) Lithium ions up-regulate mRNAs encoding dense-core vesicle proteins in nerve growth factor-differentiated PC12 cells. J Neurochem 75:2622–2625

Corsi AK, Schekman R (1997) The lumenal domain of Sec63p stimulates the ATPase activity of BiP and mediates BiP recruitment to the translocon in Saccharomyces cerevisiae. J Cell Biol 137:1483–1493

Cyr DM, Langer T, Douglas MG (1994) DnaJ-like proteins: molecular chaperones and specific regulators of Hsp70. Trends Biochem Sci 19:176–181

Cziepluch C, Kordes E, Poirey R, Grewenig A, Rommelaere J, Jauniaux JC (1998) Identification of a novel cellular TPR-containing protein, SGT, that interacts with the nonstructural protein NS1 of parvovirus H-1. J Virol 72:4149–4156

Daugaard M, Rohde M, Jaattela M (2007) The heat shock protein 70 family: Highly homologous proteins with overlapping and distinct functions. FEBS Lett 581:3702–3710

Dawson-Scully K (2003) The role of cysteine string proteins at the neuromuscular junction of Drosophila. National Library of Canada, Thesis (Ph.D.) University of Toronto

Dawson-Scully K, Bronk P, Atwood HL, Zinsmaier KE (2000) Cysteine-string protein increases the calcium sensitivity of neurotransmitter exocytosis in Drosophila. J Neurosci 20:6039–6047

Dawson-Scully K, Lin Y, Imad M, Zhang J, Marin L, Horne JA, Meinertzhagen IA, Karunanithi S, Zinsmaier KE, Atwood HL (2007) Morphological and functional effects of altered cysteine string protein at the Drosophila larval neuromuscular junction. Synapse 61:1–16

Demand J, Luders J, Hohfeld J (1998) The carboxy-terminal domain of Hsc70 provides binding sites for a distinct set of chaperone cofactors. Mol Cell Biol 18:2023–2028

Demaurex N, Distelhorst C (2003) Cell biology. Apoptosis – the calcium connection. Science 300:65–67

Dou F, Netzer WJ, Tanemura K, Li F, Hartl FU, Takashima A, Gouras GK, Greengard P, Xu H (2003) Chaperones increase association of tau protein with microtubules. Proc Natl Acad Sci USA 100:721–726

Eberle KK, Zinsmaier KE, Buchner SMG, Jenni M, Arnold C, Leibold C, Reisch D, Walter N, Hafen E, Hofbauer A, Pflugfelder GO, Buchner E (1998) Wide distribution of cysteine string protein in Drosophila tissues revealed by targeted mutagenesis. Cell Tiss Res 294:203–217

Eisenberg E, Greene LE (2007) Multiple roles of auxilin and hsc70 in clathrin-mediated endocytosis. Traffic 8:640–646

Evans GJ, Barclay JW, Prescott GR, Jo SR, Burgoyne RD, Birnbaum MJ, Morgan A (2006) Protein kinase B/Akt is a novel cysteine string protein kinase that regulates exocytosis release kinetics and quantal size. J Biol Chem 281:1564–1572

Evans GJ, Morgan A (2002) Phosphorylation-dependent interaction of the synaptic vesicle proteins cysteine string protein and synaptotagmin I. Biochem J 364:343–347

Evans GJ, Morgan A (2003) Regulation of the exocytotic machinery by cAMP-dependent protein kinase: implications for presynaptic plasticity. Biochem Soc Trans 31:824–827

Evans GJ, Morgan A (2005) Phosphorylation of cysteine string protein in the brain: developmental, regional and synaptic specificity. Eur J Neurosci 21:2671–2680

Evans GJ, Morgan A, Burgoyne RD (2003) Tying everything together: the multiple roles of cysteine string protein (CSP) in regulated exocytosis. Traffic 4:653–659

Evans GJ, Wilkinson MC, Graham ME, Turner KM, Chamberlain LH, Burgoyne RD, Morgan A (2001) Phosphorylation of cysteine string protein by protein kinase A. Implications for the modulation of exocytosis. J Biol Chem 276:47877–47885

Eybalin M, Renard N, Aure F, Safieddine S (2002) Cysteine-string protein in inner hair cells of the organ of Corti: synaptic expression and upregulation at the onset of hearing. Eur J Neurosci 15:1409–1420

Fernandez-Chacon R, Wolfel M, Nishimune H, Tabares L, Schmitz F, Castellano-Munoz M, Rosenmund C, Montesinos ML, Sanes JR, Schneggenburger R, Sudhof TC (2004) The synaptic vesicle protein CSP alpha prevents presynaptic degeneration. Neuron 42:237–251

Fonte V, Kapulkin V, Taft A, Fluet A, Friedman D, Link CD (2002) Interaction of intracellular beta amyloid peptide with chaperone proteins. Proc Natl Acad Sci USA 99:9439–9444

Frydman J, Hohfeld J (1997) Chaperones get in touch: the Hip-Hop connection. Trends Biochem Sci 22:87–92

Galli T, Haucke V (2001) Cycling of synaptic vesicles: how far? How fast! Sci STKE 2001:RE1

Galli T, Haucke V (2004) Cycling of synaptic vesicles: how far? How fast! Sci STKE 2004:RE19

Goedert M (2001) Alpha-synuclein and neurodegenerative diseases. Nat Rev Neurosci 2:492–501

Goldstein LS (2003) Do disorders of movement cause movement disorders and dementia? Neuron 40:415–425

Graham ME, Burgoyne RD (2000) Comparison of cysteine string protein (Csp) and mutant a-SNAP overexpression reveals a role for Csp in late steps of membrane fusion in Dense-Core granule exocytosis in adrenal chromaffin cells. J Neurosci 20:1281–1289

Greaves J, Chamberlain LH (2006) Dual role of the cysteine-string domain in membrane binding and palmitoylation-dependent sorting of the molecular chaperone cysteine-string protein. Mol Biol Cell 17:4748–4759

Greaves J, Salaun C, Fukata Y, Fukata M, Chamberlain LH (2008) Palmitoylation and membrane interactions of the neuroprotective chaperone cysteine-string protein. J Biol Chem 283:25014–25026

Gruschus JM, Greene LE, Eisenberg E, Ferretti JA (2004a) Experimentally biased model structure of the Hsc70/auxilin complex: substrate transfer and interdomain structural change. Protein Sci 13:2029–2044

Gruschus JM, Han CJ, Greener T, Ferretti JA, Greene LE, Eisenberg E (2004b) Structure of the functional fragment of auxilin required for catalytic uncoating of clathrin-coated vesicles. Biochemistry (Mosc) 43:3111–3119

Gunawardena S, Her LS, Brusch RG, Laymon RA, Niesman IR, Gordesky-Gold B, Sintasath L, Bonini NM, Goldstein LS (2003) Disruption of axonal transport by loss of huntingtin or expression of pathogenic polyQ proteins in Drosophila. Neuron 40:25–40

Gundersen CB, Mastrogiacomo A, Faull K, Umbach JA (1994) Extensive lipidation of a Torpedo cysteine string protein. J Biol Chem 269:19197–19199

Gundersen CB, Mastrogiacomo A, Umbach JA (1995) Cysteine-string proteins as templates for membrane fusion: models of synaptic vesicle exocytosis. J Theor Biol 172:269–277

Gundersen CB, Umbach JA (1992) Suppression cloning of the cDNA for a candidate subunit of a presynaptic calcium channel. Neuron 9:527–537

Hartl FU, Hayer-Hartl M (2002) Molecular chaperones in the cytosol: from nascent chain to folded protein. Science 295:1852–1858

Heckmann M, Adelsberger H, Dudel J (1997) Evoked transmitter release at neuromuscular junctions in wild type and cysteine string protein null mutant larvae of Drosophila. Neurosci Lett 228:167–170

Hennessy F, Nicoll WS, Zimmermann R, Cheetham ME, Blatch GL (2005) Not all J domains are created equal: implications for the specificity of Hsp40-Hsp70 interactions. Protein Sci 14:1697–1709

Hohfeld J, Jentsch S (1997) GrpE-like regulation of the hsc70 chaperone by the anti-apoptotic protein BAG-1. EMBO J 16:6209–6216

Hsu CC, Davis KM, Jin H, Foos T, Floor E, Chen WQ, Tyburski JB, Yang CY, Schloss JV, Wu JY (2000) Association of L-glutamic acid decarboxylase to the 70-kDa heat shock protein as a potential anchoring mechanism to synaptic vesicles. J Biol Chem 275: 20822–20828

Huang K, Ghose R, Flanagan JM, Prestegard JH (1999) Backbone dynamics of the N-terminal domain in E. coli DnaJ determined by 15N- and 13CO-relaxation measurements. Biochemistry (Mosc) 38:10567–10577

Irmer H, Hohfeld J (1997) Characterization of functional domains of the eukaryotic co-chaperone Hip. J Biol Chem 272:2230–2235

Jahn R, Lang T, Sudhof TC (2003) Membrane fusion. Cell 112:519–533

Jana NR, Tanaka M, Wang G, Nukina N (2000) Polyglutamine length-dependent interaction of Hsp40 and Hsp70 family chaperones with truncated N-terminal huntingtin: their role in suppression of aggregation and cellular toxicity. Hum Mol Genet 9:2009–2018

Jiang J, Ballinger CA, Wu Y, Dai Q, Cyr DM, Hohfeld J, Patterson C (2001) CHIP is a U-box-dependent E3 ubiquitin ligase: identification of Hsc70 as a target for ubiquitylation. J Biol Chem 276:42938–42944

Jiang J, Taylor AB, Prasad K, Ishikawa-Brush Y, Hart PJ, Lafer EM, Sousa R (2003) Structure-function analysis of the auxilin J-domain reveals an extended Hsc70 interaction interface. Biochemistry (Mosc) 42:5748–5753

Jin H, Wu H, Osterhaus G, Wei J, Davis K, Sha D, Floor E, Hsu CC, Kopke RD, Wu JY (2003) Demonstration of functional coupling between gamma-aminobutyric acid (GABA) synthesis and vesicular GABA transport into synaptic vesicles. Proc Natl Acad Sci USA 100:4293–4298

Kelley WL (1998) The J-domain family and the recruitment of chaperone power. TIBS 23:222–227

Kelley WL (1999) Molecular chaperones: How J domains turn on Hsp70s. Curr Biol 9:R305–R308

Klucken J, Shin Y, Masliah E, Hyman BT, McLean PJ (2004) Hsp70 reduces alpha-synuclein aggregation and toxicity. J Biol Chem 279:25497–25502

Kobayashi Y, Sobue G (2001) Protective effect of chaperones on polyglutamine diseases. Brain Res Bull 56:165–168

Kohan SA, Pescatori M, Brecha NC, Mastrogiacomo A, Umbach JA, Gundersen CB (1995) Cysteine string protein immunoreactivity in the nervous system and adrenal gland of rat. J Neurosci 15:6230–6238

Krobitsch S, Lindquist S (2000) Aggregation of huntingtin in yeast varies with the length of the polyglutamine expansion and the expression of chaperone proteins. Proc Natl Acad Sci USA 97:1589–1594

Landry SJ (2003) Structure and energetics of an allele-specific genetic interaction between dnaJ and dnaK: correlation of nuclear magnetic resonance chemical shift perturbations in the J-domain of Hsp40/DnaJ with binding affinity for the ATPase domain of Hsp70/DnaK. Biochemistry (Mosc) 42:4926–4936

Leveque C, Pupier S, Marqueze B, Geslin L, Kataoka M, Takahashi M, De Waard M, Seagar M (1998) Interaction of cysteine string proteins with the alpha1A subunit of the P/Q-type calcium channel. J Biol Chem 273:13488–13492

Levine MS, Cepeda C, Hickey MA, Fleming SM, Chesselet MF (2004) Genetic mouse models of Huntington's and Parkinson's diseases: illuminating but imperfect. Trends Neurosci 27:691–697

Liu FH, Wu SJ, Hu SM, Hsiao CD, Wang C (1999) Specific interaction of the 70-kDa heat shock cognate protein with the tetratricopeptide repeats. J Biol Chem 274:34425–34432

Lou X, Scheuss V, Schneggenburger R (2005) Allosteric modulation of the presynaptic Ca^{2+} sensor for vesicle fusion. Nature 435:497–501

Magga JM, Jarvis SE, Arnot MI, Zamponi GW, Braun JEA (2000) Cysteine string protein regulates G protein modulation of N-type calcium channels. Neuron 28:195–204

Magrane J, Smith RC, Walsh K, Querfurth HW (2004) Heat shock protein 70 participates in the neuroprotective response to intracellularly expressed beta-amyloid in neurons. J Neurosci 24:1700–1706

Marsh JL, Thompson LM (2004) Can flies help humans treat neurodegenerative diseases? Bioessays 26:485–496

Mastrogiacomo A, Gundersen CB (1995) The nucleotide and deduced amino acid sequence of a rat cysteine string protein. Brain Res Mol Brain Res 28:12–18

Mastrogiacomo A, Kohan SA, Whiteledge JP, Gundersen CB (1998a) Intrinsic membrane association of Drosophila cysteine string proteins. FEBS Lett 436:85–91

Mastrogiacomo A, Kornblum HI, Umbach JA, Gundersen CB (1998b) A Xenopus cysteine string protein with a cysteine residue in the J domain. Biochim Biophys Acta 1401:239–241

Mastrogiacomo A, Parsons SM, Zampighi GA, Jenden DJ, Umbach JA, Gundersen CB (1994) Cysteine string proteins – a potential link between synaptic vesicles and presynaptic Ca^{2+} channels. Science 263:981–982

Mayer MP, Bukau B (2005) Hsp70 chaperones: cellular functions and molecular mechanism. Cell Mol Life Sci 62:670–684

McLean PJ, Kawamata H, Shariff S, Hewett J, Sharma N, Ueda K, Breakefield XO, Hyman BT (2002) TorsinA and heat shock proteins act as molecular chaperones: suppression of alpha-synuclein aggregation. J Neurochem 83:846–854

Miller LC, Swayne LA, Chen L, Feng ZP, Wacker JL, Muchowski PJ, Zamponi GW, Braun JE (2003a) Cysteine string protein (CSP) inhibition of N-type calcium channels is blocked by mutant huntingtin. J Biol Chem 278:53072–53081

Miller LC, Swayne LA, Kay JG, Feng ZP, Jarvis SE, Zamponi GW, Braun JE (2003b) Molecular determinants of cysteine string protein modulation of N-type calcium channels. J Cell Sci 116:2967–2974

Morano KA (2007) New tricks for an old dog: the evolving world of Hsp70. Ann N Y Acad Sci 1113:1–14

Muchowski PJ (2002) Protein misfolding, amyloid formation, and neurodegeneration: a critical role for molecular chaperones? Neuron 35:9–12

Muchowski PJ, Schaffar G, Sittler A, Wanker EE, Hayer-Hartl MK, Hartl FU (2000) Hsp70 and hsp40 chaperones can inhibit self-assembly of polyglutamine proteins into amyloid-like fibrils. Proc Natl Acad Sci USA 97:7841–7846

Muchowski PJ, Wacker JL (2005) Modulation of neurodegenration by molecular chaperones. Nat Rev Neurosci 6:11–22

Natochin M, Campbell TN, Barren B, Miller LC, Hameed S, Artemyev NO, Braun JE (2005) Characterization of the G alpha(s) regulator cysteine string protein. J Biol Chem 280:30236–30241. Epub 32005 Jun 30222

Nie Z, Ranjan R, Wenniger JJ, Hong SN, Bronk P, Zinsmaier KE (1999) Overexpression of cysteine string protein in Drosophila reveals interactions with syntaxin. J Neurosci 19:10270–10279

Ohtsuka K, Hata M (2000) Mammalian HSP40/DNAJ homologs: cloning of novel cDNAs and a proposal for their classification and nomenclature. Cell Stress Chaperones 5:98–112

Ohyama T, Verstreken P, Ly CV, Rosenmund T, Rajan A, Tien AC, Haueter C, Schulze KL, Bellen HJ (2007) Huntingtin-interacting protein 14, a palmitoyl transferase required for exocytosis and targeting of CSP to synaptic vesicles. J Cell Biol 179:1481–1496

Poage RE, Meriney SD, Gundersen CB, Umbach JA (1999) Antibodies against cysteine string proteins inhibit evoked neurotransmitter release at Xenopus neuromuscular junctions. J Neurophysiol 82:50–59

Pupier S, Leveque C, Marqueze B, Kataoka M, Takahashi M, Seagar MJ (1997) Cysteine string proteins associated with secretory granules of the rat neurohypophysis. J Neurosci 17:2722–2727

Qian YQ, Patel D, Hartl FU, McColl DJ (1996) Nuclear magnetic resonance solution structure of the human Hsp40 (HDJ-1) J-domain. J Mol Biol 260:224–235

Qiu XB, Shao YM, Miao S, Wang L (2006) The diversity of the DnaJ/Hsp40 family, the crucial partners for Hsp70 chaperones. Cell Mol Life Sci 63:2560–2570

Ranjan R, Bronk P, Zinsmaier KE (1998) Cysteine string protein is required for calcium secretion coupling of evoked neurotransmission in Drosophila but not for vesicle recycling. J Neurosci 18:956–964

Redecker P, Pabst H, Grube D (1998) Munc-18-1 and cysteine string protein (csp) in pinealocytes of the gerbil pineal gland. Cell Tiss Res 293:245–252

Rhee JS, Betz A, Pyott S, Reim K, Varoqueaux F, Augustin I, Hesse D, Sudhof TC, Takahashi M, Rosenmund C, Brose N (2002) Beta phorbol ester- and diacylglycerol-induced augmentation of transmitter release is mediated by Munc13s and not by PKCs. Cell 108:121–133

Rizo J, Rosenmund C (2008) Synaptic vesicle fusion. Nat Struct Mol Biol 15:665–674

Rohrbough J, Broadie K (2005) Lipid regulation of the synaptic vesicle cycle. Nat Rev Neurosci 6:139–150

Rossi V, Banfield DK, Vacca M, Dietrich LE, Ungermann C, D'Esposito M, Galli T, Filippini F (2004) Longins and their longin domains: regulated SNAREs and multifunctional SNARE regulators. Trends Biochem Sci 29:682–688

Ruiz R, Casanas JJ, Sudhof TC, Tabares L (2008) Cysteine string protein-alpha is essential for the high calcium sensitivity of exocytosis in a vertebrate synapse. Eur J Neurosci 27:3118–3131

Sakisaka T, Meerlo T, Matteson J, Plutner H, Balch WE (2002) Rab-alphaGDI activity is regulated by a Hsp90 chaperone complex. EMBO J 21:6125–6135

Scheufler C, Brinker A, Bourenkov G, Pegoraro S, Moroder L, Bartunik H, Hartl FU, Moarefi I (2000) Structure of TPR domain-peptide complexes: critical elements in the assembly of the Hsp70-Hsp90 multichaperone machine. Cell 101:199–210

Schiavo G, Matteoli M, Montecucco C (2000) Neurotoxins affecting neuroexocytosis. Physiol Rev 80:717–766

Schmitz F, Tabares L, Khimich D, Strenzke N, de la Villa-Polo P, Castellano-Munoz M, Bulankina A, Moser T, Fernandez-Chacon R, Sudhof TC (2006) CSPalpha-deficiency causes massive and rapid photoreceptor degeneration. Proc Natl Acad Sci USA 103:2926–2931

Selkoe DJ (2001) Alzheimer's disease: genes, proteins, and therapy. Physiol Rev 81:741–766

Shapira R, Silberberg SD, Ginsburg S, Rahamimoff R (1987) Activation of protein kinase C augments evoked transmitter release. Nature 325:58–60

Sherman MY, Goldberg AL (2001) Cellular defenses against unfolded proteins: a cell biologist thinks about neurodegenerative diseases. Neuron 29:15–32

Slepnev VI, De Camilli P (2000) Accessory factors in clathrin-dependent synaptic vesicle endocytosis. Nat Rev Neurosci 1:161–172

Sondermann H, Scheufler C, Schneider C, Hohfeld J, Hartl FU, Moarefi I (2001) Structure of a Bag/Hsc70 complex: convergent functional evolution of Hsp70 nucleotide exchange factors. Science 291:1553–1557

Stevens CF, Sullivan JM (1998) Regulation of the readily releasable vesicle pool by protein kinase C. Neuron 21:885–893

Stowers RS, Isacoff EY (2007) Drosophila huntingtin-interacting protein 14 is a presynaptic protein required for photoreceptor synaptic transmission and expression of the palmitoylated proteins synaptosome-associated protein 25 and cysteine string protein. J Neurosci 27:12874–12883

Sudhof TC (2004) The synaptic vesicle cycle. Annu Rev Neurosci 27:509–547

Swayne LA, Beck KE, Braun JE (2006) The cysteine string protein multimeric complex. Biochem Biophys Res Commun 348:83–91

Swayne LA, Blattler C, Kay JG, Braun JE (2003) Oligomerization characteristics of cysteine string protein. Biochem Biophys Res Commun 300:921–926

Szyperski T, Pellecchia M, Wall D, Georgopoulos C, Wuthrich K (1994) NMR structure determination of the Escherichia coli DnaJ molecular chaperone: secondary structure and backbone fold of the N-terminal region (residues 2–108) containing the highly conserved J domain. Proc Natl Acad Sci USA 91:11343–11347

Takayama S, Reed JC (2001) Molecular chaperone targeting and regulation by BAG family proteins. Nat Cell Biol 3:E237–E241

Tobaben S, Thakur P, Fernandez-Chacon R, Sudhof TC, Rettig J, Stahl B (2001) A trimeric protein complex functions as a synaptic chaperone machine. Neuron 31:987–999

Tobaben S, Varoqueaux F, Brose N, Stahl B, Meyer G (2003) A brain-specific isoform of small glutamine-rich tetratricopeptide repeat-containing protein binds to Hsc70 and the cysteine string protein. J Biol Chem 278:38376–38383

Tsai J, Douglas MG (1996) A conserved HPD sequence of the J-domain is necessary for YDJ1 stimulation of Hsp70 ATPase activity at a site distinct from substrate binding. J Biol Chem 271:9347–9354

Umbach JA, Gundersen CB (1997) Evidence that cysteine string proteins regulate an early step in the Ca^{2+}-dependent secretion of neurotransmitter at Drosophila neuromuscular junctions. J Neurosci 17:7203–7209

Umbach JA, Mastrogiacomo A, Gundersen CB (1995) Cysteine string proteins and presynaptic function. J Physiol Paris 89:95–101

Umbach JA, Saitoe M, Kidokoro Y, Gundersen CB (1998) Attenuated influx of calcium ions at nerve endings of csp and shibire mutant Drosophila. J Neurosci 18:3233–3240

Umbach JA, Zinsmaier KE, Eberle KK, Buchner E, Benzer S, Gundersen CB (1994) Presynaptic dysfunction in Drosophila csp mutants. Neuron 13:899–907

Ungewickell E, Ungewickell H, Holstein SE, Lindner R, Prasad K, Barouch W, Martin B, Greene LE, Eisenberg E (1995) Role of auxilin in uncoating clathrin-coated vesicles. Nature 378:632–635

Ungewickell EJ, Hinrichsen L (2007) Endocytosis: clathrin-mediated membrane budding. Curr Opin Cell Biol 19:417–425

van deGoor J, Kelly RB (1996) Association of Drosophila cysteine string proteins with membranes. FEBS Lett 380:251–256

van deGoor J, Ramaswami M, Kelly R (1995) Redistribution of synaptic vesicles and their proteins in temperature-sensitive shibire(ts1) mutant Drosophila. Proc Natl Acad Sci USA 92:5739–5743

Virmani T, Ertunc M, Sara Y, Mozhayeva M, Kavalali ET (2005) Phorbol esters target the activity-dependent recycling pool and spare spontaneous vesicle recycling. J Neurosci 25:10922–10929

Walsh DM, Selkoe DJ (2004) Deciphering the molecular basis of memory failure in Alzheimer's disease. Neuron 44:181–193

Walsh P, Bursac D, Law YC, Cyr D, Lithgow T (2004) The J-protein family: modulating protein assembly, disassembly and translocation. EMBO Rep 5:567–571

Warrick JM, Chan HY, Gray-Board GL, Chai Y, Paulson HL, Bonini NM (1999) Suppression of polyglutamine-mediated neurodegeneration in Drosophila by the molecular chaperone HSP70. Nat Genet 23:425–428

Warrick JM, Paulson HL, Gray-Board GL, Bui QT, Fischbeck KH, Pittman RN, Bonini NM (1998) Expanded polyglutamine protein forms nuclear inclusions and causes neural degeneration in Drosophila. Cell 93:939–949

Waters J, Smith SJ (2000) Phorbol esters potentiate evoked and spontaneous release by different presynaptic mechanisms. J Neurosci 20:7863–7870

Wilbur JD, Hwang PK, Brodsky FM (2005) New faces of the familiar clathrin lattice. Traffic 6:346–350

Wu MN, Fergestad T, Lloyd TE, He Y, Broadie K, Bellen HJ (1999) Syntaxin 1A interacts with multiple exocytic proteins to regulate neurotransmitter release in vivo. Neuron 23:593–605

Wu SJ, Liu FH, Hu SM, Wang C (2001) Different combinations of the heat-shock cognate protein 70 (hsc70) C-terminal functional groups are utilized to interact with distinct tetratricopeptide repeat-containing proteins. Biochem J 359:419–426

Wu XS, Wu LG (2001) Protein kinase c increases the apparent affinity of the release machinery to Ca^{2+} by enhancing the release machinery downstream of the Ca^{2+} sensor. J Neurosci 21:7928–7936

Yamada M, Yamazaki S, Takahashi K, Nara K, Ozawa H, Yamada S, Kiuchi Y, Oguchi K, Kamijima K, Higuchi T, Momose K (2001) Induction of cysteine string protein after chronic antidepressant treatment in rat frontal cortex. Neurosci Lett 301:183–186

Yawo H (1999) Protein kinase C potentiates transmitter release from the chick ciliary presynaptic terminal by increasing the exocytotic fusion probability. J Physiol (Lond) 515:169–180

Young JC, Barral JM, Ulrich Hartl F (2003) More than folding: localized functions of cytosolic chaperones. Trends Biochem Sci 28:541–547

Zhang H, Kelley WL, Chamberlain LH, Burgoyne RD, Lang J (1999) Mutational analysis of cysteine-string protein function in insulin exocytosis. J Cell Sci 112:1345–1351

Zhang H, Kelley WL, Chamberlain LH, Burgoyne RD, Wollheim CB, Lang J (1998) Cysteine-string proteins regulate exocytosis of insulin independent from transmembrane ion fluxes. FEBS Lett 437:267–272

Zhang W, Khan A, Ostenson CG, Berggren PO, Efendic S, Meister B (2002) Down-regulated expression of exocytotic proteins in pancreatic islets of diabetic GK rats. Biochem Biophys Res Commun 291:1038–1044

Zhao CM, Jacobsson G, Chen D, Hakanson R, Meister B (1997) Exocytotic proteins in enterochromaffin-like (ECL) cells of the rat stomach. Cell Tiss Res 290:539–551

Zhao X, Braun AP, Braun JE (2008) Biological roles of neural J proteins. Cell Mol Life Sci 65:2385–2396

Zinsmaier KE (1997) Cysteine String Proteins. In Guidebook to Molecular Chaperones and Protein Folding Catalysts, Marey-Jane Gething, ed. (Oxford: Sambrook & Tooze, Oxford University Press), pp. 115–117.

Zinsmaier KE, Bronk P (2001) Molecular chaperones and the regulation of neurotransmitter exocytosis. Biochem Pharm 62:1–11

Zinsmaier KE, Eberle KK, Buchner E, Walter N, Benzer S (1994) Paralysis and early death in cysteine string protein mutants of Drosophila. Science 263:977–980

Zinsmaier KE, Hofbauer A, Heimbeck G, Pflugfelder GO, Buchner S, Buchner E (1990) A cysteine-string protein is expressed in retina and brain of Drosophila. J Neurogenet 7:15–29

Zoghbi HY, Orr HT (2000) Glutamine repeats and neurodegeneration. Annu Rev Neurosci 23:217–247

Chapter 9
The Role of Protein SUMOylation in Neuronal Function

Kevin A. Wilkinson and Jeremy M. Henley

Abstract The post-translational modification of proteins by members of the Small Ubiquitin-like MOdifier (SUMO) family is beginning to emerge as a key regulator of neuronal function. SUMO conjugation modifies the interaction of target proteins with protein partners, and thereby alters their subcellular localization, activity and stability. Importantly, SUMOylation is readily reversible, allowing cells to respond rapidly to varying cellular demands. SUMO has already been implicated in the regulation of multiple neuronal signalling pathways, mitochondrial dynamics, spine formation and synaptogenesis, as well as the direct control of neuronal excitability via its modulation of cell surface receptors and ion channels. Here, we outline the basic mechanics of the SUMO pathway, review major recent advances in the field and discuss the far-reaching implications of neuronal SUMOylation in health and disease.

9.1 Introduction

The post-translational modification of proteins is critical to their dynamic and reversible regulation of nearly all cellular functions. In neurons, where the temporal control of protein function is particularly crucial, post-translational regulation of proteins by the addition of phosphate groups, lipid groups (as in palmitoylation) and other proteins (such as ubiquitin) have been shown to control a vast number of signalling pathways. Recently, it has emerged that modification of proteins with a member of the (SUMO) Small Ubiquitin-like MOdifier family of proteins also plays key roles in neuronal function.

SUMOylation involves the covalent conjugation of a member of the SUMO family to lysine residues in target proteins. While the consequences of an analogous

---

J.M. Henley (✉)
Department of Anatomy MRC Centre for Synaptic Plasticity,
School of Biochemistry, Medical Sciences Building,
University of Bristol, University Walk, Bristol, BS8 1TD, UK
e-mail: J.M.Henley@bristol.ac.uk

A. Wyttenbach and V. O'Connor (eds.), *Folding for the Synapse*,
DOI 10.1007/978-1-4419-7061-9_9,

protein modification, ubiquitination, generally (but not always) lead to protein degradation (see Chap. 3 by J. Höhfeld and colleagues for details on protein degradation), the roles of SUMOylation are much more diverse, with roles in nuclear function, protein trafficking, and even the regulation of proteins at the cell surface.

9.2 The SUMO Family

SUMO-1 was first reported in 1996 as an 11 kDa ubiquitin-like protein, which was conjugated to the nuclear pore complex protein RanGAP (Matunis et al. 1996; Mahajan et al. 1997), regulating its localisation. Since then several hundred targets of SUMOylation have been reported. Initially, SUMO conjugation was shown to occur predominantly in the nucleus (Kamitani et al. 1997) and the majority of identified SUMOylated proteins played roles in nuclear processes. However, more recently multiple non-nuclear SUMO targets have been reported indicating that SUMO modification is not solely a nuclear phenomenon.

Deletion of the single SUMO protein in yeast, Smt3, is fatal for *Saccharomyces cerevisiae* (Meluh and Koshland 1995). Further, ablation of the SUMO pathway by either knockdown or deletion of Ubc9, a critical enzyme in the SUMOylation pathway, results in a loss of viability of mammalian cells (Hayashi et al. 2002; Nacerddine et al. 2005), highlighting the critical cellular roles played by SUMOylation.

In mammals, there are four SUMO paralogues, designated SUMO-1 to -4. SUMO-2 and -3 differ only in three N-terminal residues and are often referred to collectively as SUMO-2/3; however, they share approximately 50% homology with SUMO-1. SUMO-1 to -3 have been shown to conjugate substrate proteins. SUMO-4 has been reported to be a diabetes susceptibility gene (Bohren et al. 2004; Guo et al. 2004), but despite high expression of the mRNA in kidney and immune tissues, there are no reports currently documenting the expression of the protein. Further, it remains to be determined if SUMO-4 can conjugate proteins due to a proline residue unique to the SUMO-4 paralogue that appears to prevent the maturation of the immature polypeptide in vitro by proteins known to mature other SUMO proteins (Owerbach et al. 2005). This review therefore only discusses SUMO-1 to -3.

While some substrates are modified exclusively by one SUMO paralogue, many can be modified by both SUMO-1 and SUMO-2/3. It is currently unclear in most cases how the SUMOylation machinery distinguishes between SUMO-1 and SUMO-2/3 for specific conjugation of one paralogue or the other to proteins. Furthermore, for most substrates that can be modified by either paralogue, the functional differences of SUMO-1 or SUMO-2/3 conjugation have yet to be defined.

Interestingly, SUMO-2/3 has been reported to form SUMO chains, whereas SUMO-1 does not (Tatham et al. 2001; Bencsath et al. 2002; Bylebyl et al. 2003). However, SUMO-1 may act as a chain-terminator of elongating SUMO-2/3 chains (Matic et al. 2008). As with many aspects of SUMO biology, the function of chain formation on most substrates is unclear. However, the exciting discovery of SUMO-targeted ubiquitin ligases (StUbLs) highlights an elegant example of interplay between the SUMO and ubiquitin systems (for review see (Perry et al. 2008)) and adds an increasing level of sophistication to mechanisms highlighted elsewhere in this book. For example, the protein PML can be modified by both SUMO-1 and SUMO-2/3. SUMO-1 modification is required for the formation of nuclear domains known as PML nuclear bodies (Heun 2007); however, SUMO-2/3 chain formation leads to the degradation of PML, via the recruitment of the SUMO-binding protein RNF4, which mediates the ubiquitination of the SUMO chains, targeting the whole complex for degradation (Lallemand-Breitenbach et al. 2008; Tatham et al. 2008; Weisshaar et al. 2008).

SUMO-1 and SUMO-2/3 also appear to differ in their conjugation dynamics. Whereas there is a large free pool of cellular SUMO-2/3, there is very little unconjugated SUMO-1, suggesting SUMO-2/3 may act as a cellular reserve of SUMO (Saitoh and Hinchey 2000). Indeed, upon various cellular stresses, there is a massive increase in SUMO-2/3 conjugation in many cell types (for reviews see Bossis and Melchior 2006; Agbor and Taylor 2008; Tempe et al. 2008; Yang et al. 2008a)

9.3 SUMO Conjugation

SUMO proteins are first synthesised as inactive precursors that are cleaved at the C-terminus by a family of SUMO-specific protease (SENP) enzymes. This cleavage reveals a di-glycine motif that renders SUMO capable of conjugation to lysine residues in target proteins (Fig. 9.1). An elegant aspect of the SUMO pathway is that the same SENP enzymes are also responsible for the removal of SUMO from substrates once it has been conjugated, suggesting that the SENPs regulate both the levels of free SUMO in addition to the abundance of SUMO conjugates (for review see Mukhopadhyay and Dasso 2007). Indeed, it will be interesting to determine how SENP activities regarding these opposing processes are regulated in vivo.

SUMO proteins are first activated in an ATP-dependent manner by an E1 "activating" enzyme, a heterodimer of SUMO activating enzyme (SAE)1 and SAE2 in mammals (Gong et al. 1999). Once activated, SUMO is passed to the active-site cysteine of the sole SUMO conjugating enzyme Ubc9 (Desterro et al. 1997; Gong et al. 1997; Johnson and Blobel 1997; Lee et al. 1998; Saitoh et al. 1998; Schwarz et al. 1998).

By virtue of Ubc9 being the only SUMO conjugating enzyme, and its ability to directly recognise substrates (Okuma et al. 1999), SUMOylation predominantly occurs at a consensus motif in substrate proteins (Rodriguez et al. 2001; Sampson et al. 2001) to which Ubc9 binds directly (Sampson et al. 2001). This consensus motif can be defined as ψKxD/E, where ψ is a large hydrophobic residue, K is the target lysine, x can be any residue, and D/E are aspartate or glutamic acid (acidic residues). However, it is important to emphasise that while ~75% of known SUMO

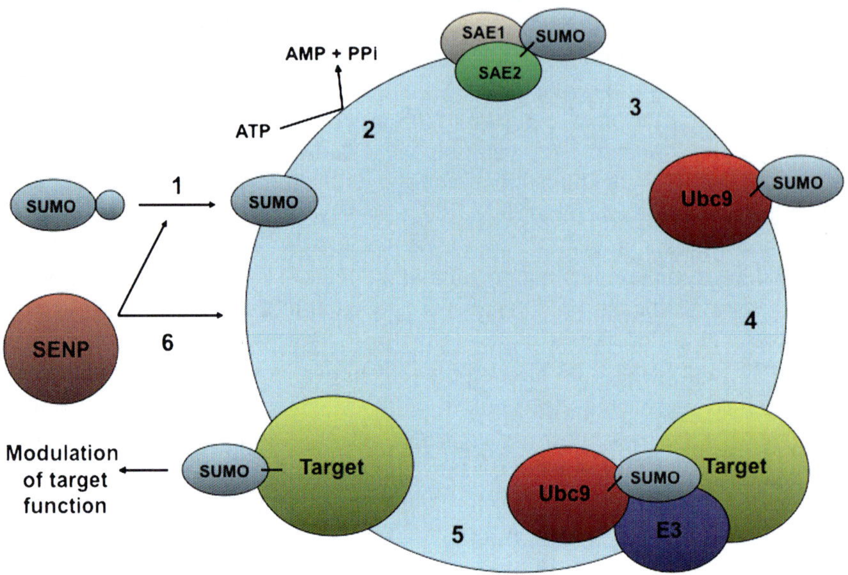

Fig. 9.1 The SUMOylation cycle. (1) SUMO proteins are first synthesised as precursors that must be cleaved via the hydrolase activity of the SENP enzymes before they can be conjugated. (2) Mature SUMO proteins are first activated in an ATP-dependent manner, resulting in the formation of a thioester bond between SUMO and the active-site cysteine of the E1 subunit SAE2. (3) Activated SUMO is then passed to the active-site cysteine of the E2 conjugating enzyme Ubc9. (4) Ubc9, often in conjunction with E3 enzymes, mediates target protein recognition. (5) SUMO is passed from the active-site cysteine of Ubc9 to a lysine residue in the target protein, resulting in changes in the function of the target protein. (6) SUMO can be removed from substrates via the isopeptidase activity of the SENP enzymes, allowing the cycle to be repeated

substrates are modified within a consensus motif (Xu et al. 2008), many are not and not all ψKxD/E motifs are SUMOylated. Thus, although knowledge of a general SUMOylation motif can be very useful in the bioinformatic identification of potential SUMO substrates it is by no means a definitive indicator that a protein is SUMOylated.

Because Ubc9 is capable of directly recognising (Gong et al. 1997; Bencsath et al. 2002; Tatham et al. 2003) and conjugating SUMO (Okuma et al. 1999) to substrate proteins, there was initially some debate as to whether E3 "ligase" proteins were required for SUMOylation, as is generally the case in the ubiquitin pathway. However, a number of proteins have subsequently been discovered which possess E3 activity in the SUMOylation pathway. These proteins appear to enhance the specificity of SUMOylation in vivo, generally through binding Ubc9 and the substrate protein to act as an adaptor for SUMOylation, or by holding the SUMO-Ubc9 intermediate or the substrate in a conformation more conducive to SUMO transfer.

9.4 Non-covalent SUMO Binding

In addition to conjugation of SUMO to substrate proteins (SUMOylation), various proteins have also been reported to bind SUMO non-covalently (for review see Kerscher 2007). For ubiquitination, many proteins have been reported to bind ubiquitin, and these non-covalent interactions appear to mediate the effects of ubiquitination by binding ubiquitin once it has been conjugated to substrate proteins. A similar situation is likely to exist for the SUMOylation pathway.

Currently, more than 16 ubiquitin-binding motifs (Hurley et al. 2006), and 4 SUMO-interacting motifs (SIMs) have been reported (Minty et al. 2000; Song et al. 2004; Hannich et al. 2005; Hecker et al. 2006). SIMs comprise a hydrophobic core surrounded by acidic flanking residues or phosphorylatable serines. Each of the reported SIMs appears to bind both SUMO-1 and SUMO-2/3; however, the affinity of the interaction varies depending on the context. Interestingly, numerous neuronal and synaptic proteins appear to contain these SUMO-interacting motifs and future research will shed light on how these potential SUMO-interacting proteins mediate the consequences of protein SUMOylation at the synapse.

9.5 Extranuclear SUMOylation and Neuronal Function

The majority of SUMOylation occurs in the nucleus or in the nuclear trafficking of cytosolic proteins (for reviews see Johnson 2004; Hay 2005; Geiss-Friedlander and Melchior 2007). Nonetheless, a growing number of extranuclear SUMO substrates have been identified and many of these may have important implications for neuronal function (Fig. 9.2). For a recent review, see Martin et al. (2007b).

9.6 SUMO as a Regulator of Multiple Signalling Pathways in Neurons

9.6.1 Regulation of G-Protein Signalling

G protein-coupled receptors (GPCRs) are integral membrane proteins characterised by seven transmembrane domains. More than 90% of known GPCRs are expressed in the brain, and together participate in all functions of the nervous system (Vassilatis et al. 2003). In their inactive state, GPCRs are bound to heterotrimeric G proteins comprising a GDP-bound Gα subunit and G$\beta\gamma$ subunits. Upon stimulation, the Gα subunits exchange their bound GDP for GTP and dissociate from the G$\beta\gamma$ subunit to perform multiple effector functions, depending on the bound Gα subunit (for a review see Pierce et al. 2002). The bound G$\beta\gamma$ subunit also dissociates from the receptor

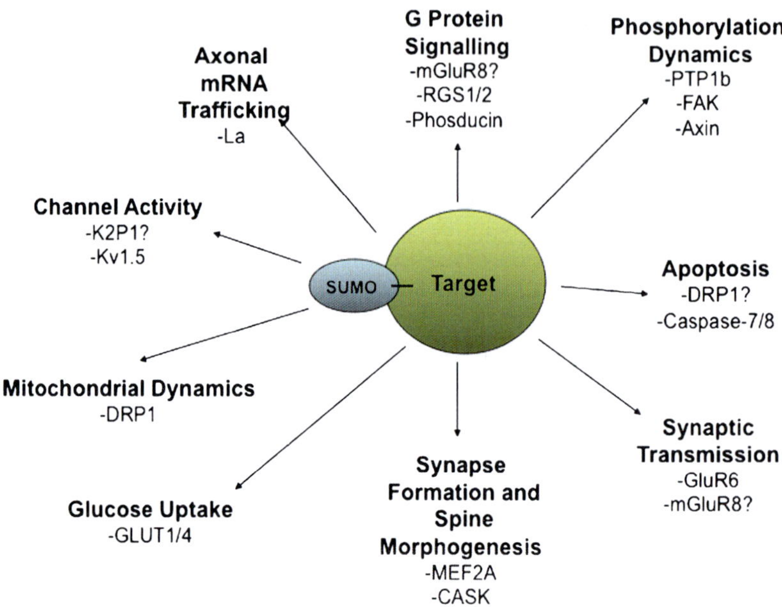

Fig. 9.2 SUMOylation has been associated with various neuronal functions. Neuronal functions shown to be modulated by protein SUMOylation are shown, along with SUMO substrates involved in these processes. Putative or unconfirmed targets are shown with a question mark. Abbreviations: *mGluR* metabotropic glutamate receptor, *RGS* regulator of G protein signalling, *PTP1B* protein tyrosine phosphatase 1B, *FAK* focal adhesion kinase, *DRP* dynamin-related protein, *MEF* myocyte enhancer factor, *CASK* calcium/calmodulin-dependent serine protein kinase, *GLUT* glucose transporter

to regulate further downstream pathways. Intriguingly, SUMOylation has been reported to potentially modify GPCRs themselves (see below), in addition to the regulation of proteins that modulate the activity of the released Gα and G$\beta\gamma$ subunits.

Regulator of G protein signalling (RGS) proteins are GTPase activating proteins (GAPs) that stimulate the GTPase activity of G protein Gα subunits, facilitating their return to their GDP-bound "off" state (Berman and Gilman 1998). RGS proteins are almost exclusively expressed in the brain (Mao et al. 2004), and a role for the SUMOylation of RGS1 and RGS2 has recently been reported in the regulation of μ-opioid receptor signalling (Rodriguez-Munoz et al. 2007). RGS1 and RGS2 can be modified by SUMO-1 and SUMO-2/3, and upon μ-opioid receptor activation, SUMOylated RGS proteins bind strongly both to the receptor and to Gα subunits, mediating desensitisation of the receptors (Rodriguez-Munoz et al. 2007).

The protein phosducin is highly expressed in the retina and pineal gland and functions in the regulation of G$\beta\gamma$ subunit signalling via sequestering G$\beta\gamma$ subunits and inhibiting their ability to activate effector pathways (Bauer et al. 1992). SUMOylation of phosducin appears to function both in promoting the stability of phosducin via reducing its ubiquitination, and also by decreasing its capability to bind Gβγ subunits (Klenk et al. 2006), favouring prolonged Gβγ signalling.

9.6.2 Regulation of Kinase and Phosphatase Signalling

Phosphorylation dynamics play critical roles in various cellular signalling pathways. Interestingly, kinases, phosphatases and scaffolding proteins involved in the activation of these pathways have been reported to be targets of SUMO conjugation.

Focal adhesion kinase (FAK) regulates the maturation and turnover of focal adhesions (Mitra et al. 2005), contacts between the extracellular matrix and the cytoskeleton. FAK has been reported to be a SUMO substrate, and SUMOylation was associated with an increase in autophosphorylated FAK (Kadare et al. 2003). Autophosphorylation of FAK leads to binding of Src, which can phosphorylate FAK and leads to its full activation (Mitra et al. 2005). Interestingly, SUMO-FAK was enriched in the nuclear fraction, indicating that SUMOylation may regulate nuclear shuttling of FAK in addition to enhancing its activation.

Protein tyrosine phosphatase 1B (PTP1B) is an endoplasmic reticulum-associated protein phosphatase involved in the downregulation of multiple signalling pathways via dephosphorylation of various receptor tyrosine kinases (Dube and Tremblay 2005). PTP1B is highly expressed in the brain, and functions as a critical negative regulator of leptin signalling in the hypothalamus (Cheng et al. 2002; Zabolotny et al. 2002; Bence et al. 2006). Leptin acts on the hypothalamus to decrease feeding behaviour and promote energy expenditure (Ahima and Flier 2000). PTP1B SUMOylation was recently reported in fibroblasts in response to insulin signalling and appears to act both in the downregulation of catalytic activity and enzyme expression levels (Dadke et al. 2007). Although SUMOylation of PTP1B in the brain has yet to be shown, it remains possible that SUMOylation of PTP1B may play a role in the hypothalamic regulation of feeding behaviour and body mass.

The scaffolding protein axin acts to cluster proteins required for the activation of both c-Jun N-terminal kinase (JNK) and WNT signalling in neurons. JNK is activated by various surface receptors and signalling pathways while WNT activation plays a major role in axon guidance. Interestingly, axin is SUMOylated at two atypical SUMO consensus sites at its extreme C-terminus (Rui et al. 2002). A SUMOylation-deficient form of axin is unable to interact with MEKK, a critical enzyme in the activation of JNK, resulting in incapacity to activate the JNK pathway. Intriguingly, however, its roles in the activation of the WNT pathway are unaffected (Rui et al. 2002), indicating a selective effect of SUMOylation upon JNK signalling.

Thus, it seems SUMOylation can have profound effects on phosphorylation cascades, modifying both the enzymes directly involved in these pathways, as well as selectively affecting pathways via its effects on multifunctional scaffolding proteins. These examples raise the possibility that SUMOylation may function as a general regulator of cellular phosphorylation cascades.

9.7 SUMOylation at Mitochondria

Mitochondrial dynamics are critical to meeting the energy needs of all eukaryotic cells. Indeed, in both cultured neurons and hippocampal slices, neuronal activity is linked to an increase in mitochondrial number (Li et al. 2004). Inhibiting this increase in mitochondrial number by preventing mitochondrial fission severely reduces the number of dendritic protrusions, highlighting a role for mitochondrial dynamics in the maintenance of synapse number (Li et al. 2004). A role for SUMOylation at mitochondria was first reported for the mitochondrial GTPase DRP1. This protein is required for the membrane scission step during mitochondrial fission (Smirnova et al. 2001). DRP1 has been shown to be SUMOylated, and this modification appears to stabilise DRP1, resulting in an increased active pool of the protein and enhanced mitochondrial fission (Harder et al. 2004). Further work has also lead to the identification of a mitochondrial SUMO E3 enzyme, MAPL, that appears to function as a positive regulator of mitochondrial fission via the stimulation of DRP1 SUMOylation (Braschi et al. 2009). Interestingly, the SUMOylation of DRP1 has also been reported to be enhanced by the pro-apoptotic protein Bax (Wasiak et al. 2007), potentially implicating SUMOylation in the induction of apoptosis. In support of this, the apoptotic proteases caspase 7 and 8 have also been reported to be targets of SUMOylation (Besnault-Mascard et al. 2005; Hayashi et al. 2006).

Intriguingly, anti-SUMO western blotting of mitochondrial fractions or neuronal fractions enriched for this organelle indicate that there are multiple mitochondrial SUMO substrates, which are yet to be defined (Harder et al. 2004; Martin et al. 2007a; Braschi et al. 2009).

9.7.1 SUMOylation and Axonal mRNA Trafficking

Local protein synthesis in axons is critical for growth-cone guidance in developing neurons, and for axonal regeneration and synaptic plasticity in adult neurons (Giuditta et al. 2002; Martin 2004). An elegant example of the effect SUMOylation can have on substrate protein interaction profiles has recently been reported for the mRNA-binding protein La (van Niekerk et al. 2007). La binds numerous axonal mRNAs and facilitates their axonal trafficking from the cell soma and their protection from exonucleases (Wolin and Cedervall 2002). Fast axonal transport relies on the motor proteins dynein (for retrograde transport) and kinesin (generally for anterograde transport) (Hirokawa and Takemura 2005). Interestingly, non-SUMOylated La binds kinesin, but not dynein (van Niekerk et al. 2007). However, when SUMOylated, La binds only to dynein. Thus, SUMOylation of La dictates the direction of its axonal transport, and likely allows it to undergo multiple rounds of mRNA transport.

9.8 SUMOylation at the Cell Surface in Neurons

Given the extensive role of SUMOylation in the nucleus, the discovery that multiple classes of plasma membrane proteins are SUMO substrates was an exciting advancement to the field. Indeed, four classes of neuronal membrane protein have now been reported to be SUMOylated, heavily implicating SUMOylation in the control of neuronal excitability, synaptic transmission and glucose transport.

9.8.1 The K2P1 and Kv1.5 Potassium Channels

K2P channels are potassium-selective plasma membrane leak channels (Plant et al. 2005). K2P1 mRNA has been shown to be expressed strongly in heart, brain and kidney (Orias et al. 1997); however, this channel had been reported to be inactive, even when expressed at the cell surface (Plant et al. 2005). This inactivity has now been proposed to be due to SUMO conjugation to K274 of the channel, which acts to block the pore (Rajan et al. 2005). Immunoprecipitation experiments indicated that K2P1 could be SUMO modified in *Xenopus* oocytes, and overexpression of a lysine point mutant that cannot be SUMOylated generated an active, pH-sensitive potassium channel in COS-7 cells. Indeed, overexpression of wild-type K2P1 along with SENP-1, enhancing the number of deSUMOylated channels, produced more active channels, as determined by single channel recordings (Rajan et al. 2005).

However, it must be noted that SUMOylation of native K2P1 channels has not been shown, and the trigger(s) for SUMOylation or deSUMOylation of the channel remain elusive. Indeed, a recent report has cast doubt over whether K2P1 channels are bona fide SUMO substrates (Feliciangeli et al. 2007). Feliciangeli et al. were unable to detect SUMOylation of K2P1 channels and reported that mutation of the proposed SUMOylatable lysine to arginine (instead of a negatively-charged glutamic acid) had no effect on channel functions. The observed effects of the K274E mutation may thus be due to changes in ion flux dependent on charge, and not the prevention of SUMOylation. Further work will be required to determine whether K2P1 is a genuine SUMO substrate and to isolate the functional consequences of this modification.

Another potassium channel, Kv1.5, has also been reported to be a SUMO substrate (Benson et al. 2007). Kv1.5 could be modified by all three SUMO paralogues at two cytosolic lysine residues both in vivo and in vitro, and disruption of SUMOylation either by mutation of the acceptor lysines or overexpression of the deSUMOylating enzyme SENP-2 caused a change in the inactivation kinetics of the channel, indicating SUMOylation has a role in fine-tuning channel function.

These reports of changes in potassium channel function via SUMOylation, coupled with the observation that multiple potassium channel subunits present in the brain also contain SUMOylation consensus motifs, strongly implicates SUMOylation in the regulation of ion fluxes and neuronal excitability.

9.8.2 Glucose Transporters

Glucose is the predominant energy providing metabolite in the brain, and enters cells from the blood via the actions of a family of specific glucose transporters. The glucose transporters GLUT1 and GLUT4 were the first membrane proteins to be shown to be SUMOylated (Giorgino et al. 2000), and are highly expressed in the brain. Interestingly, overexpression of Ubc9 caused a 65% decrease in GLUT1 levels, but an eightfold increase in GLUT4 protein levels. As GLUT1 is responsible for the majority of basal glucose transport, and GLUT4 is transported to the surface in response to insulin (causing an increase in glucose uptake in response to insulin) (Bryant et al. 2002), these results suggest that SUMO decreases the basal transport of glucose, but enhances the insulin-responsiveness of glucose transport into adipocytes. However, data on how SUMOylation differentially affects GLUT1 versus GLUT4 has not been published, nor has any explanation been offered as to how SUMOylation of these two transporters is regulated and under what conditions SUMO modification occurs. Indeed, recent evidence suggests that the increase in GLUT4 stability versus GLUT1 can be reproduced with an inactive point mutant of Ubc9, suggesting this phenomenon may be due to a scaffolding role of Ubc9 rather than by SUMO conjugation (Liu et al. 2007). Further work will therefore be required to determine whether these transporters are genuine SUMO substrates. In addition, the majority of work on these transporters has been carried out in adipose tissue; however, their abundant expression in neurons suggests that SUMO may be a major regulator of glucose transport in the brain.

9.8.3 Ionotropic Glutamate Receptors

Recently, the kainate receptor subunit GluR6 has been shown to be SUMOylated (Martin et al. 2007b). Kainate receptors are tetrameric glutamate-gated ion channels composed of combinations of the glutamate receptor subunits KA1, KA2 and GluRs 5–7 (Pinheiro and Mulle 2006). Kainate receptor subunits are expressed widely in the nervous system and have various roles in regulating neuronal responses – either acting presynaptically to modulate neurotransmitter release, acting postsynaptically to contribute to the postsynaptic response, or by regulating neuronal excitability via the inhibition of K^+ channels (for reviews, see (Jaskolski et al. 2005; Pinheiro and Mulle 2006)).

GluR6 was shown to bind both Ubc9 and the E3 enzyme PIAS3, and was identified as a SUMO substrate (Martin et al. 2007b). Modification with SUMO appears to lead to endocytosis of GluR6 in response to kainate, but not in response to NMDA. Thus, SUMOylation appears to be a trigger for agonist-induced endocytosis of GluR6-containing kainate receptors. Indeed, at mossy fibre synapses, the infusion of SUMO-1 led to a run down in kainate receptor-mediated excitatory postsynaptic currents (EPSCs), but had no effect on AMPA receptor-mediated EPCS. This study also reported that anti-SUMO-1 western blotting of synaptosomes reveals multiple synaptic SUMO-1 substrates. Although these substrates remain unidentified, it highlights the potential significance of SUMOylation in regulating synaptic function.

9.8.4 Metabotropic Glutamate Receptors

The metabotropic glutamate receptor mGluR8 has been described as a potential SUMO substrate (Tang et al. 2005). mGluR8 belongs to the Group III mGluRs, which are expressed widely throughout the brain and predominantly function as presynaptic autoreceptors for glutamate, acting to modulate presynaptic release. The C-terminus of mGluR8 was shown to bind Ubc9 and the E3 enzyme PIAS1, and can be modified with SUMO-1 when expressed as a GST-fusion protein in HEK293 cells (Tang et al. 2005), although evidence for SUMOylation of full-length mGluR8 is lacking. Interestingly, all of the group III mGluRs have been shown to interact with PIAS1 (Tang et al. 2005), and the C-terminus of each of these proteins can be modified by SUMO-1 in vitro (Wilkinson et al. 2008). Further work will be required to confirm these proteins as genuine SUMO substrates, and to determine what role SUMOylation plays in mGluR function.

9.9 SUMO Regulation of Presynaptic Glutamate Release

In addition to directly regulating synaptic transmission postsynaptically via modulation of the kainate receptor subunit GluR6, recent evidence also suggests a direct role for SUMOylation in presynaptic glutamate release. Feligioni et al. (2009) used synaptosomes as a model system to measure glutamate release from presynaptic terminals. During the process of synaptosome production, the inside of the sheared presynaptic terminal becomes accessible to the external media before resealing, making it possible to entrap non-membrane permeable compounds within the synaptosome. Feligioni et al. studied the effects of entrapping either conjugatable or non-conjugatable SUMO, or wild-type or inactive SENP-1 catalytic domains, with a view to investigating a role for SUMOylation in evoked neurotransmitter release. Interestingly, enhancing presynaptic SUMOylation via entrapment of conjugatable SUMO-1 leads to a decrease in glutamate release evoked by potassium chloride (Feligioni et al. 2009). Consistent with this, the infusion of SENP-1 catalytic domain enhanced neurotransmitter release, suggesting that SUMOylation may be playing a negative role in activity-dependent glutamate release. In addition, it appears that this effect may be stimulus dependent – entrapment of conjugatable SUMO appeared to enhance glutamate release evoked by kainate, whereas the infusion of SENP-1 decreased kainate-evoked glutamate release (Feligioni et al. 2009). These findings suggest that SUMOylation can act to either enhance or reduce glutamate release depending on the stimulus. It is possible that different stimuli induce the SUMOylation or deSUMOylation of distinct subsets of proteins that can lead to different effects on neurotransmitter release. However, the nature of the SUMO substrates that mediate this effect are currently unclear, as is whether SUMO plays a similar role in the release of other neurotransmitters.

9.10 SUMOylation in Spine Morphogenesis and Synapse Formation

In neurons, synapse formation is usually initiated by contact between pre- and post-synaptic membranes. In the cerebellum, granule neuron dendrites form dendritic claws onto which mossy fibre terminals form synapses. The family of MEF2 transcription factors are critical regulators of dendritic claw formation. MEF2 family members suppress synapse number in an activity- and calcineurin-dependent manner – neuronal activation leads to calcineurin activation and dephosphorylation of MEF2. Dephosphorylated (active) MEF2 then leads to the transcription of various genes that restrict synapse number, such as ARC and synGAP (Flavell et al. 2006).

Recently, SUMOylation of one of the MEF2 family members, MEF2A, has been reported to strongly influence synapse formation (Shalizi et al. 2006). Early in differentiation, MEF2A is repressed by SUMOylation at lysine 403. Neuronal activity leads to the activation of calcineurin, and dephosphorylation of the nearby serine S408. This favours deSUMOylation of K403 and the replacement of SUMO with an acetyl group, leading to MEF2A activation and the inhibition of synapse formation. This phosphorylation-dependent switch between an inhibitory modification (SUMOylation) and an activating modification (acetylation) highlights MEF2A as an elegant example of the interplay between multiple post-translational modifications tightly regulating a complex physiological process. The role of MEF2A SUMOylation in synapse development has been further strengthened by the observation that the SUMO E3 PIASx functions as an E3 for MEF2A in vivo (Shalizi et al. 2007). In addition, overexpression or knockdown of PIASx were shown to significantly enhance or inhibit dendritic claw formation, respectively, and this effect was dependent on the presence of MEF2A, consistent with its role in the suppression of MEF2A (Shalizi et al. 2007).

The protein CASK binds various postsynaptic transmembrane receptors and has been shown to be required for the formation of dendritic spines, morphological protrusions onto which more than 90% of excitatory synapses form (Nimchinsky et al. 2002). Interestingly, CASK has been reported to be a SUMO substrate (Chao et al. 2008). The function of CASK in spine morphogenesis requires its interaction with protein 4.1. Chao et al. (2008) have demonstrated that fusion of SUMO to CASK significantly reduces its interaction with protein 4.1, and expression of this fusion protein in neurons causes a decrease in spine size and density.

These examples indicate that through both activity-dependent modification of transcription factors and modification of synaptic proteins, SUMOylation regulates various aspects of synapse formation and neuronal morphology.

9.11 SUMOylation and Neurodegenerative Disorders

As may be expected of a pathway that performs such critical roles in neuronal signalling, excitability and development, dysregulation of the SUMO pathway is being increasingly implicated in various neurodegenerative disorders (for recent reviews see Dorval and Fraser 2007; Martin et al. 2007b; Anderson et al. 2009). Many of these disorders are characterised by the formation of nuclear and/or cytosolic inclusions, thought to arise from aberrant protein trafficking or degradation. In a growing number of disorders, these inclusions have been shown to stain positive for SUMO, and a growing number of disease-associated SUMO substrates have been identified (Table 9.1).

9.11.1 Nuclear Inclusion Disorder

Nuclear inclusion disorder (NIID) is a rare disease of both the central and peripheral nervous systems characterised by extensive neuronal intranuclear inclusions (Sung et al. 1980). The disease predominantly manifests as ataxia and movement disorders in younger patients and as dementia in adult-onset cases, suggesting NIID may comprise

Table 9.1 SUMOylation has been implicated in various neurodegenerative disorders

		Implicated SUMO substrates
Nuclear inclusion disease		Unknown
Polyglutamine diseases	Huntington's	Huntingtin
	SBMA	Androgen Receptor
	DRPLA	Atrophin-1
	SCA Type I	Ataxin-1
Alzheimer's disease		APP
		Tau
α-Synucleinopathies	Parkinson's	α-Synuclein
		DJ-1
		Parkin (non-covalent)
	MSA	α-Synuclein?
	DLB	α-Synuclein?
Hypoxia, ischemia and cellular stress		Various

Neurodegenerative disorders in which SUMOylation has been implicated are shown, alongside known SUMO substrates involved in the pathogenesis of each disorder

SBMA spinal and bulbar muscular atrophy, *DRPLA* dentatorubral-pallidoluysian atrophy, *SCA* spinocerebellar ataxia, *MSA* multiple system atrophy, *DLB* dementia with Lewy bodies, *APP* amyloid precursor protein

a family of pathologically distinct neuopathies. However, inclusions have been reported to stain strongly for SUMO reactivity in familial (Pountney et al. 2003), juvenile (McFadden et al. 2005) and sporadic (Takahashi-Fujigasaki et al. 2006) forms of the disease, indicating the various forms of the disease may share a similar mechanism. In an attempt to identify SUMO-associated proteins in NIID inclusions, Pountney et al. (2008) performed mass spectrometry following anti-SUMO-1 immunoprecipitation from inclusions isolated from a patient with familial NIID. Interestingly, they identified NSF, mUnc18, dynamin-1 and HSP90 in these SUMO immunoprecipitations. Each of these proteins is heavily involved in cellular vesicle trafficking, suggesting impairment in these pathways may contribute to the pathogenesis of NIID. Further work will now be required in order to determine whether these proteins are bona fide SUMO substrates. In addition, it is possible that given the well-characterised role of SUMOylation in nuclear trafficking dysregulation of neuronal SUMOylation in this disease may lead to the accumulation and aggregation of proteins in the nucleus.

9.11.2 Polyglutamine Disorders

Polyglutamine (polyQ) disorders (discussed in Chap. 12) are characterised by the presence of a toxic stretch of polyQ (CAG) repeats in the disease-specific genes, resulting in neuronal inclusions composed of the affected protein (for reviews see Zoghbi and Orr 2000; Gatchel and Zoghbi 2005). Interestingly, inclusions associated with polyQ disorders such as Huntington's disease, spinal and bulbar muscular atrophy (SBMA), dentatorubral-pallidoluysian atrophy (DRPLA), and spinocerebellar ataxias (SCA), have all been reported to stain strongly for SUMO reactivity (for review see Ueda et al. 2002). In addition, Huntingtin (Htt) protein has been shown to be both SUMOylated and ubiquitinated at lysine residues in its N-terminus (Steffan et al. 2004). In a *Drosophila* model of Huntington's, SUMOylation has been shown to exacerbate disease pathology, possibly by leading to a decrease in mutant Huntingtin aggregation resulting in an increase in the cytotoxic soluble form, whereas ubiquitination decreases pathogenicity (Steffan et al. 2004), suggesting a likely interplay between SUMO and ubiquitin in the regulation of Htt stability. Recent work has also highlighted the role of Htt SUMOylation in mutant Htt-mediated cytotoxicity, and has offered explanation for the localised neurodegeneration characteristic of Huntington's disease. Subramaniam et al. (2009) identified Rhes (Ras homolog enriched in striatum) as a SUMO E3 which bound to and enhanced SUMOylation of mutant Htt, but not wild-type Htt. Interestingly, in both heterologous cells and striatal neurons, overexpression of mutant Htt did not lead to cytotoxicity in the absence of Rhes, suggesting that Rhes-mediated SUMOylation of mutant Htt was leading to Htt-mediated toxicity (Subramaniam et al. 2009). Indeed, while Htt is ubiquitously expressed, Rhes is located only in the striatum, seemingly explaining part of the striatal-specific nature of the neurodegeneration associated with Huntington's disease.

SBMA is an X-linked neuromuscular disorder characterised by a polyQ repeat in the androgen receptor (AR), a transcription factor which responds to the presence of androgenic hormones (Katsuno et al. 2006). PolyQ AR can also be modified by both ubiquitin and SUMO at the same lysine residue (Poukka et al. 2000), and SUMOylation has been reported to repress AR-dependent transcription (Takahashi et al. 2001; Nishida and Yasuda 2002). In a *Drosophila* model, polyQ AR aggregates to form both nuclear and cytosolic inclusions that are enhanced upon the prevention of both SUMOylation and ubiquitination (Chan et al. 2002) indicating that, like Huntington's, the SUMO and ubiquitin pathways are both involved in the advancement of the disease.

A polyQ expansion in the atrophin-1 protein leads to DRPLA, a neuropathological condition characterised by dementia, epilepsy and involuntary movement (Yazawa et al. 1995; Schilling et al. 1999). While the function of atrophin-1 is currently unknown, co-expression of polyQ atrophin-1 with SUMO-1 leads to a significant increase in nuclear inclusions and subsequent cell death, whereas co-expression of a non-conjugatable form of SUMO-1 actually decreases inclusion formation (Terashima et al. 2002). These data indicate that SUMO conjugation plays a significant role in the formation of nuclear inclusions associated with DRPLA.

SCAs are progressive neurodegenerative disorders, which result in atrophy of cerebellar Purkinje cells. A polyQ expansion in the ataxin-1 protein results in type I SCA. Ataxin-1 can be SUMOylated on at least five lysine residues, and the presence of the polyQ expansion significantly decreases ataxin-1 SUMOylation (Riley et al. 2005). Interestingly, a mutant polyQ ataxin-1 that cannot be phosphorylated shows normal levels of SUMOylation, indicating an inhibitory interplay between phosphorylation and SUMOylation may exist in the pathogenesis of type I SCA (Riley et al. 2005).

9.11.3 Alzheimer's Disease

Alzheimer's disease is an age-dependent neurodegenerative disorder characterised by progressive dementia. Alzheimer's patients exhibit neuronal plaques composed of amyloid-β (Aβ), a cleavage product of amyloid precursor protein (APP), and neurofibrillary plaques composed of hyperphosphorylated tau (see Chap. 14 for an exhaustive description on Alzheimer's disease). Interestingly, both APP and tau may be SUMO substrates (Gocke et al. 2005; Dorval and Fraser 2006). Exogenous SUMO-3 has been reported to regulate Aβ formation (Li et al. 2003; Dorval et al. 2007); however, a direct role for SUMO-3 is currently unclear – groups have reported SUMO-3 to both increase (Dorval et al. 2007) or decrease (Li et al. 2003) Aβ formation. Indeed, the SUMO-3-mediated enhancement of Aβ formation could be replicated with a non-conjugatable form of SUMO-3 (Dorval et al. 2007), indicating that a non-covalent effect of SUMO-3 may be regulating Aβ levels.

Tau is a microtubule-associated protein that can be both SUMOylated and ubiquitinated in cell culture models (Dorval and Fraser 2006). Inhibition of proteasomal degradation leads to an increase in tau ubiquitination and decreased SUMOylation, highlighting a role for both SUMOylation and the ubiquitin-proteasome system in tau clearance, which may be perturbed in Alzheimer's disease, leading to the accumulation of tau protein.

9.11.4 Parkinson's Disease

Parkinson's disease (PD) results in the specific loss of dopaminergic neurons projecting from the substantia nigra pars compacta to the striatum, along with the formation of inclusions known as Lewy bodies in the cytosol of affected neurons (also see Chap. 13). Lewy bodies have been reported to stain strongly for α-synuclein, a protein reported to be SUMOylated in cultured cells (Dorval and Fraser 2006). However, the existence of SUMO modified α-synuclein in neurons has not yet been reported.

Numerous other neurological disorders have also been reported to contain Lewy bodies, including multiple system atrophy (MSA) and dementia with Lewy bodies (DLB). Inclusions from patients with MSA have been reported to stain positively for SUMO-1 (Pountney et al. 2005). In addition, while α-synuclein has been reported to be a SUMO substrate, non-covalent SUMO interactors have also been implicated in synucleinopathies. Mutations in the ubiquitin E3 ligase parkin have been associated with familial Parkinsonism (Tan and Skipper 2007). Parkin has been reported to interact non-covalently with SUMO-1 both in vitro and in vivo, and this interaction appears to enhance its E3 ligase activity (Um and Chung 2006). Interestingly, a substrate of Parkin is RanBP2 (Um et al. 2006), a SUMO E3 ligase. Thus, SUMO binding to Parkin may enhance its ubiquitination activity, leading to the degradation of RanBP2, an intriguing example of negative interplay between the ubiquitin and SUMO systems.

Another PD-associated protein is the transcriptional regulator DJ-1. DJ-1 controls the expression of many genes associated with the cellular response to oxidative stress (Taira et al. 2004). Loss of function of DJ-1 has been reported to lead to PD. Interestingly, DJ-1 has been reported to be a SUMO substrate (Shinbo et al. 2006). While the physiological role of DJ-1 SUMOylation is unclear, SUMOylation of a PD-associated DJ-1 mutation (L166P) is enhanced (Shinbo et al. 2006), suggesting that inappropriate levels of SUMOylation may contribute to the pathogenicity of this mutant DJ-1.

9.11.5 Hypoxia, Ischemia and Cellular Stress

Numerous studies have reported a rapid global increase in protein SUMOylation resulting from cellular stress. Indeed, while there are relatively low levels of free SUMO-1 under resting conditions, there is a large pool of free SUMO-2/3 (Saitoh and Hinchey 2000). Upon various stresses, such as heat shock, osmotic shock or oxidative stress, there is a massive increase in SUMO-2/3 conjugation (for reviews see Bossis and Melchior 2006; Agbor and Taylor 2008; Tempe et al. 2008). It is therefore possible that various neurological disorders which result in neuronal stress may also be characterised by a rapid increase in SUMO-2/3 conjugation, which may constitute a protective cellular response to stress. Indeed, a massive increase in protein SUMOylation has recently been reported in ischemia (for reviews see Cimarosti and Henley 2008; Yang et al. 2008a). Ischemia is characterised by oxygen and glucose deprivation, and results in cell death surrounding the infarct region due to extensive glutamate release leading to excitotoxic cell death. Transient global ischemia in mice leads to significant increases in SUMO-2/3 conjugation in the cortex and hippocampus (Yang et al. 2008b). Another study also reported an increase in SUMO-1 conjugation in the cortical infarct region and the non-ischemic hippocampus in a mouse model of ischemia with reperfusion (Cimarosti et al. 2008). In this model, increases in SUMO-2/3 conjugation were restricted to the non-infarct regions. Interestingly, levels of AMPA and kainate receptors were decreased by the ischemic event, suggesting the possibility that SUMOylation may play a role in decreasing glutamatergic signalling that may lead to excitotoxic cell death. Indeed, the observation that kainate receptor EPSCs are decreased as a result of SUMOylation of the GluR6 subunit adds credence to this possibility (Martin et al. 2007a).

In addition, direct evidence for a role of SUMOylation in the protection against ischemic cell death has been provided recently by the observation that overexpression of SUMO proteins makes various cell types more resistant to in vitro ischemia stress (Ja Lee et al. 2009). Overexpression of SUMO-1 and SUMO-2 were shown to make cells more resistant to oxygen and glucose deprivation (OGD) in the SHSY5Y human neuroblastoma cell line (Ja Lee et al. 2009). Conversely, knockdown of SUMO-1, but not SUMO-2 made SHSY5Y cells more susceptible to OGD-mediated cell death. In addition, overexpression of SUMO-1 made rat cortical neurons more resistant to OGD (Ja Lee et al. 2009). While this work appears to suggest a protective role for SUMO in OGD, further work will be required to define the targets responsible for the effects observed.

9.12 Concluding Remarks

There has been rapid progress in defining the roles of SUMOylation in neurons. There remain, however, many outstanding questions – what are the identities of the multiple SUMO substrates present at synapses (Martin et al. 2007a)? How is the SUMOylation of these targets regulated? What further roles does SUMOylation play in neurons? Given the distribution of the many synaptic proteins that contain SUMOylation consensus motifs, it is tempting to speculate roles in processes such as vesicle recycling and synaptic plasticity. In addition, how does the SUMO pathway become dysregulated in so many neurodegenerative disorders? Understanding of these pathways may potentially lead to novel therapeutics targeting the SUMO system in these processes (for review see Anderson et al. 2009).

Future studies will shed light on these exciting possibilities, and we anticipate that, like phosphorylation and ubiquitination, SUMOylation will emerge as a major regulator of neuronal function.

Acknowledgements Work in the lab of JMH is funded by the MRC, the Wellcome Trust, the ERC and the BBSRC. We thank Atsushi Nishimune for invaluable guidance, discussion and advice.

References

Agbor TA and Taylor CT (2008) SUMO, hypoxia and the regulation of metabolism. Biochem Soc Trans 36(Pt 3): 445–8

Ahima RS and Flier JS (2000) Leptin. Annu Rev Physiol 62: 413–37

Anderson DB, Wilkinson KA and Henley JM (2009) Protein SUMOylation in neuropathological conditions. Drug News Perspect 22(5): 255–65

Bauer PH, Muller S, Puzicha M et al (1992) Phosducin is a protein kinase A-regulated G-protein regulator. Nature 358(6381): 73–6

Bence KK, Delibegovic M, Xue B et al (2006) Neuronal PTP1B regulates body weight, adiposity and leptin action. Nat Med 12(8): 917–24

Bencsath KP, Podgorski MS, Pagala VR et al (2002) Identification of a multifunctional binding site on Ubc9p required for Smt3p conjugation. J Biol Chem 277(49): 47938–45

Benson MD, Li QJ, Kieckhafer K et al (2007) SUMO modification regulates inactivation of the voltage-gated potassium channel Kv1.5. Proc Natl Acad Sci U S A 104(6): 1805–10

Berman DM and Gilman AG (1998) Mammalian RGS proteins: barbarians at the gate. J Biol Chem 273(3): 1269–72

Besnault-Mascard L, Leprince C, Auffredou MT et al (2005) Caspase-8 SUMOylation is associated with nuclear localization. Oncogene 24(20): 3268–73

Bohren KM, Nadkarni V, Song JH et al (2004) A M55V polymorphism in a novel SUMO gene (SUMO-4) differentially activates heat shock transcription factors and is associated with susceptibility to type I diabetes mellitus. J Biol Chem 279(26): 27233–8

Bossis G and Melchior F (2006) SUMO: regulating the regulator. Cell Div 1: 13

Braschi E, Zunino R and McBride HM (2009) MAPL is a new mitochondrial SUMO E3 ligase that regulates mitochondrial fission. EMBO Rep 10(7): 748–54

Bryant NJ, Govers R and James DE (2002) Regulated transport of the glucose transporter GLUT4. Nat Rev Mol Cell Biol 3(4): 267–77

Bylebyl GR, Belichenko I and Johnson ES (2003) The SUMO isopeptidase Ulp2 prevents accumulation of SUMO chains in yeast. J Biol Chem 278(45): 44113–20

Chan HY, Warrick JM, Andriola I et al (2002) Genetic modulation of polyglutamine toxicity by protein conjugation pathways in *Drosophila*. Hum Mol Genet 11(23): 2895–904

Chao HW, Hong CJ, Huang TN et al (2008) SUMOylation of the MAGUK protein CASK regulates dendritic spinogenesis. J Cell Biol 182(1): 141–55

Cheng A, Uetani N, Simoncic PD et al (2002) Attenuation of leptin action and regulation of obesity by protein tyrosine phosphatase 1B. Dev Cell 2(4): 497–503

Cimarosti H and Henley JM (2008) Investigating the mechanisms underlying neuronal death in ischemia using in vitro oxygen-glucose deprivation: potential involvement of protein SUMOylation. Neuroscientist 14(6): 626–36

Cimarosti H, Lindberg C, Bomholt SF et al (2008) Increased protein SUMOylation following focal cerebral ischemia. Neuropharmacology 54(2): 280–9

Dadke S, Cotteret S, Yip SC et al (2007) Regulation of protein tyrosine phosphatase 1B by SUMOylation. Nat Cell Biol 9(1): 80–5

Desterro JM, Thomson J and Hay RT (1997) Ubch9 conjugates SUMO but not ubiquitin. FEBS Lett 417(3): 297–300

Dorval V and Fraser PE (2006) Small ubiquitin-like modifier (SUMO) modification of natively unfolded proteins tau and alpha-synuclein. J Biol Chem 281(15): 9919–24

Dorval V and Fraser PE (2007) SUMO on the road to neurodegeneration. Biochim Biophys Acta 1773(6): 694–706

Dorval V, Mazzella MJ, Mathews PM et al (2007) Modulation of Abeta generation by small ubiquitin-like modifiers does not require conjugation to target proteins. Biochem J 404(2): 309–16

Dube N and Tremblay ML (2005) Involvement of the small protein tyrosine phosphatases TC-PTP and PTP1B in signal transduction and diseases: from diabetes, obesity to cell cycle, and cancer. Biochim Biophys Acta 1754(1–2): 108–17

Feliciangeli S, Bendahhou S, Sandoz G et al (2007) Does SUMOylation control K2P1/TWIK1 background K+ channels? Cell 130(3): 563–9

Feligioni M, Nishimune A and Henley JM (2009) Protein SUMOylation modulates calcium influx and glutamate release from presynaptic terminals. Eur J Neurosci 29(7): 1348–56

Flavell SW, Cowan CW, Kim TK et al (2006) Activity-dependent regulation of MEF2 transcription factors suppresses excitatory synapse number. Science 311(5763): 1008–12

Gatchel JR and Zoghbi HY (2005) Diseases of unstable repeat expansion: mechanisms and common principles. Nat Rev Genet 6(10): 743–55

Geiss-Friedlander R and Melchior F (2007) Concepts in SUMOylation: a decade on. Nat Rev Mol Cell Biol 8(12): 947–56

Giorgino F, de Robertis O, Laviola L et al (2000) The sentrin-conjugating enzyme mUbc9 interacts with GLUT4 and GLUT1 glucose transporters and regulates transporter levels in skeletal muscle cells. Proc Natl Acad Sci U S A 97(3): 1125–30

Giuditta A, Kaplan BB, van Minnen J et al (2002) Axonal and presynaptic protein synthesis: new insights into the biology of the neuron. Trends Neurosci 25(8): 400–4

Gocke CB, Yu H and Kang J (2005) Systematic identification and analysis of mammalian small ubiquitin-like modifier substrates. J Biol Chem 280(6): 5004–12

Gong L, Kamitani T, Fujise K et al (1997) Preferential interaction of sentrin with a ubiquitin-conjugating enzyme, Ubc9. J Biol Chem 272(45): 28198–201

Gong L, Li B, Millas S et al (1999) Molecular cloning and characterization of human AOS1 and UBA2, components of the sentrin-activating enzyme complex. FEBS Lett 448(1): 185–9

Guo D, Li M, Zhang Y et al (2004) A functional variant of SUMO4, a new I kappa B alpha modifier, is associated with type 1 diabetes. Nat Genet 36(8): 837–41

Hannich JT, Lewis A, Kroetz MB et al (2005) Defining the SUMO-modified proteome by multiple approaches in *Saccharomyces cerevisiae*. J Biol Chem 280(6): 4102–10

Harder Z, Zunino R and McBride H (2004) Sumo1 conjugates mitochondrial substrates and participates in mitochondrial fission. Curr Biol 14(4): 340–5

Hay RT (2005) SUMO: a history of modification. Mol Cell 18(1): 1–12

Hayashi N, Shirakura H, Uehara T et al (2006) Relationship between SUMO-1 modification of caspase-7 and its nuclear localization in human neuronal cells. Neurosci Lett 397(1–2): 5–9

Hayashi T, Seki M, Maeda D et al (2002) Ubc9 is essential for viability of higher eukaryotic cells. Exp Cell Res 280(2): 212–21

Hecker CM, Rabiller M, Haglund K et al (2006) Specification of SUMO1- and SUMO2-interacting motifs. J Biol Chem 281(23): 16117–27

Heun P (2007) SUMOrganization of the nucleus. Curr Opin Cell Biol 19(3): 350–5

Hirokawa N and Takemura R (2005) Molecular motors and mechanisms of directional transport in neurons. Nat Rev Neurosci 6(3): 201–14

Hurley JH, Lee S and Prag G (2006) Ubiquitin-binding domains. Biochem J 399(3): 361–72

Ja Lee Y, Castri P, Bembry J et al (2009) SUMOylation participates in induction of ischemic tolerance. J Neurochem 109(1): 257–67

Jaskolski F, Coussen F and Mulle C (2005) Subcellular localization and trafficking of kainate receptors. Trends Pharmacol Sci 26(1): 20–6

Johnson ES (2004) Protein modification by SUMO. Annu Rev Biochem 73: 355–82

Johnson ES and Blobel G (1997) Ubc9p is the conjugating enzyme for the ubiquitin-like protein Smt3p. J Biol Chem 272(43): 26799–802

Kadare G, Toutant M, Formstecher E et al (2003) PIAS1-mediated SUMOYlation of focal adhesion kinase activates its autophosphorylation. J Biol Chem 278(48): 47434–40

Kamitani T, Nguyen HP and Yeh ET (1997) Preferential modification of nuclear proteins by a novel ubiquitin-like molecule. J Biol Chem 272(22): 14001–4

Katsuno M, Adachi H, Waza M et al (2006) Pathogenesis, animal models and therapeutics in spinal and bulbar muscular atrophy (SBMA). Exp Neurol 200(1): 8–18

Kerscher O (2007) SUMO junction-what's your function? New insights through SUMO-interacting motifs. EMBO Rep 8(6): 550–5

Klenk C, Humrich J, Quitterer U et al (2006) SUMO-1 controls the protein stability and the biological function of phosducin. J Biol Chem 281(13): 8357–64

Lallemand-Breitenbach V, Jeanne M, Benhenda S et al (2008) Arsenic degrades PML or PML-RARalpha through a SUMO-triggered RNF4/ubiquitin-mediated pathway. Nat Cell Biol 10(5): 547–55

Lee GW, Melchior F, Matunis MJ et al (1998) Modification of Ran GTPase-activating protein by the small ubiquitin-related modifier SUMO-1 requires Ubc9, an E2-type ubiquitin-conjugating enzyme homologue. J Biol Chem 273(11): 6503–7

Li Y, Wang H, Wang S et al (2003) Positive and negative regulation of APP amyloidogenesis by SUMOYlation. Proc Natl Acad Sci U S A 100(1): 259–64

Li Z, Okamoto K, Hayashi Y et al (2004) The importance of dendritic mitochondria in the morphogenesis and plasticity of spines and synapses. Cell 119(6): 873–87

Liu LB, Omata W, Kojima I et al (2007) The SUMO conjugating enzyme Ubc9 is a regulator of GLUT4 turnover and targeting to the insulin-responsive storage compartment in 3T3-L1 adipocytes. Diabetes 56: 1977–85

Mahajan R, Delphin C, Guan T et al (1997) A small ubiquitin-related polypeptide involved in targeting RanGAP1 to nuclear pore complex protein RanBP2. Cell 88(1): 97–107

Mao H, Zhao Q, Daigle M et al (2004) RGS17/RGSZ2, a novel regulator of Gi/o, Gz, and Gq signaling. J Biol Chem 279(25): 26314–22

Martin KC (2004) Local protein synthesis during axon guidance and synaptic plasticity. Curr Opin Neurobiol 14(3): 305–10

Martin S, Nishimune A, Mellor JR et al (2007a) SUMOylation regulates kainate-receptor-mediated synaptic transmission. Nature 447: 321–5

Martin S, Wilkinson KA, Nishimune A et al (2007b) Emerging extranuclear roles of protein SUMOylation in neuronal function and dysfunction. Nat Rev Neurosci 8(12): 948–59

Matic I, van Hagen M, Schimmel J et al (2008) In vivo identification of human small ubiquitin-like modifier polymerization sites by high accuracy mass spectrometry and an in vitro to in vivo strategy. Mol Cell Proteomics 7(1): 132–44

Matunis MJ, Coutavas E and Blobel G (1996) A novel ubiquitin-like modification modulates the partitioning of the Ran-GTPase-activating protein RanGAP1 between the cytosol and the nuclear pore complex. J Cell Biol 135(6 Pt 1): 1457–70

McFadden K, Hamilton RL, Insalaco SJ et al (2005) Neuronal intranuclear inclusion disease without polyglutamine inclusions in a child. J Neuropathol Exp Neurol 64(6): 545–52

Meluh PB and Koshland D (1995) Evidence that the MIF2 gene of *Saccharomyces cerevisiae* encodes a centromere protein with homology to the mammalian centromere protein CENP-C. Mol Biol Cell 6(7): 793–807

Minty A, Dumont X, Kaghad M et al (2000) Covalent modification of p73alpha by SUMO-1. Two-hybrid screening with p73 identifies novel SUMO-1-interacting proteins and a SUMO-1 interaction motif. J Biol Chem 275(46): 36316–23

Mitra SK, Hanson DA and Schlaepfer DD (2005) Focal adhesion kinase: in command and control of cell motility. Nat Rev Mol Cell Biol 6(1): 56–68

Mukhopadhyay D and Dasso M (2007) Modification in reverse: the SUMO proteases. Trends Biochem Sci 32(6): 286–95

Nacerddine K, Lehembre F, Bhaumik M et al (2005) The SUMO pathway is essential for nuclear integrity and chromosome segregation in mice. Dev Cell 9(6): 769–79

Nimchinsky EA, Sabatini BL and Svoboda K (2002) Structure and function of dendritic spines. Annu Rev Physiol 64: 313–53

Nishida T and Yasuda H (2002) PIAS1 and PIASxalpha function as SUMO-E3 ligases toward androgen receptor and repress androgen receptor-dependent transcription. J Biol Chem 277(44): 41311–7

Okuma T, Honda R, Ichikawa G et al (1999) In vitro SUMO-1 modification requires two enzymatic steps, E1 and E2. Biochem Biophys Res Commun 254(3): 693–8

Orias M, Velazquez H, Tung F et al (1997) Cloning and localization of a double-pore K channel, KCNK1: exclusive expression in distal nephron segments. Am J Physiol 273(4 Pt 2): F663–6

Owerbach D, McKay EM, Yeh ET et al (2005) A proline-90 residue unique to SUMO-4 prevents maturation and SUMOYlation. Biochem Biophys Res Commun 337(2): 517–20

Perry JJ, Tainer JA and Boddy MN (2008) A SIM-ultaneous role for SUMO and ubiquitin. Trends Biochem Sci 33(5): 201–8

Pierce KL, Premont RT and Lefkowitz RJ (2002) Seven-transmembrane receptors. Nat Rev Mol Cell Biol 3(9): 639–50

Pinheiro P and Mulle C (2006) Kainate receptors. Cell Tissue Res 326(2): 457–82

Plant LD, Rajan S and Goldstein SA (2005) K2P channels and their protein partners. Curr Opin Neurobiol 15(3): 326–33

Poukka H, Karvonen U, Janne OA et al (2000) Covalent modification of the androgen receptor by small ubiquitin-like modifier 1 (SUMO-1). Proc Natl Acad Sci U S A 97(26): 14145–50

Pountney DL, Chegini F, Shen X et al (2005) SUMO-1 marks subdomains within glial cytoplasmic inclusions of multiple system atrophy. Neurosci Lett 381(1–2): 74–9

Pountney DL, Huang Y, Burns RJ et al (2003) SUMO-1 marks the nuclear inclusions in familial neuronal intranuclear inclusion disease. Exp Neurol 184(1): 436–46

Pountney DL, Raftery MJ, Chegini F et al (2008) NSF, Unc-18–1, dynamin-1 and HSP90 are inclusion body components in neuronal intranuclear inclusion disease identified by anti-SUMO-1-immunocapture. Acta Neuropathol 116(6): 603–14

Rajan S, Plant LD, Rabin ML et al (2005) SUMOylation silences the plasma membrane leak K+ channel K2P1. Cell 121(1): 37–47

Riley BE, Zoghbi HY and Orr HT (2005) SUMOylation of the polyglutamine repeat protein, ataxin-1, is dependent on a functional nuclear localization signal. J Biol Chem 280(23): 21942–8

Rodriguez MS, Dargemont C and Hay RT (2001) SUMO-1 conjugation in vivo requires both a consensus modification motif and nuclear targeting. J Biol Chem 276(16): 12654–9

Rodriguez-Munoz M, Bermudez D, Sanchez-Blazquez P et al (2007) Sumoylated RGS-Rz proteins act as scaffolds for Mu-opioid receptors and G-protein complexes in mouse brain. Neuropsychopharmacology 32(4): 842–50

Rui HL, Fan E, Zhou HM et al (2002) SUMO-1 modification of the C-terminal KVEKVD of Axin is required for JNK activation but has no effect on Wnt signaling. J Biol Chem 277(45): 42981–6

Saitoh H and Hinchey J (2000) Functional heterogeneity of small ubiquitin-related protein modifiers SUMO-1 versus SUMO-2/3. J Biol Chem 275(9): 6252–8

Saitoh H, Sparrow DB, Shiomi T et al (1998) Ubc9p and the conjugation of SUMO-1 to RanGAP1 and RanBP2. Curr Biol 8(2): 121–4

Sampson DA, Wang M and Matunis MJ (2001) The small ubiquitin-like modifier-1 (SUMO-1) consensus sequence mediates Ubc9 binding and is essential for SUMO-1 modification. J Biol Chem 276(24): 21664–9

Schilling G, Wood JD, Duan K et al (1999) Nuclear accumulation of truncated atrophin-1 fragments in a transgenic mouse model of DRPLA. Neuron 24(1): 275–86

Schwarz SE, Matuschewski K, Liakopoulos D et al (1998) The ubiquitin-like proteins SMT3 and SUMO-1 are conjugated by the UBC9 E2 enzyme. Proc Natl Acad Sci U S A 95(2): 560–4

Shalizi A, Bilimoria PM, Stegmuller J et al (2007) PIASx is a MEF2 SUMO E3 ligase that promotes postsynaptic dendritic morphogenesis. J Neurosci 27(37): 10037–46
Shalizi A, Gaudilliere B, Yuan Z et al (2006) A calcium-regulated MEF2 SUMOylation switch controls postsynaptic differentiation. Science 311(5763): 1012–7
Shinbo Y, Niki T, Taira T et al (2006) Proper SUMO-1 conjugation is essential to DJ-1 to exert its full activities. Cell Death Differ 13(1): 96–108
Smirnova E, Griparic L, Shurland DL et al (2001) Dynamin-related protein Drp1 is required for mitochondrial division in mammalian cells. Mol Biol Cell 12(8): 2245–56
Song J, Durrin LK, Wilkinson TA et al (2004) Identification of a SUMO-binding motif that recognizes SUMO-modified proteins. Proc Natl Acad Sci U S A 101(40): 14373–8
Steffan JS, Agrawal N, Pallos J et al (2004) SUMO modification of Huntingtin and Huntington's disease pathology. Science 304(5667): 100–4
Subramaniam S, Sixt KM, Barrow R et al (2009) Rhes, a striatal specific protein, mediates mutant-huntingtin cytotoxicity. Science 324(5932): 1327–30
Sung JH, Ramirez-Lassepas M, Mastri AR et al (1980) An unusual degenerative disorder of neurons associated with a novel intranuclear hyaline inclusion (neuronal intranuclear hyaline inclusion disease). A clinicopathological study of a case. J Neuropathol Exp Neurol 39(2): 107–30
Taira T, Saito Y, Niki T et al (2004) DJ-1 has a role in antioxidative stress to prevent cell death. EMBO Rep 5(2): 213–8
Takahashi K, Taira T, Niki T et al (2001) DJ-1 positively regulates the androgen receptor by impairing the binding of PIASx alpha to the receptor. J Biol Chem 276(40): 37556–63
Takahashi-Fujigasaki J, Arai K, Funata N et al (2006) SUMOylation substrates in neuronal intranuclear inclusion disease. Neuropathol Appl Neurobiol 32(1): 92–100
Tan EK and Skipper LM (2007) Pathogenic mutations in Parkinson disease. Hum Mutat 28(7): 641–53
Tang Z, El Far O, Betz H et al (2005) Pias1 interaction and SUMOylation of metabotropic glutamate receptor 8. J Biol Chem 280(46): 38153–9
Tatham MH, Geoffroy MC, Shen L et al (2008) RNF4 is a poly-SUMO-specific E3 ubiquitin ligase required for arsenic-induced PML degradation. Nat Cell Biol 10(5): 538–46
Tatham MH, Jaffray E, Vaughan OA et al (2001) Polymeric chains of SUMO-2 and SUMO-3 are conjugated to protein substrates by SAE1/SAE2 and Ubc9. J Biol Chem 276(38): 35368–74
Tatham MH, Kim S, Yu B et al (2003) Role of an N-terminal site of Ubc9 in SUMO-1, -2, and -3 binding and conjugation. Biochemistry 42(33): 9959–69
Tempe D, Piechaczyk M and Bossis G (2008) SUMO under stress. Biochem Soc Trans 36(Pt 5): 874–8
Terashima T, Kawai H, Fujitani M et al (2002) SUMO-1 co-localized with mutant atrophin-1 with expanded polyglutamines accelerates intranuclear aggregation and cell death. Neuroreport 13(17): 2359–64
Ueda H, Goto J, Hashida H et al (2002) Enhanced SUMOylation in polyglutamine diseases. Biochem Biophys Res Commun 293(1): 307–13
Um JW and Chung KC (2006) Functional modulation of parkin through physical interaction with SUMO-1. J Neurosci Res 84(7): 1543–54
Um JW, Min DS, Rhim H et al (2006) Parkin ubiquitinates and promotes the degradation of RanBP2. J Biol Chem 281(6): 3595–603
van Niekerk EA, Willis DE, Chang JH et al (2007) SUMOylation in axons triggers retrograde transport of the RNA-binding protein La. Proc Natl Acad Sci U S A 104(31): 12913–8
Vassilatis DK, Hohmann JG, Zeng H et al (2003) The G protein-coupled receptor repertoires of human and mouse. Proc Natl Acad Sci U S A 100(8): 4903–8
Wasiak S, Zunino R and McBride HM (2007) Bax/Bak promote SUMOYlation of DRP1 and its stable association with mitochondria during apoptotic cell death. J Cell Biol 177(3): 439–50
Weisshaar SR, Keusekotten K, Krause A et al (2008) Arsenic trioxide stimulates SUMO-2/3 modification leading to RNF4-dependent proteolytic targeting of PML. FEBS Lett 582 (21–22): 3174–8
Wilkinson KA, Nishimune A and Henley JM (2008) Analysis of SUMO-1 modification of neuronal proteins containing consensus SUMOylation motifs. Neurosci Lett 436(2): 239–44

Wolin SL and Cedervall T (2002) The La protein. Annu Rev Biochem 71: 375–403
Xu J, He Y, Qiang B et al (2008) A novel method for high accuracy SUMOylation site prediction from protein sequences. BMC Bioinformatics 9: 8
Yang W, Sheng H, Homi HM et al (2008a) Cerebral ischemia/stroke and small ubiquitin-like modifier (SUMO) conjugation – a new target for therapeutic intervention? J Neurochem 106: 989–99
Yang W, Sheng H, Warner DS et al (2008b) Transient global cerebral ischemia induces a massive increase in protein SUMOylation. J Cereb Blood Flow Metab 28(2): 269–79
Yazawa I, Nukina N, Hashida H et al (1995) Abnormal gene product identified in hereditary dentatorubral-pallidoluysian atrophy (DRPLA) brain. Nat Genet 10(1): 99–103
Zabolotny JM, Bence-Hanulec KK, Stricker-Krongrad A et al (2002) PTP1B regulates leptin signal transduction in vivo. Dev Cell 2(4): 489–95
Zoghbi HY and Orr HT (2000) Glutamine repeats and neurodegeneration. Annu Rev Neurosci 23: 217–47

Chapter 10
The Ubiquitin–Proteasome System in Synapses

Suzanne Tydlacka, Shi-Hua Li, and Xiao-Jiang Li

Abstract The ubiquitin–proteasome system (UPS) is responsible for clearing most soluble proteins in the cytoplasm and nucleus. Recent studies reveal that the UPS function is critical for maintaining synaptic plasticity and transmission and that the UPS dysfunction is associated with axonal degeneration and impaired synaptic transmission. In this chapter, we will focus on the role of the UPS in synapses and the association of UPS impairment with neurological disorders. Since protein misfolding causes several neurological disorders that show synaptic dysfunction during the early stages of disease, understanding the involvement of the synaptic UPS in neurological disorders may help determine effective strategies for treating neurological disorders caused by the accumulation of misfolded proteins.

10.1 The Ubiquitin–Proteasome System Pathway

The ubiquitin–proteasome system (UPS) plays a key role in degrading short-lived and misfolded proteins.

Protein clearance by the UPS involves two sequential reactions: a ubiquitination reaction and a subsequent degradation of the ubiquitinated protein by the proteasome (Ciechanover 2005). The ubiquitination reaction requires a three-step covalent attachment of ubiquitin, a small and highly conserved peptide (76 residues), to the substrate. First, a ubiquitin monomer is activated by forming an intermolecular thiol ester in its C-terminal Gly with the ubiquitin-activating enzyme (E1) in an ATP-dependent reaction (Ciechanover 1998). Next, the activated ubiquitin is transferred to a Cys residue in the active site of a ubiquitin-conjugating enzyme (E2). Then the ubiquitin is linked by its C-terminus through an amide isopeptide linkage to the ε-amino group of a Lys residue of the substrate protein. This step is catalyzed by a series of distinct ubiquitin-protein ligases (E3) that confer specificity to the process by selectively binding to target proteins. Finally, activated ubiquitin molecules are

X.-J. Li (✉)
Department of Human Genetics, Emory University School of Medicine, Atlanta, GA, USA
e-mail: Xli2@emory.edu

A. Wyttenbach and V. O'Connor (eds.), *Folding for the Synapse*,
DOI 10.1007/978-1-4419-7061-9_10,

sequentially added at Lys48 to the previously ubiquitin-conjugated protein, forming a polyubiquitin chain. Proteins tagged with chains of four or more ubiquitins are recognized and unfolded by the 19S regulatory cap of the 26S proteasome. The unfolded protein can then enter the 20S catalytic core of the proteasome for degradation (Chau et al. 1989). Ubiquitin monomers are released after proteasome degradation or are actively removed by the ubiquitin carboxyl-terminal hydrolases (Mayer and Wilkinson 1989).

The 26S proteasome is known as an energy-dependent multicatalytic protease localized both in the nucleus and cytoplasm that degrades polyubiquitylated proteins. It consists of three major subunits: one 20S catalytic core flanked by two 19S regulatory caps. The catalytic core is a barrel-shaped compartment made up of four stacked rings of seven different α_{1-7} (the outer two rings) and β_{1-7} subunits (the inner two rings). Three of the two inner subunits ($\beta_{1,2,5}$) conduct the proteolytic activities. The β_1 subunit is the site of postglutamyl activity, the β_2 subunit is the site of trypsin-like activity, and the β_5 subunit is the site of chymotrypsin-like activity (Layfield et al. 2003). The outer α subunits stabilize the holoenzyme (Wolf and Hilt 2004). The 19S subunits participate in substrate recognition, untagging, and unfolding. They also induce conformational modifications toward an open state of the α-ring and provide the force that drives the substrate into the catalytic core (Pickart and Cohen 2004).

10.2 The UPS at the Synapse

In the central nervous system (CNS), the synapse is the major site of communication between neurons. The synapse consists of the presynaptic and postsynaptic terminals, as well as the space between them. The presynaptic terminal contains specialized structures necessary for vesicular neurotransmitter release. Once the neurotransmitter is released, the vesicle membrane will be retrieved into the presynaptic terminal by clathrin-mediated endocytosis and can be used again. The presynaptic terminal is separated by a cleft from the postsynaptic terminal, which is enriched in neurotransmitter receptors and downstream signaling molecules. In the synapse, ubiquitination is an important mechanism modulating synapse formation and function (Yi and Ehlers 2005), whereas polyubiquitination targets proteins for degradation through the UPS. Proteomic approaches using mass spectroscopy have confirmed the presence of UPS components at synapses (Li et al. 2004). In light of these findings, the UPS has been examined as a regulator of presynaptic and postsynaptic compartments in the developing and mature nervous system (Patrick 2006).

10.2.1 The Regulation of the UPS in the Presynaptic Terminal

In the presynaptic terminal, the UPS regulates presynaptic function through multiubiquitination and protein turnover, thereby altering protein activity and vesicle dynamics (Chen et al. 2003). Studies in *Drosophila* have shown that the UPS is involved in presynaptic development. Gain-of-function mutations in the deubiquitinating enzyme fat facets (faf) cause overgrowth of the presynaptic terminals in *Drosophila* motor neurons (DiAntonio et al. 2001). Loss of function of highwire, an E3 enzyme, has the same phenotype as the faf gain-of-function mutants (Huang et al. 1995; Wan et al. 2000), suggesting that the balance between ubiquitination and deubiquitination is important for presynaptic development.

In addition to its role in presynaptic development, the UPS is also found to regulate neurotransmitter release. For example, the ataxia mouse, which exhibits spontaneous tremors, is defective in the ubiquitin-specific protease Usp14 (Wilson et al. 2002). Usp14 is believed to play a role in the recycling of ubiquitin from multiubiquitinated proteins, which maintains cellular levels of free ubiquitin. Ataxia mice have 53% less quantal content and hippocampal long-term plasticity, both hallmarks of a presynaptic deficit, underscoring the importance of ubiquitin recycling. Additionally, acute proteasome inhibition results in a larger recycling pool of synaptic vesicles in cultured hippocampal neurons (Willeumier et al. 2006). This effect was enhanced with greater synaptic activity and reduced by less synaptic activity, suggesting an important role for the UPS in the activity-mediated regulation of presynaptic neurotransmitter release.

In addition, it has been found that the protein RIM1α, located in the presynaptic terminal, forms a scaffold that links synaptic vesicles with the fusion machinery and priming vesicles for release under the control of a E3 ubiquitin ligase called SCRAPPER (Yao et al. 2007). SCRAPPER is selectively enriched in the presynaptic terminal and binds to and polyubiquitinates RIM1α in a complex with accessory E3 ubiquitin ligase subunits. Mice lacking SCRAPPER have an increase in half-life and steady-state levels of RIM1α. Additionally, these mice have enhanced neurotransmission, and their cultured neurons, when treated with a proteasome inhibitor, show decreased vesicle release, in contrast to wild-type neurons, which have a higher frequency of vesicle release. Thus, SCRAPPER appears to be a major mediator of proteasomal regulation in the presynaptic terminal. These findings are consistent with those from *Drosophila* studies conducted by Willeumier et al. (2006), indicating that the spatial regulation of UPS-mediated degradation would allow for precise tuning of neuronal networks (Dobie and Craig 2007). Given these findings, it may be important to investigate the role of UPS-mediated protein degradation in regulating synaptic neurotransmission.

10.2.2 The Regulation of the UPS in the Postsynaptic Terminal

Direct immunofluorescence and immunoblotting of the postsynaptic density (PSD) fraction have revealed that there are two main components of the UPS: the ubiquitin and proteasome subunits (Ehlers 2003). Further, immunofluorescent labeling in cultured hippocampal neurons indicated that dendrites near synapses contain both proteasome and ubiquitin (Patrick et al. 2003); in fact, the proteasome and the synaptic marker synaptophysin were found to colocalize. When hippocampal proteasomes were pharmacologically inhibited, AMPA-induced glutamate receptor (GluR) endocytosis was blocked. Interestingly, there was no effect on the free ubiquitin pool, suggesting a direct role of proteasome-dependent degradation in the postsynaptic terminal. In light of these findings, polyubiquitination and degradation of proteins may be required for ligand-induced internalization of GluR receptors. The requirement for proteasome activity in GluR endocytosis suggests that there are UPS targets that can regulate GluR trafficking at the synapse. In principle, proteins that stabilize the AMPA-type glutamate receptor at the surface would be perfect candidates for such UPS targets (Bingol and Schuman 2005). Studies have found that the major scaffolding protein PSD-95 is regulated by the UPS. PSD-95 tethers NMDA- and AMPA-type glutamate receptors to signaling proteins and neuronal cytoskeleton. When NMDA receptors are activated, PSD-95 is ubiquitinated by the E3 ligase Mdm2 and removed from the synapse in a UPS-dependent manner (Colledge et al. 2003). Another postsynaptic UPS substrate is the spine-associated Rap GTPase activating protein (SPAR) (Pak and Sheng 2003). SPAR is phosphorylated by the serum-inducible kinase known as SNK, the expression of which is upregulated in the soma, but not in the dendrites. Upon phosphorylation of SPAR, there is an immediate degradation of SPAR that leads to loss of spines as well as a decrease in PSD-95 levels.

In line with this, studies of the *Drosophila* neuromuscular junction (NMJ) glutamatergic synapse were performed via a UAS/GAL4 transgenic method to drive the postsynaptic expression of proteasome β_2 and β_6 subunit mutant proteins, which operate through a dominant-negative mechanism to block proteasome function (Haas et al. 2007). This method allowed for the identification of specific postsynaptic effects of impaired UPS function on glutamatergic synapses located at the NMJ. The studies demonstrated that impaired postsynaptic proteasome function caused the B-class glutamate receptor and its interacting scaffold to be selectively misregulated, resulting in altered synaptic transmission strength and activity-dependent plasticity. The UPS inhibition was specific to B-class glutamate receptors and had no effect on A-class glutamate receptors.

10.3 The UPS and Axonal Growth

At the tip of outgrowing axons, growth cones sense guidance cues and transfer this information into cytoskeletal reorganizations to orient growth in a specific direction (Ciechanover and Brundin 2003). The belief is that the UPS activity is stimulated by guidance cues, and the resulting protein degradation could play a role in the structural rearrangements that mediate axonal outgrowth (Campbell and Holt 2001). For example, in growing axons netrin-1 is expressed in regions of the embryonic CNS, where it acts to attract the growth of receptor-bearing axons (Colamarino and Tessier-Lavigne 1995). Netrin-1 stimulates the production of ubiquitin-protein conjugates in growth cones, and the inhibition of proteasome-mediated proteolysis blocks chemotropic responses to growth guidance cues.

Additionally, genetic screens in *Drosophila* have identified mutations that result in axon guidance defects in the genes encoding a ubiquitin ligase (E3) and ubiquitin-specific protease enzymes, such as bendless (Muralidhar and Thomas 1993; Uthaman et al. 2008), non-stop (Martin et al. 1995; Poeck et al. 2001), and ariadne (Aguilera et al. 2000). Furthermore, genetic screens in *Caenorhabditis elegans* for regulators of GABAergic and glutamatergic synaptic growth identified the presynaptic E3 ligase, RPM-1 (Schaefer et al. 2000). RPM-1 acts as a positive regulator of growth, and RPM-1 mutants have a reduced number of synapses. Lending support to these findings are additional studies that found RPM-1 as a negative regulator of the p38 MAPK (mitogen-activated protein kinase) signaling pathway, mediating the ubiquitination and degradation of the MAPK kinase kinase (MAPKKK) DLK.

In mice, the knockout of Phr1, a mammalian ortholog of RPM-1, impaired phrenic nerve development and disrupted the morphology of phrenic nerve terminals (Bloom et al. 2007). Phr1 knockout mice have severe defects in axon tracts of the CNS, including failure of the corticofugal and thalamocortical axons to enter the subcortical telencephalon and form an internal capsule. This axonal tract is the predominant path for information flow between the cerebral cortex and the rest of the nervous system. These findings offer independent evidence that protein synthesis and degradation could be involved in growth cone guidance.

10.4 The UPS and Axonal Degeneration

Wallerian degeneration, which is degeneration of the distal segment of the nerve, occurs in many neuropathies and neurodegenerative diseases. Wallerian degeneration is distinct from death of the nerve cell body, because it requires a nerve fiber containing an axon to be cut and separated from the cell body, resulting in degeneration. The mechanism behind the Wallerian degeneration is unclear; however, in a mouse model of Wallerian degeneration (Wld^s), a fusion protein containing a truncated region of UFD2, an E4 protein involved in polyubiquitination, was found to be overexpressed (Conforti et al. 2000). Despite this result, it remains unclear whether the UPS contributed to the phenotype of Wld^s mice. To examine the role of the UPS in axonal degeneration, additional studies were performed using explanted superior cervical ganglion neurons whose axons were transected (Zhai et al. 2003). Normally, transected axons undergo complete degeneration within hours; however, when axons were treated with pharmacological inhibitors of the proteasome or when the deubiquitinating enzyme UBP2 was expressed to prevent polyubiquitination, the transected axons survived. Inhibitors of caspases and serine proteases, on the other hand, had no protective effect, indicating that the effect was selective for proteasome-mediated proteolysis.

Studies conducted on the gracile axonal dystrophy (gad) mouse, which has sensory and motor ataxia, illustrated that the gad mutation transmitted by a gene on chromosome 5 results in an in-frame deletion of exons 7 and 8 of Uchl1 (Saigoh et al. 1999). Uchl1 encodes the ubiquitin carboxy-terminal hydrolase isozyme (Uch-l1), which is selectively expressed in the nervous system and is thought to stimulate protein degradation by generating free monomeric ubiquitin. The gad mutation negatively affected protein turnover, suggesting that altered function of the ubiquitin system led directly to neurodegeneration.

In addition to injury-associated pathology, the role of the UPS in axon retraction and loss suggests its potential relevance to late-onset neurodegenerative diseases. Decreases in brain proteasome activity are known to occur during normal brain aging (Zhou et al. 2003; Tydlacka et al. 2008). Similarly, synaptic UPS activity is also decreased with age (Wang et al. 2008a), suggesting that aging has a negative impact on the UPS in synapses. Consistent with this idea, the age-dependent accumulation of ubiquitin-conjugated proteins and ubiquitin-associated inclusion bodies are present in the brains of patients with many neurodegenerative diseases. For example, Alzheimer's disease is associated with neurofibrillary tangles in the brain, Parkinson's disease with Lewy bodies, and polyglutamine diseases with polyglutamine-containing inclusions that are linked to ubiquitin. Mutations in the Ub-modifying enzymes and aberrant UPS function have both been implicated in these diseases as well as in other neurological syndromes, like gracile axonal dystrophy. Nonetheless, most efforts have been focused on understanding neuronal cell death, and little is known about synaptic UPS activity in neurodegenerative diseases. Since axonal degeneration and synaptic dysfunction often precede the death of the cell body, investigating the function of the UPS in synapses could provide a novel target for treating these neurological disorders.

10.5 Regulation of the UPS in Polyglutamine Diseases

Protein misfolding and aggregation are common pathological changes seen in age-dependent neurodegenerative disorders (Kopito 2000; Goldberg 2003). Polyglutamine (polyQ) expansion in various proteins causes nine known inherited neurodegenerative disorders, including Huntington's disease (HD), six types of spinocerebellar ataxia (SCA) diseases, dentatorubral-pallidoluysian atrophy (DRPLA), and spinal and bulbar muscular atrophy (SBMA) (Orr and Zoghbi 2007). In HD, polyQ expansion (>37 glutamines) in the N-terminal region of huntingtin (htt) causes progressive neurological symptoms and neurodegeneration. The feature common to all polyQ diseases is the presence of polyQ protein aggregates or inclusions due to the accumulation of misfolded proteins in selective brain regions. These misfolded proteins accumulate predominantly in neurons, despite the fact that polyQ disease proteins are widely expressed throughout the brain and body. Thus, polyQ diseases make an ideal model for investigating how protein misfolding affects neuronal function and whether polyQ expansion impairs UPS function in neurons.

Early studies using cell models of in vitro systems revealed the inhibitory effects of expanded polyQ proteins on UPS activity (Bence et al. 2001; Waelter et al. 2001; Jana et al. 2001). Two models have since been proposed to account for this UPS impairment seen in polyQ diseases. The first model emerged from immunocytochemistry staining of samples from patients with polyQ diseases. In the brains of HD and spinocerebellar ataxia type 1 (SCA1) patients, there was colocalization of polyQ inclusions with ubiquitin and proteasome subunits (DiFiglia et al. 1997; Cummings et al. 1998). These findings led to the idea that the proteasome subunits were sequestered into inclusions, thereby impairing their normal function. However, subsequent studies found that just a relatively small percentage of the total UPS elements were trapped in the polyQ aggregates (Bennett et al. 2005). In addition, polyQ protein-mediated impairment of the UPS occurred in the absence of detectable aggregates (Bennett et al. 2005). In SCA-1, proteasomal degradation of the disease protein ataxin-1 was delayed by the presence of an expanded polyQ tract without impaired ubiquitination (Cummings et al. 1999). In vitro assays also illustrated that the proteasome could not digest expanded polyQ sequences efficiently, but only cut the flanking basic residues (Venkatraman et al. 2004). These findings led to the second model, wherein soluble proteins containing expanded polyQ repeats may block the proteasome, preventing the entry of substrates and their subsequent degradation.

Further evidence supporting this theory came to light when proteasomal function was found to be reduced by the expression of N-terminal mutant huntingtin (htt) with an expanded polyQ tract (Jana et al. 2001). However, our own lab has found that the accumulation of N-terminal htt fragments in HD mouse brains increases with age, while proteasome activity decreases (Zhou et al. 2003). This is also consistent with the fact that the striatal region accumulates abundant intranuclear inclusions (NII), and proteasome activity in the striatum is lower than in the cortex (Wang et al. 2008b).

Nevertheless, transgenic mouse models of polyQ disease have not revealed that UPS activity is significantly reduced in their brains compared with wild-type mice. For example, a transgenic approach was used to express a fluorescent UPS reporter (Ub^{G76V}GFP) in mice (Lindsten et al. 2003). This transgenic mouse model was crossed with a mouse model of SCA7 to determine whether UPS activity was responsible for the NII formation in vulnerable retinal neurons (Bowman et al. 2005); however, the crossed mice yielded no evidence that mutant SCA-7 proteins could impair UPS activity in the retina. Likewise, biochemical assays of brain homogenates from HD mice that express exon 1 mutant huntingtin showed no reduction in proteasomal activity or levels of proteasomal subunits LMP2 and LMP7 (Díaz-Hernández et al. 2003; Bett et al. 2006). Rather, increased chymotrypsin-like activity was seen in whole brain homogenates of 13-week-old R6/2 mice (Bett et al. 2006) and in conditional HD (HD94) mice (Díaz-Hernández et al. 2003), which could be an indirect consequence of htt toxicity or cell stress. Similarly, studies of HD CAG150 repeat knock-in mice revealed no significant difference in UPS activities in brain lysates compared with wild-type mouse brain samples (Wang et al. 2008a; Tydlacka et al. 2008). A different assay, which measures Lys 48-linked polyubiquitin chain accumulation, does suggest global changes in polyubiquitination in HD mouse brains (Bennett et al. 2007). It is possible that the effect of polyQ proteins on UPS activity depends on their accumulation and subcellular localization, which cannot be detected by examining whole cell homogenates. Because UPS activity varies in different types of cells and subcellular regions, it is important to evaluate UPS activity in different subcellular compartments of neurons, especially in synapses, where mutant htt accumulates (Li et al. 1999, 2000a; Yu et al. 2003). Furthermore, it is important to point out that aging can significantly reduce UPS function in the brain (Zhou et al. 2003; Wang et al. 2008a), which may play an important role in the age-dependent accumulation of misfolded polyQ proteins and formation of polyQ inclusions in the brain.

10.5.1 The Role of the UPS in the Synaptic Dysfunction During Huntington's Disease

Recently, mounting evidence suggests that alterations in synaptic function in HD and other polyQ diseases could underlie early symptoms (Smith et al. 2005). Mice expressing htt with 80 glutamine repeats are found to have reduced synaptic plasticity and impairment of LTP induction (Usdin et al. 1999). In SCA-1, there is altered trafficking of glutamate receptor subunits and PKCγ in Purkinje cells (Skinner et al. 2001). Further, huntingtin is known to interact with cytoskeletal and synaptic vesicle proteins that are essential for exocytosis and endocytosis (Li and Li 2004). For example, huntingtin interacts with HAP1, which is involved in vesicle trafficking, and HIP1, which is involved in clathrin-mediated endocytosis of presynaptic nerve terminals (Li et al. 1995; Gutekunst et al. 1998; Kalchman et al. 1997). HAP1 is also distributed in synapses, and cotransfection of mutant huntingtin with HAP1 in PC12 cells inhibits the neurite outgrowth promoted by HAP1 (Li et al. 2000b).

Mutant htt could affect the UPS function in synapses. Using adenoviral vectors that express fluorescent UPS reporters fused with the presynaptic (SNAP25) or postsynaptic (PSD95) protein, we were able to target the UPS reporters to presynaptic or postsynaptic terminals (Wang et al. 2008a). Targeting of these reporters to defined synaptic structures allows for sensitive detection of changes in synaptic UPS activity. Accordingly, these reporters allowed us for the first time to detect synaptic UPS activity in the brain, revealing decreased UPS activity in the synapses of HD mouse brains (Wang et al. 2008a); this decrease was confirmed by biochemical assays of proteasomal activity in isolated synaptosomes. Chymotrypsin-like activity in synaptosomes isolated from the cortex and striatum revealed a decrease in activity with age. This decrease in proteasome activity is enhanced by the presence of mutant htt. It has been shown that mutant htt can affect intracellular vesicle trafficking (Gauthier et al. 2004), and indeed, the intracellular transport of mitochondria was also shown to be decreased in the presence of mutant htt (Trushina et al. 2004; Orr et al. 2008). Such impaired trafficking could lower the number of functional mitochondria available in nerve terminals and the ATP supply to the synaptic UPS (Orr et al. 2008; Wang et al. 2008a), thereby decreasing UPS activity in synapses (see Chap. 6 for details on mitochondrial transport). Thus, although assays of brain homogenates could not establish a difference in brain UPS activity between wild-type and HD transgenic mice, targeting of a UPS reporter to the synaptic compartment and biochemical analysis on synaptosomes enabled detection of an inhibitory effect of mutant htt on synaptic UPS function.

The finding that mutant htt can indeed affect synaptic UPS function gives us important insight into the mechanism by which misfolded proteins affect synaptic function and neurotransmitter release. It is also possible that UPS function is impaired in other neurodegenerative diseases in which misfolded proteins accumulate in synapses.

10.6 Summary

Protein degradation mediated by the UPS is now known to be as important as protein synthesis in regulating the synaptic proteins critical for neurotransmission. In the synapse, the UPS has a role in regulating synapse development, neurotransmission strength, and pre- and postsynaptic function. Synaptic dysfunction is thought to occur in the early stages of many neurological disorders that are caused by the accumulation of misfolded proteins. Recent studies suggest that synaptic UPS function may be impaired in HD; whether UPS function in synapses in other neurological disorders is also affected remains to be investigated. Given the important roles of the synaptic UPS and the synaptic dysfunction seen in the early stages of several neurological disorders, elucidating the role of the UPS in the synapse will give us a better understanding of the pathogenesis of these diseases, and may eventually lead to the development of effective treatments for these neurodegenerative disorders.

References

Aguilera M, Oliveros M, Martinez-Padron M, Barbas JA, and Ferrus A (2000) Aridne-1: a vital Drosophila gene is required in development and defines a new conserved family of ring-finger proteins. Genetics 155:1231–1244

Bence NF, Sampat RM, and Kopito RR (2001) Impairment of the ubiquitin–proteasome system by protein aggregation. Science 292:1552–1555

Bennett EJ, Bence NF, Jayakumar R, and Kopito RR (2005) Global impairment of the ubiquitin–proteasome system by nuclear or cytoplasmic protein aggregates precedes inclusion body formation. Mol Cell 17:351–365

Bennett EJ, Shaler TA, Woodman B, Ryu KY, Zaitseva TS, Becker CH, Bates GP, Schulman H, and Kopito RR (2007) Global changes to the ubiquitin system in Huntington's disease. Nature 448:704–708

Bett JS, Goellner GM, Woodman B, Pratt G, Rechsteiner M, and Bates GP (2006) Proteasome impairment does not contribute to pathogenesis in R6/2 Huntington's disease mouse: exclusion of proteasome activator REGγ as a therapeutic target. Hum Mol Genet 15:33–44

Bingol B and Schuman EM (2005) Synaptic protein degradation by the ubiquitin proteasome system. Curr Opin Neurobiol 15:536–541

Bloom AJ, Miller BR, Sanes JR, and DiAntonio A (2007) The requirement for Phr1 in CNS axon tract formation reveals the corticostriatal boundary as a choice point for cortical axons. Genes Dev 21:2593–2606

Bowman AB, Yoo SY, Dantuma, NP and Zoghbi HY (2005) Neuronal dysfunction in a polyglutamine disease model occurs in the absence of ubiquitin–proteasome system impairment and inversely correlates with the degree of nuclear inclusion formation. Hum Mol Genet 14:679–691

Campbell DS and Holt CE (2001) Chemotropic responses of retinal growth cones mediated by rapid local protein synthesis and degradation. Neuron 32:1013–1026

Chau V, Tobias JW, Bachmar A, Marriott D, Ecker DJ, Gonda DK, and Varshavsky A (1989) A multiubiquitin chain is confined to specific lysine in a targeted short-lived protein. Science 243:1576–1583

Chen H, Polo S, Fiore PP, and Camilli PV (2003) Rapid Ca^{2+}-dependent decrease of protein ubiquitination at synapses. Proc Natl Acad Sci U S A 100:14908–14913

Ciechanover A (1998) The ubiquitin–proteasome pathway: on protein death and cell life. EMBO 17:7151–7160

Ciechanover A (2005) Proteolysis: from the lysosome to ubiquitin and the proteasome. Nat Rev Mol Cell Biol 6:79–87

Ciechanover A and Brundin P (2003) The ubiquitin proteasome system in neurodegenerative diseases: sometimes the chicken, sometimes the egg. Neuron 40:427–446

Colamarino CY and Tessier-Lavigne M (1995) The axonal chemoattractant netrin-1 is also a chemorepellent for trochlear motor axons. Cell 81:621–629

Colledge M, Snyder EM, Crozier RA, Soderling JA, Jin Y, Langeberg LK, Lu H, Bear MF, and Scott JD (2003) Ubiquitination regulates PSD-95 degradation and AMPA receptor surface expression. Neuron 40:595–607

Conforti L, Tarlton A, Mack TG, Mi W, Buckmaster EA, Wagner D, Perry VH, and Coleman MP (2000) A Ufd2/D4cole1e chimeric protein and overexpersion of Rbp7 in the slow Wallerian degeneration (WldS) mouse. Proc Natl Acad Sci U S A 97:11377–11382

Cummings CJ, Mancini MA, Antalffy B, DeFranco DB, Orr HT, and Zoghbi HY (1998) Chaperone suppression of aggregation and altered subcellular proteasome localization imply protein misfolding in SCA1. Nat Genet 19:148–154

Cummings CJ, Reinstein E, Sun Y, Antalffy B, Jiang Y, Ciehanoer A, Orr HT, Beaudet AL, and Zoghbi HY (1999) Mutation of the E6-AP ubiquitin ligase reduces nuclear inclusion frequency while accelerating polyglutamine-induced pathology in SCA1 mice. Neuron 24:879–892

DiAntonio A, Haghighi AP, Portman SL, Lee JD, Amaranto AM, and Goodman CS (2001) Ubiquitination-dependent mechanisms regulate synaptic growth and function. Nature 412:449–452

Díaz-Hernández M, Hernández F, Martín-Aparicio E, Gomez-Ramos P, Moran MA, Castano JG, Ferrer I, Avila J, and Lucas JJ (2003) Neuronal induction of the immunoproteasome in Huntington's disease. J Neurosci 23:11653–11661

DiFiglia M, Sapp E, Chase KO, Davies SW, Bates GP, Vonsattel JP, and Aronin N (1997) Aggregation of huntingtin in neuronal intranuclear inclusions and dystrophic neurites in brain. Science 277:1990–1993

Dobie F and Craig AM (2007) A fight for neurotransmission: SCRAPPER trashes RIM. Cell 130:775–777

Ehlers MD (2003) Activity levels controls postsynaptic composition and signaling via the ubiquitin–proteasome system. Nat Neurosci 6:231–242

Gauthier LR, Charrin BC, Borrell-Pagès M, Dompierre JP, Rangone H, Cordelières FP, De Mey J, MacDonald ME, Lessmann V, Humbert S, Saudou F. (2004) Huntingtin controls neurotrophic support and survival of neurons by enhancing BDNF vesicular transport along microtubules. Cell 118:127–138

Goldberg AL (2003) Protein degradation and protection against misfolded or damaged proteins. Nature 426:895–899

Gutekunst CA, Li SH, Yi H, Ferrante RJ, Li XJ, and Hersch SM (1998) The cellular and subcellular localization of huntingtin-associated protein 1 (HAP1): comparison with huntingtin in rat and human. J Neurosci 18:7674–7686

Haas KF, Miller SLH, Friedman DB, and Broadie K (2007) The ubiquitin–proteasome system postsynaptically regulates glutamatergic synaptic function. Mol Cell Neurosci 35:64–75

Huang Y, Baker RT, and Fischer-Vize JA (1995) Control of cell fate by a deubiquitinating enzyme encoded by the fat facets gene. Science 270:1828–1831

Jana NR, Zemskov EA, Wang G, and Nukina N (2001) Altered proteasomal function due to the expression of polyglutamine-expanded truncated N-terminal huntingtin induces apoptosis by caspase activation through mitochondrial cytochrome c release. Hum Mol Genet 10:1049–1059

Kalchman MA, Koide HB, McCutcheon K, Graham RK, Nichol K, Nishiyama K, Kazemi-Esfarjani P, Lynn FC, Wellington C, Metzler M, Goldberg YP, Kanazawa I, Gietz RD, and Hayden MR (1997) HIP, a human homologue of S. cerevisiae Sla2p, interacts with membrain-associated huntingtin in the brain. Nat Genet 16:44–53

Kopito RR (2000) Aggregosmes, inclusion bodies and protein aggregation. Trends Cell Biol 10:524–530

Layfield R, Cavey JR, and Lowe J (2003) Role of ubiquitin-mediated proteolysis in the pathogenesis of neurodegenerative disorders. Ageing Res Rev 2:343–356

Li SH and Li XJ (2004) Huntingtin-protein interactions and the pathogenesis of Huntington's disease. Trends Genet 20:146–154

Li XJ, Li SH, Sharp AH, Nucifora FC Jr, Schilling G, Lanahan A, Worley P, Snyder SH, and Ross CA (1995) A huntingtin-associated protein enriched in brain with implications for pathology. Nature 378:398–402

Li H, Li SH, Cheng AL, Mangiarini L, Bates GP, and Li XJ (1999) Ultrastructural localization and progressive formation of neuropil aggregates in Huntington's disease transgenic mice. Hum Mol Genet 8:1227–1236

Li H, Li SH, Johnston H, Shelbourne PF, and Li XJ (2000a) Amino-terminal fragments of mutant huntingtin show selective accumulation in striatal neurons and synaptic toxicity. Nat Genet 25:385–389

Li SH, Li H, Torre ER, and Li XJ (2000b) Expression of huntingtin-associated protein-1 in neuronal cells implicates a role in neurite growth. Mol Cell Neuroscience 16:168–183

Li KW, Hornshaw MP, Van Der Schors RC, Watson R, Tate S, Casetta B, Jimenez CR, Gouwenberg Y, Gundelfinger ED, and Smalla KW (2004) Proteomics analysis of rat brain postsynaptic density. Implications of the diverse protein functional groups for the integration of synaptic physiology. J Biol Chem 279:987–1002

Lindsten K, Menendez-Benito V, Masucci MG, and Dantuma NP (2003) A transgenic mouse model of the ubiquitin/proteasome system. Nat Biotechnol 21:897–902

Martin KA, Poeck B, Roth H, Ebens AJ, Ballard LC, and Zipursky SL (1995) Mutations disrupting neuronal connectivity in the Drosophila visual system. Neuron 14:229–240

Mayer AN and Wilkinson KD (1989) Detection, resolution, and nomenclature of multiple ubiquitin carboxyl-terminal esterases from bovine calf thymus. Biochemistry 28:166–172

Muralidhar MG and Thomas JB (1993) The Drosophila bendless gene encodes a neural protein related to ubiquitin-conjugating enzymes. Neuron 11:253–266

Orr AL, Li S, Wang CE, Li H, Wang J, Rong J, Xu X, Mastroberardino PG, Greenamyre JT, Li XJ. (2008) N-terminal mutant huntingtin associates with mitochondria and impairs mitochondrial trafficking. J Neurosci. 28:2783–2792

Orr HT and Zoghbi HY (2007) Trinucleotide repeat disorders. Annu Rev Neurosci 30:575–621

Pak DT and Sheng M (2003) Targeted protein degradation and synpase remodeling by an inducible protein kinase. Science 302:1368–1373

Patrick GN (2006) Synapse formation and plasticity: recent insights from the perspective of the ubiquitin proteasome system. Curr Opin Neurobiol 16:90–94

Patrick GN, Bingol B, Weld HA, and Schuman EM (2003) Ubiquitin-mediated proteasome activity is required for agonist-induced endocytosis of GluRs. Curr Biol 13:2073–2081

Pickart CM and Cohen RE (2004) Proteasomes and their kin: proteases in the machine age. Nat Rev Mol Cell Dev Biol 5:177–187

Poeck B, Fischer S, Gunning D, Zipursky SL, and Salecker I (2001) Glial cells mediate target layer selection of retinal axons in the developing visual system of Drosophila. Neuron 29:99–113

Saigoh K, Wang YL, Suh T, Yamanishi Y, Sakai H, Kiyosawa T, Harada N, Ichihara S, Wakana T, Kikuchi, and Wada K (1999) Intragenic deletion in the gene encoding ubiquitin carboxy-terminal hydrolase in gad mice. Nat Genet 23:47–51

Schaefer AM, Hadwiger GD, and Nonet ML (2000) RPM-1, a conserved neuronal gene that regulates targeting and synaptogenesis in *C. elegans*. Neuron 26:345–356

Skinner PJ, Vierra-Green CA, Clark HB, Zoghbi HY, and Orr HT (2001) Altered trafficking of membrane proteins in purkinje cells of SCA1 transgenic mice. Am J Pathol 159:905–913

Smith R, Brudin P, and Li JY (2005) Synaptic dysfunction in Huntington's disease: a new perspective. Cell Mol Life Sci 62:1901–1912

Trushina E, Dyer RB, Badger JD 2nd, Ure D, Eide L, Tran DD, Vrieze BT, Legendre-Guillemin V, McPherson PS, Mandavilli BS, Van Houten B, Zeitlin S, McNiven M, Aebersold R, Hayden M, Parisi JE, Seeberg E, Dragatsis I, Doyle K, Bender A, Chacko C, McMurray CT (2004) Mutant huntingtin impairs axonal trafficking in mammalian neurons in vivo and in vitro. Mol Cell Biol 24:8195–8209

Tydlacka S, Wang CE, Wang X, Li SH, and Li XJ (2008) Differential activities of the ubiquitin–proteasome system in neurons versus glia may account for the preferential accumulation of misfolded proteins in neurons. J Neurosci 28:13285–13295

Usdin MT, Shelbourne PF, Myers RM, and Madison DV (1999) Impaired synaptic plasticity in mice carrying the Huntington's disease mutation. Hum Mol Genet 8:839–946

Uthaman SB, Godenschwege TA, and Murphey RK (2008) A mechanism distinct from highwire for the drosophila ubiquitin conjugase bendless in synaptic growth and maturation. J Neursci 28:8615–8623

Venkatraman P, Wetzel R, Tanaka M, Nukina N, Goldberg AL. (2004) Eukaryotic proteasomes cannot digest polyglutamine sequences and release them during degradation of polyglutamine-containing proteins. Mol Cell 14:95–104

Waelter S, Boeddrich A, Lurz R, Scherzinger E, Lueder G, Lehrach H, and Wanker EE (2001) Accumulation of mutant huntingtin fragments in aggresome-like inclusion bodies as a result of insufficient protein degradation. Mol Biol Cell 12:1393–1407

Wan HI, DiAntonio A, Fetter RD, Bergstrom K, Strauss R, and Goodman CS (2000) Highwire regulates synaptic growth in Drosophila. Neuron 26:313–329

Wang JJ, Wang CE, Orr A, Tydlacka S, Li Sh, and Li XJ (2008a) Impaired ubiquitin–proteasome system activity in the synapses of Huntington's disease mice. J Cell Biol 180:1177–1189

Wang CE, Tydlacka S, Orr AL, Yang SH, Graham RK, Hayden MR, Li S, Chan AW, and Li XJ (2008b) Accumulation of N-terminal mutant huntingtin in mouse and monkey models implicated as a pathogenic mechanism in Huntington's disease. Hum Mol Genet 17:2738–2751

Willeumier K, Pulst SM, and Schweizer FE (2006) Proteasome inhibition triggers activity-dependent increase in size of the recycling vesicle pool in cultured hippocampal neurons. 26:11333–11341

Wilson SM, Bhattacharyya B, Rachel RA, Coppola V, Tessarollo L, Householder DB, Fletcher CF, Miller RJ, Copeland NG, and Jenkins NA (2002) Synaptic defects in ataxia mice result from a mutation in Usp14, encoding a ubiquitin-specific protease. Nat Genet 32:420–425

Wolf DH and Hilt W (2004) The proteasome: a proteolytic nanomachine of cell regulation and waste disposal. Biochim Biophys Acta 1695:19–31

Yao I, Takagi H, Ageta H, Kahyo T, Sato S, Hatanaka K, Fukauda Y, Chiba T, Morone N, Yuasa S, Inokuchi K, Ohtsuka T, MacGregor GR, Tanaka K, and Setou M (2007) SCRAPPER-dependent ubiquitination of active zone protein RIM1 regulates synaptic vesicle release. Cell 130:943–957

Yu ZX, Li SH, Evans J, Pillarisetti A, Li H, Li XJ. (2003) Mutant huntingtin causes context-dependent neurodegeneration in mice with Huntington's disease. J Neurosci. 23:2193-202

Yi JJ and Ehlers MD (2005) Ubiquitin and protein turnover in synapse formation. Neuron 47:629–632

Zhai Q, Wang J, Kim A, Liu Q, Watts R, Hoopfer E, Mitchison T, Luo L, and He Z (2003) Involvement of the ubiquitin–proteasome system in the early stages of wallerian degeneration. Neuron 39:217–225

Zhou H, Cao F, Wang Z, Yu ZX, Nguyen HP, Evans J, Li SH, and Li XJ (2003) Huntingtin forms toxic NH2-terminal fragment complexes that are promoted by the age-dependent decrease in proteasome activity. J Cell Biol 163:109–118

Part IV
Chronic Neurodegeneration Associated with Protein Misfolding and Synaptic Dysfunction

Chapter 11
VAPB Aggregates and Neurodegeneration

P. Skehel

Abstract VAP proteins are a small family of type II membrane proteins enriched on the endoplasmic reticulum that have been conserved from yeast to mammals. The N-terminal half of the proteins consists of a domain highly homologous to a polypeptide found in the motile sperm of nematodes known as the major sperm protein (MSP). A mis-sense mutation in human *vapB* that changes a proline residue to a serine in the most highly conserved region of the MSP domain causes a rare form of motor neuron disease, amyotrophic lateral sclerosis type 8. Whether $vapB^{P56S}$ is a gain or loss of function mutation is not yet clear, however, it causes the protein to aggregate and may disrupt the normal function and regulation of the ER.

11.1 Introduction

Motor neuron disease describes a range of degenerative conditions in which neurons of the motor system are selectively, although not exclusively, lost. These diseases are invariably fatal, and there are currently no effective therapies or treatments (Talbot 2002).

The majority of MND cases appear to arise spontaneously, however, in approximately 5–10% of patients the disease is associated with an inherited genetic mutation (Hand and Rouleau 2002; Schymick et al. 2007). The similarity in symptoms suffered in both forms of disease suggests that they share similar pathological mechanisms. The first human mutation identified in familial motor neuron disease was in the Zn^{2+}/Cu^{2+} superoxide dismutase type 1 gene, *sod1* (Rosen et al. 1993). To date, there are now more than 100 different mutations in *sod1* associated with MND (Andersen 2006). Within a family most patients exhibit similar clinical phenotypes, although in some cases variability between siblings has been observed (Ito et al. 2002). Different

P. Skehel (✉)
Centre for Integrative Physiology, University of Edinburgh, Hugh Robson Building, George Square, Edinburgh EH8 9XD, UK
e-mail: Paul.Skehel@ed.ac.uk

A. Wyttenbach and V. O'Connor (eds.), *Folding for the Synapse*,
DOI 10.1007/978-1-4419-7061-9_11, © Springer Science+Business Media, LLC 2011

sod1 mutations, however, produce diseases with significant differences in age at onset, severity and progression, and which may also involve other functions such as the sensory system (Abe et al. 1996). The molecular characterization of inherited mutations associated with motor neuron diseases may, therefore, provide information relevant to the pathophysiological processes of many neurodegenerative conditions and to motor neuron degeneration in particular. Mutations identified in genes previously not associated with disease can implicate different cellular and molecular pathways with the disease process.

11.1.1 Amyotrophic Lateral Sclerosis Type 8

In 2004, a mis-sense mutation in human *vapB* was identified as the causative genotype associated with familial motor neuron disease in a large Brazilian family (Nishimura et al. 2004a, b). The disease is inherited in a dominant autosomal fashion, and clinical phenotypes vary from that of classical amyotrophic lateral sclerosis (ALS), to a slow progression lower motor neuron selective spinal muscular atrophy (SMA). In some instances, there are also autonomic dysfunctions (Nishimura et al. 2004a, b; Marques et al. 2006). The disease associated with this mutation was classified as ALS type VIII or ALS 8. Understanding the molecular properties and biological activities of VAPB, and how these are disrupted by the mis-sense mutation, could provide information on the pathophysiological mechanisms associated with ALS 8. ALS 8 is extremely rare but understanding the disease at the molecular level may provide important information on the mechanisms of, and selective motor system sensitivity to, neuronal degeneration.

11.2 VAP Proteins

The first metazoan VAP protein was identified in the marine mollusc *Aplysia californica* by a yeast two hybrid screen for VAMP/Synaptobrevin interacting proteins, and hence the nomenclature VAMP/Synaptobrevin-associated proteins or "VAP" (Skehel et al. 1995). VAP homologues are found in yeast, *Drosophila*, mice, rats and humans (Nikawa et al. 1995; Weir et al. 1998; Nishimura et al. 1999; Skehel et al. 2000; Pennetta et al. 2002). These proteins are expressed ubiquitously, but at different levels depending on cell and tissue type (Weir et al. 1998; Nishimura et al. 1999; Skehel et al. 2000; Gkogkas et al. 2008). Vertebrates have two closely related VAP genes, *vapA* and *vapB*. The proteins share the same general structural features (Fig. 11.1a). The amino terminal halves of the proteins contain a structural motif highly homologous to a nematode sperm protein termed the "major sperm protein" or MSP (Sepsenwol et al. 1989; King et al. 1992). The remaining C-terminal domains contain a region predicted to form an amphipathic coiled/coil, and a short hydrophobic membrane anchor at the carboxy-terminus.

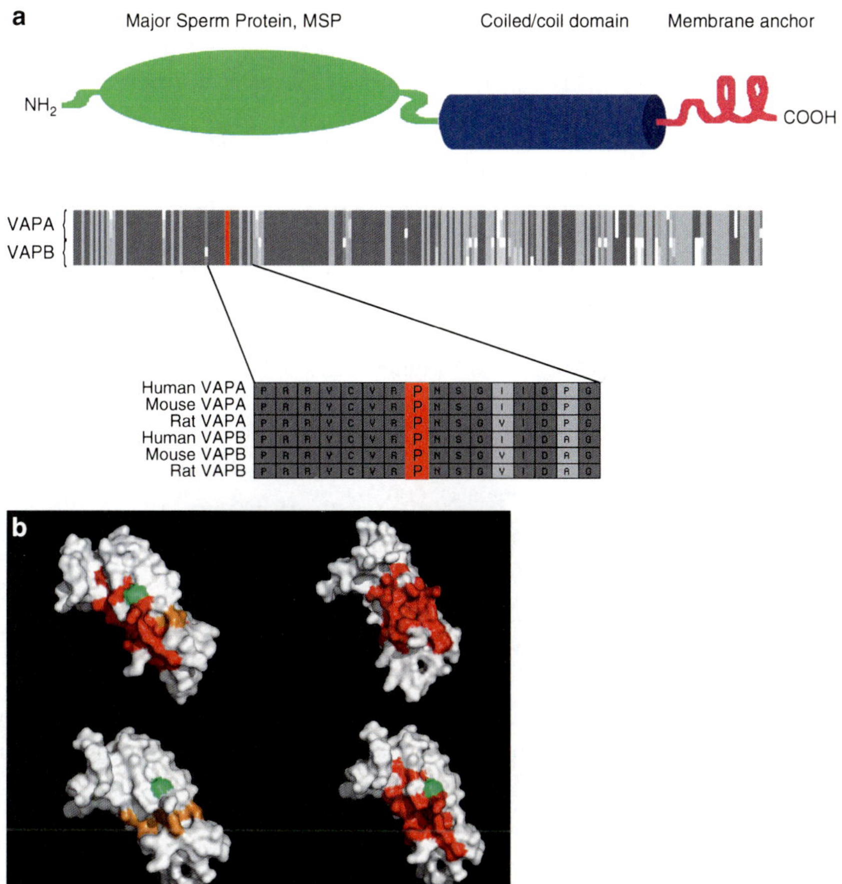

Fig. 11.1 General domain structure of VAP proteins. (**a**) The amino terminal half of the protein contains a structural motif highly homologous to the nematode "major sperm protein", or MSP (Sepsenwol et al. 1989; King et al. 1992). The C-terminal domains of the proteins contain a region predicted to form an amphipathic coiled/coil, and a short hydrophobic membrane anchor at the carboxy-terminus. Sequences shown are from human, mouse and rat, *darker shading* indicates higher levels of conservation. The conserved proline 56 is highlighted in *red*. (**b**) Evolutionary trance analysis identified a patch on the surface of the MSP domain, which is therefore predicted to be involved in protein–protein interactions (shown in *red*). Residues involved in FFAT binding are shown in *orange*, and the Pro56 residue is shown in *green*. The *panels* depict four different perspectives of the rat VAPA MSP domain structure as determined by Kaiser et al. (2005)

Injection of antisera against *Aplysia* VAP (ApVAP33) into the presynaptic cell of an *Aplysia* sensory–motor neuron co-culture had the effect of reducing evoked synaptic transmission (Skehel et al. 1995). This suggested that VAP proteins might function in the release or trafficking of synaptic vesicles. Analysis of miniature EPSPs in these cultures indicates that the total number of vesicle available for the release does not change, but the efficiency with which they are released is decreased

(Ghirardi and Skehel, unpublished). Subsequent studies have shown that VAP proteins accumulate on intracellular membranes, such as the endoplasmic reticulum and not on synaptic vesicles (Lapierre et al. 1999; Soussan et al. 1999; Skehel et al. 2000). The influence of VAP proteins on synaptic vesicle exocytosis, therefore, is most likely due to a function earlier in the secretory pathway, rather than at the point of synaptic transmission.

Several studies have indicated that VAP proteins can act in membrane trafficking between the ER and Golgi apparatus, and the plasma membrane. Soussman et al., showed that antibodies to VAPB (termed ERG30 in the report) inhibited intra-Golgi transport in an *in vitro* system, and resulted in the accumulation of COPI coated vesicles (Soussan et al. 1999). More recently, Prosser et al., have suggested that over expression of VAPA but not VAPB inhibits ER-Golgi transport (Prosser et al. 2008). At the plasma membrane, VAPA (VAP-33) has been implicated in GLUT4 incorporation (Foster et al. 2000), and is also found associated with occludin at tight junctions (Lapierre et al. 1999). *Drosophila* DVAP-33 seems to localise also to the plasma membrane (Pennetta et al. 2002).

11.3 VAP Proteins and Microtubules

Ultrastructural analysis of hippocampal neurons detected VAPA immunoreactivity associated with microtubules (Skehel et al. 2000). Pennetta et. al., subsequently showed that hypomorphic and hypermorphic mutations in one of the *Drosophila* VAP genes, DVAP-33A, resulted in the rearrangement of microtubule networks in motor neuron boutons (Pennetta et al. 2002). Later, Amarillo et. al., identified interactions between VAPB and the microtubule associated proteins Nir1, 2 and 3. The interactions between VAPB and Nir2 and Nir3 are required to maintain the structure of the ER (Amarilio et al. 2005). Over expression of VAPB together with Nir2 induced the formation of granular and multilamellar structures, whereas over expression of VAPB and Nir3 resulted in a rearrangement of the microtubule network and the ER (Amarilio et al. 2005).

11.4 VAPs as ER Docking Proteins

VAP proteins associate with intracellular membranes, such as the ER via a short hydrophobic C-terminal anchor, with the large majority of the protein facing the cytoplasm (Soussan et al. 1999). A number of cytoplasmic proteins associate with the ER membrane via interactions with VAPA and/or VAPB (Wyles et al. 2002; Loewen and Levine 2005; Kawano et al. 2006). Oxysterol-binding proteins (OSBP), the ceramide binding protein CERT, and the yeast transcriptional regulator Opi1, all bind to VAPA via their "FFAT" motif (two phenylalanines in an acid tract) (Wyles et al. 2002; Loewen et al. 2003). The same motif is present in Nir1, 2 and

3 (Amarilio et al. 2005). The structural basis of the FFAT interaction with the rat VAPA MSP domain has been determined, demonstrating that the MSP domain constitutes a FFAT binding activity (Kaiser et al. 2005). Thus, through the MSP domain VAP proteins may mediate the interaction of cytoplasmic proteins and cytoskeletal elements with intracellular membranes, such as the ER. Pathogens may also have exploited this property of VAP proteins. Hepatitis C virus replication takes place on the surface of the ER and has been shown to be dependent on the interaction of VAPA or VAPB, with the viral proteins NS5A and NS5B (Tu et al. 1999; Gao et al. 2004; Hamamoto et al. 2005).

The ALS 8 mis-sense mutation lies within the MSP domain of VAPB and converts an absolutely conserved proline to a serine (Nishimura et al. 2004a, b) (Fig. 11.1a). An evolutionary trace (ET) analysis of 348 MSP domains from the PFAM database, constrained by the secondary structure of the rat VAPA MSP domain (Kaiser et al. 2005), identified a surface patch of 29 amino acids predicted to participate in protein–protein interactions (Fig. 11.1b). These residues include, and extend beyond, those responsible for the VAPA-FFAT interaction. Pro56 lies within this area, but does not take part directly in binding the FFAT motif. How the P56S mutation would disrupt the over all structure of the MSP domain is not clear. However, Teuling et al., have argued that FFAT binding is perturbed in VAPB^{P56S} (Teuling et al. 2007).

11.5 VAP Aggregates

Like many degenerative diseases ALS is associated with the presence of intracellular protein aggregates in affected cells. It is unclear whether these aggregates are cytotoxic, or the product of a cytotoxic processes. Three distinct types of aggregate have been identified by histological analysis of autopsy samples from ALS patients and transgenic animals. Thread-like ubiquinated aggregates have been found in all ALS patients, and the abundance of these structures correlates with the severity of the disease (Welsem et al. 2002). Eosin-staining "Bunina bodies" are also found in a large majority of ALS cases. The third class of aggregate are found in somatic and proximal regions of affected neurons and are predominantly composed of neurofilament. The presence of these aggregates could be related to cellular responses to oxidative damage (see (Williamson et al. 2000; Menzies et al. 2002). It has also been proposed that aggregates are themselves cytotoxic, leading to the blockage of intracellular transport, sequestration of cellular components or activation of stress responses (Soto 2003). Studies on other neurodegenerative diseases have suggested the formation of aggregates may in fact be protective. Neuropathology studies on tissue from Parkinson's disease patients indicated that cells containing protein aggregates in the form of Lewy body inclusions appeared to be healthier than the surrounding cells that contained no aggregates (Tompkins and Hill 1997). Furthermore, in a cell culture model of Huntington's disease, cell survival was associated with inclusion body formation, whereas accumulation of less aggregated huntingtin protein correlated with cell death (Arrasate et al. 2004).

When expressed in cultured primary neurons or cell lines, the ALS8 mutant VAPB^{P56S} forms intracellular aggregates (Nishimura et al. 2004a, b). These aggregates appear to be absent from the nucleoplasm, but present in all other regions of the cell body and neuronal processes (Fig. 11.2a, b). At moderate levels of recombinant protein expression the ER does not appear to be disrupted (see Fig. 11.2b). However, other reports have suggested that over expression of VAPA, VAPB or VAPB^{P56S} can

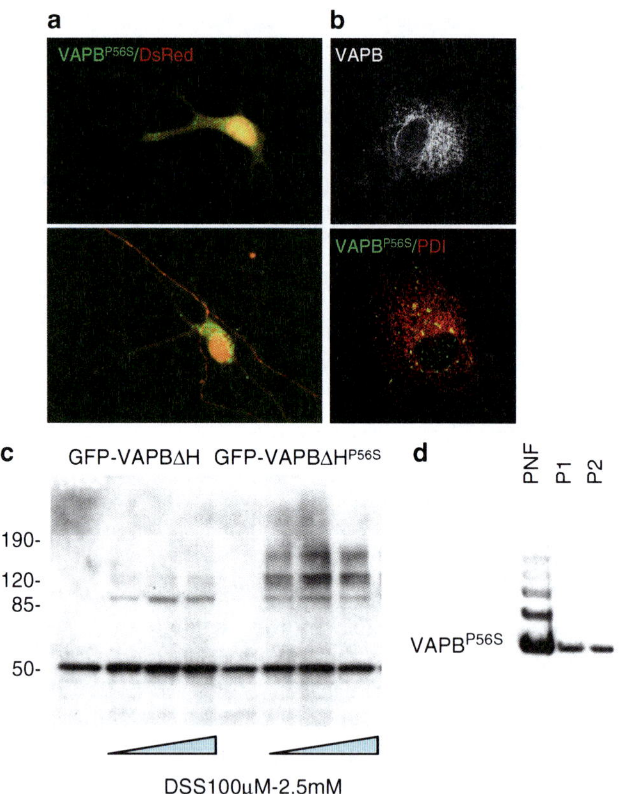

Fig. 11.2 Aggregation of the ALS8 mutantVAPB^{P56S}. (**a**) GFP-VAPB^{P56S} was co-expressed with DsRed in primary hippocampal cell culture. Fluorescent puncta are clearly visible throughout the cell, but not within the nucleus. (**b**) Myc-epitope-tagged VAPB or VAPB^{P56S} was expressed in COS7 cells, which were subsequently analysed by indirect immunocytochemical analysis to detect the myc-epitope (*green*) or the ER marker PDI (*red*). (**c**) Soluble forms of VAPB and VAPB^{P56S} lacking the C-terminal hydophobic domain (ΔH) were expressed as GFP-fusion proteins in HEK293 cells. Cells were incubated for 30 min with increasing concentrations of the cell permeable cross-linking reagent disuccinimidyl suberate (DSS). Cell lysates were then analysed by immunoblot with GFP antisera. Approximate molecular weights are indicated in kiloDaltons. (**d**) myc-tagged VAPB^{P56S} is present in insoluble material in higher molecular forms. Significant levels of the monomeric protein are present in lighter membrane fractions

cause changes to ER morphology (Pennetta et al. 2002; Amarilio et al. 2005; Kaiser et al. 2005; Teuling et al. 2007; Prosser et al. 2008).

Two groups have examined the mutation in *Drosophila*. Both studies identified P56S-induced aggregations of the VAP protein and an ensuing (associated) toxicity (Chai et al. 2008; Ratnaparkhi et al. 2008). Whether similar aggregations form in the cells of ALS8 patients has yet to be confirmed by post-mortem histological analysis.

When expressed at higher levels, GFP-fusion proteins of both VAPA and VAPB, but VAPA in particular, induce the formation of multilamellar membrane structures that have been termed karmellae (Fig. 11.3 and (Snapp et al. 2003)). In the original report of Snapp et al., karmellae formation was shown to be an artefact produced by the expression of some GFP-fusion proteins resulting from the low affinity dimer formed between two GFP molecules. Recombinant proteins containing only a short C-terminal membrane anchor fused to GFP readily induced karmellae (Snapp et al. 2003). This is not the case with the membrane anchor of VAPA. However, the effect is clearly due to GFP, since epitope-tagged versions of VAPA or VAPB do not form karmellae. Importantly, myc epitope-tagged forms of $VAPB^{P56S}$ still aggregate which dissociate aggregation from the experimentally induced karmella formation (Fig. 11.2). Nevertheless, increasing the level of interaction between VAP proteins can clearly induce the formation of these distinctive membrane structures. Therefore, aggregations induced by the P56S mutation may also induce the formation of karmellae or karmellae-like structures at some level. These structures are extensions of the ER. Thus, co-localisation of proteins with $VAPB^{P56S}$-induced aggregates must be viewed with caution, since an ER-resident protein within a small karmellae may appear to be part of an aggregate, when it is in fact simply co-expressed in the ER membrane.

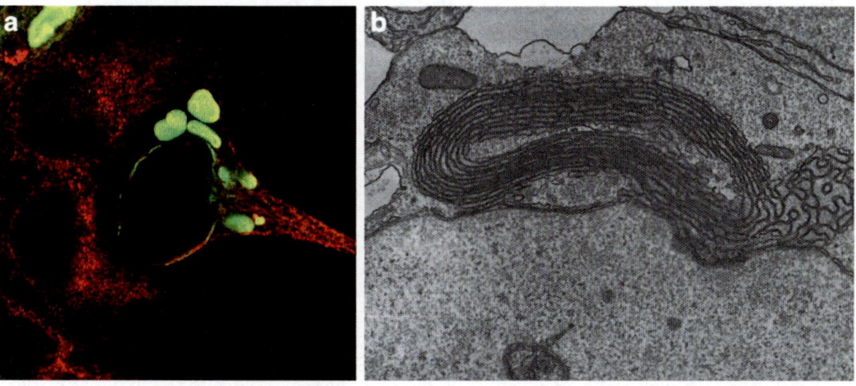

Fig. 11.3 Induction of multilameller bodies. (**a**) GFP-VAPB fusion proteins can induce the formation of karemellea. HEK293 cells co-transfected with EGFP-VAPB and the ER marker ER-DsRed contain these distinctive large membranous structures. Co-expressed fluorescent forms of ER markers are clearly visible in karemellea. Immunofluorescent methods are less efficient in identifying karemellea due to the poor antibody accessibility of inner membrane layers. (**b**) An electron micrograph of GFP-VAPB-induced multilameller bodies. Similar structures are seen with high levels of expression if GFP-VAPA fusion proteins

The aggregates formed by VAPB^{P56S} may be accumulations of unstructured misfolded protein. However, this is not necessarily the case. A large proportion of heterologously expressed VAPB^{P56S} is associated with ubiquitin (Kanekura et al. 2006), and multiple higher molecular weight species are detected by immunoblot of lysates from cultured cells expressing the mutant protein. However, a significant proportion of the protein remains unmodified and associated with lighter membrane fractions (Fig. 11.2d). Protein cross-linking in cells expressing wild type VAPB and VAPB^{P56S} as soluble GFP-fusion proteins lacking their C-terminal membrane anchors, generates similar higher molecular weight species (Fig. 11.2c). The P56S mutation is converted to the larger molecular weight complexes at lower concentrations of the cross-linker DSS, but these are discrete species rather than the random pattern one would expect from an amorphous aggregate formed by an unstructured protein. Teuling et al., have reported VAPB^{P56S} does not bind the FFAT motif from Nir2 (Teuling et al. 2007). Prosser et al., however, have shown recently that over expression of a fusion protein containing an isolated FFAT motif can dissociate VAPB^{P56S}-induced aggregates (Prosser et al. 2008).

Deletion of the hydrophobic C-terminus renders VAPA and VAPB cytosolic, in which state they do not appear to aggregate. However, further deletion to remove the coiled/coil domain causes the MSP domain of VAPA to aggregate dramatically (Fig. 11.4). The presence of these large cytosolic aggregates arrests the growth of cell lines and is toxic to primary cultured neurons (Fig. 11.4b, c). The mechanism of this toxicity or its relationship to ALS 8 is not known. One might speculate that the oligomeric state of the MSP domain is influenced by other proteins or by the other VAP domains, and that this is in some way important for VAP function. Perturbation of such oligomerisation, either by the P56S substitution or deletion of the C-terminal structures from the protein, might then lead to some form of cellular toxicity.

11.5.1 Molecular Interactions Mediated by the MSP Domain

A yeast two-hybrid screen identified seven MSP domain interacting polypeptides. Consistent with previous reports, several clones were identified for the Good pasture antigen-binding protein or CERT (GPBP), oxysterol binding protein 1 (OSBP1),

Fig.11.4 (continued) in which the expression of the VAPA MSP domain GFP fusion protein was under the control of a tetracycline inducible promoter. Induction of GFP-MSP rapidly arrests cell division. Initially, the protein is soluble, but aggregates are clearly visible after 48 h. Uninduced cells or T-REx 293 cells treated with tetracycline (Trex), continue to proliferate over the same time course. Full length VAPA and the MSP domain of VAPA were transiently co-expressed with DsRed in dissociated rat primary hippocampal neuron cultures. The full length VAPA is expressed throughout the neurons, but in comparison to DsRed it does not enter the dentritic spines or terminal growth cones (images of live neurons). In contrast, the MSP-GFP protein accumulates in large aggregates in the soma of the neurons, which rapidly degenerate and exhibit fragmented processes. The nucleus of these neurons reacts positively in a TUNEL assay (not shown)

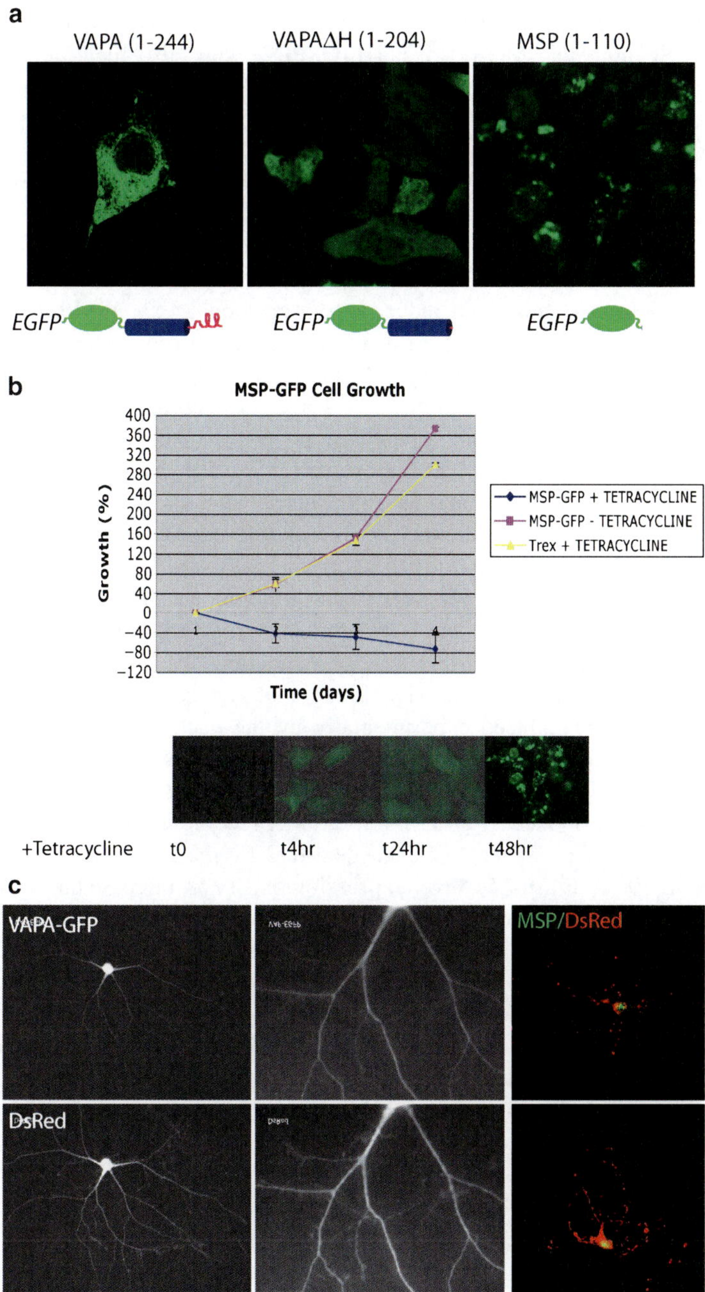

Fig. 11.4 Properties of the VAPA MSP domain. Truncations of VAPA were expressed as EGFP fusion proteins in T-REx 293 cells. Deletion of the hydrophobic C-terminus of the protein releases it from the ER membrane, and it distributes evenly throughout the cell, including the nucleus. In contrast, the MSP domain GFP-fusion protein lacking both the hydrophobic C-terminus and the coiled/coil structure forms large protein aggregates. A stable cell line was derived from T-REx 293

COG7 and OSBPL9, all of which contain an FFAT motif. Two proteins containing internal MSP domains, MOSPD4 and MOSPD2, were also identified but their function is unknown. A strong interaction was also detected with the transcription factor ATF6α. ATF6α is part of a regulatory pathway that controls gene transcription in response to biosynthetic demands on the ER, and ER stress (Yoshida et al. 1998). Increased demand for secreted or membrane proteins, or inadequate chaperone capacity, leads to the accumulation of unfolded proteins in the lumen of the ER. This induces a homeostatic response called the unfolded protein response or UPR (Schroder and Kaufman 2005). Three canonical pathways act co-ordinately to reduced overall levels of translation and induce a transcriptional response to increase the size and capacity of the ER. ATF6α is an integral ER membrane protein that acts in one of these regulatory pathways. Accumulation of unfolded proteins in the lumen of the ER, or ER stress induced by Ca^{2+} store depletion, causes ATF6α to move from the ER to the Golgi, where it is proteolysed sequentially by S1P and S2P (Ye et al. 2000). These cleavage events release the cytoplasmic N-terminal domain containing the DNA-binding and transcription activation domains of the protein into the cytoplasm from where it moves to the nucleus.

The MSP-ATF6α interaction identified in the two-hybrid assay was confirmed by fluorescence complementation assay using full length ATF6α and VAPB, indicating the interaction was not simply an artefact of the yeast screen. The potential functional significance of the MSP-ATF6α interaction was suggested by transient transfection assays for ATF6α activity (Wang et al. 2000; Gkogkas et al. 2008).

The UPR can be induced experimentally by the bacterial toxin tunicamycin, which inhibits N-linked glycosylation in the ER/Golgi, resulting in an accumulation of unglycosylated unfolded proteins in the lumen of the ER. Tunicamycin treatment robustly induced transcription from a recombinant ER-stress reporter plasmid containing promoter elements identified by the interaction with ATF6α (Wang et al. 2000) (Fig. 11.5). Over expression of ATF6α increase the basal levels of activity from this reporter plasmid (Wang et al. 2000; Gkogkas et al. 2008). In contrast, co-expression of either VAPB or $VAPB^{P56S}$ attenuates tunicamycin-induced transcription. $VAPB^{P56S}$ appears to be a slightly more potent inhibitor than the wild type protein. If VAPB acts as a negative regulator of ATF6α activity, it might be expected that a reduction in VAPB levels would lead to increased ATF6α-dependent transcription. This appears to be the case. Following siRNA mediated reduction of endogenous VAPB in HEK293 cells, basal and induced levels of transcription from the ATF6α-regulated reporter are enhanced (Fig. 11.5b). Thus, VAPB levels appear to be inhibitory to ATF6α activity. This seems contradictory to other reports that indicate increase VAP levels activate the UPR (Kanekura et al. 2006). Indeed, over expression of VAPA can induce BiP mRNA (Fig. 11.5c). It might be expected that increasing the level of an ER membrane protein, such as VAPA, could induce a luminal protein chaperone, such as BiP. However, the MSP domain of VAPA that lacks the C-terminal membrane anchor also induces BiP to similar levels as full length VAPA. Thus, in addition to their structural role, VAP proteins may also influence the regulatory pathways of the ER, perhaps by direct interaction with ATF6α. Disrupting the MSP domain by the P56S mutation

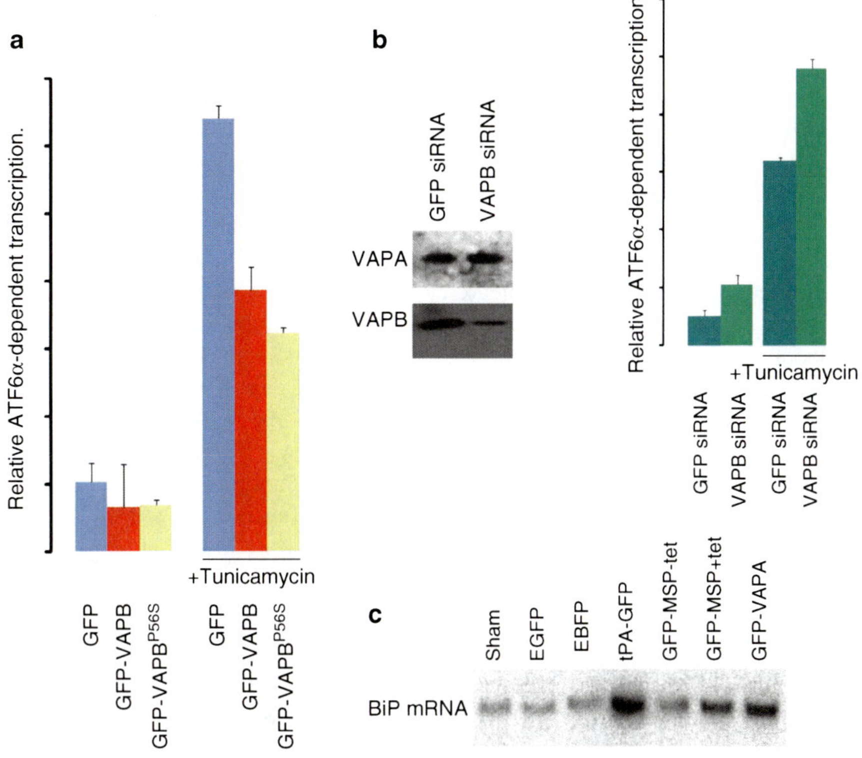

Fig. 11.5 VAPB levels reduce the activity from a UPR-regulated promoter. (**a**) Tunicamycin-induced transcription from the UPR-regulated promoter pGL3(5XATF6) is reduced by co-expression of VAPB and VAPB^{P56S}. HEK293 cells were transfected with luciferase-based transient transcription reporters for the UPR and expression vectors for GFP or GFP fusions of VAPB and VAPB^{P56S}. Both VAPB and VAPB^{P56S} reduced UPR-regulated transcription relative to GFP alone, and the ALS8 mutation has a slightly enhanced inhibitory activity. (**b**) siRNA against human VAPB effectively reduces endogenous levels of VAPB in HEK293 cells. Control siRNA against GFP has no effect on VAPB levels, and neither siRNA affects VAPA. Basal and tunicamycin-induced ATF6α-regulated transcription is increased following the reduction of VAPB levels. (**c**) HEK293 cells were transfected with expression constructs for EGFP, EBFP, Tissue plasminogen activating protein GFP fusion protein (tPA-GFP), GFP-VAPA or a tetracycline inducible GFP-MSP expression vector. mRNA for the ER chaperone BiP is induced by the secreted tPA-GFP fusion protein. BiP mRNA is also induced, but to a lesser extent by GFP-VAPA expression, even though this protein has no luminal structure. When expression of MSP-GFP is induced by tetracycline (± tet), a concomitant increase in BiP mRNA is also detected. GFP alone (EGFP, EBFP) or transfection reagent does not induce BiP mRNA

of ALS 8 may therefore lead to inappropriate responses to ER stress that, in someway, ultimately results in cellular degeneration. ER stress responses have been implicated in disease processes previously (Katayama et al. 2001; Zhang and Kaufman 2005; Atkin et al. 2006; Marciniak and Ron 2006). Furthermore, Hep C infection that, as noted before, requires virus protein interaction with VAPA and B, also leads to inappropriate activation of UPR transcriptional activity (Tardif et al. 2004; Zheng et al. 2005).

As VAPB provides a link between the ER and the microtuble network, reduced synaptic ER in both the pre- and post-synaptic compartments might be predicted as a consequence of the P56S mutation (Fig. 11.6). This would have compromised the delivery of membrane proteins and constituents to the synapse and could also reduce the capacity for local intracellular Ca^{2+} signalling. However, VAPB may also have a more direct role in maintaining synapses. VAP proteins are subject to limited proteolytic cleavage in a tissue-specific fashion (Gkogkas et al. 2008). VAPB is only proteolysed to detectable levels in neuronal tissue. The cleavage reaction releases the MSP domain from the ER membrane. Tsuda et al., have proposed that, similar to the MSP polypeptide in *Caenorhabditis elegans*, the MSP domain is released from cells and acts as an Eph receptor ligand and regulates glutamate receptor function (Tsuda et al. 2008). Although identified more in the regulation of developmental processes, the Eph/Ephrin pathway is increasingly linked to synaptic function and disease (Tomita et al. 2006; Edwards et al. 2008; Klein 2009; Simon et al. 2009). In such a scenario, the disruption of either the formation or the function of the MSP domain from VAPB, might lead to synaptic dysfunction or loss.

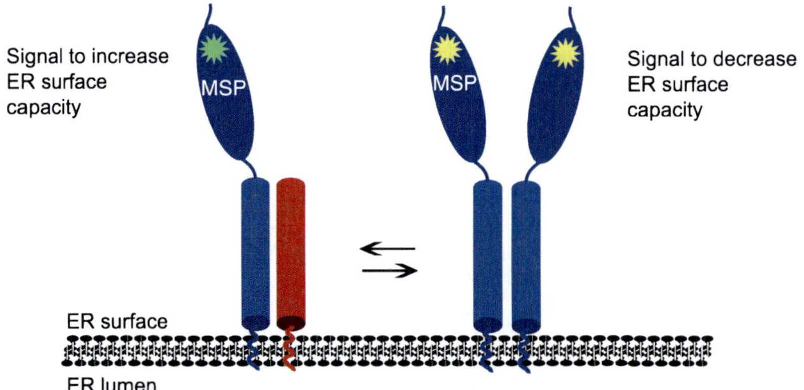

Fig. 11.6 Possible model for VAP protein-mediated signalling of ER surface capacity. VAP proteins are suggested to exist as homodimers or heterodimers; although larger multimeric forms are not inconsistent with the proposed mechanism. The balance between these two states would be influenced by other type II membrane proteins, such as VAMP/Synaptobrevin (indicated in *red*), possibly mediated by interactions with the promiscuous coiled-coil and C-terminal domains. The monomeric MSP domain in the heterodimer could have different binding properties compared to the paired MSP domains of the homodimer. Increased synthesis of type II membrane proteins would push the balance towards the heteromeric forms inducing or facilitating ER expansion, while decreased levels would favour homodimers and be inhibitory to expansion. This is one simple form of such a model. Depending on relative binding affinities involved it could equally function in the opposite direction, with the homomeric form of VAP proteins acting to increase ER capacity. In either case, the principle feature of the proposed mechanism would be the same; the nature of the multimers formed by the VAP proteins would influence the properties of the associated MSP domains (indicated by the different *coloured stars*)

11.6 Conclusion

The ALS8 mis-sense *vapB* mutation affects the structure of the protein causing it to aggregate. The dominant inheritance pattern of the disease indicates that either the mutant protein is non-functional and the disease results from haploinsufficiency or the protein retains some of its properties but acts in a dominant fashion. Protein aggregations could have dominant affects by physically disrupting normal cellular processes, such as transport, or they may sequester cellular constituents into non-functional states. The wild type protein seems to be also present in the VAPB^{P56S}-induced aggregates. Thus, *vapB*^{P56S} may act as a null mutation by causing the remaining wild type protein to be rendered inactive. Such a mechanism has been proposed by Kanekura, who showed that the ability of wild type VAPB to activate XBP1 mRNA processing was reduced in the presence of elevated VAPB^{P56S} levels (Kanekura et al. 2006).

The large intracellular aggregates formed by VAPB^{P56S} are conspicuous under the microscope. However, there is no direct evidence that these structures are deleterious to the cell. Studies of other degenerative conditions, such as Parkinson's, Huntington's, and Alzheimer's diseases, indicated that the more toxic forms of protein aggregates are in fact small multimers or oligomers rather than the larger more obvious aggregations (for review see (Dominic and Walsh 2007; Haass and Selkoe 2007)). As presented in Fig. 11.2, induced expression of VAPA-MSP rapidly arrests cell growth, although protein aggregates are only visible after 24 h. Similarly, the VAPB^{P56S}-induced aggregates may be a dramatic end point of an aggregative process that builds from smaller constituents. Our observations that the mutant protein cross-links into multimers of a similar size distribution to the wild type protein is consistent with a more ordered assembly of protein aggregates rather than an amorphous assembly of denatured protein. Oligomerization could, therefore, be a normal feature of VAP protein function or regulation, and the P56S mutation may simply enhance or exaggerate this process. This suggestion is further supported by the observations of Prosser et al., who showed elevated levels of FFAT-domain containing protein could dissociate VAPB^{P56S}-induced aggregates (Prosser et al. 2008). The level of VAP oligomerization may, therefore, be regulated or influenced by the levels of other proteins with which it interacts, such as FFAT-domain containing proteins. The assembly of large protein aggregations could be a general regulatory process of membrane proteins in the ER. In resting cells, for instance, ATF6α is found predominantly in a 600kD protein complex, which redistributes to lower molecular weight following the treatment with DTT (Shen et al. 2002).

The luminal domains of the membrane proteins ATF6α, IRE1 and PERK sense the biosynthetic capacity and luminal environment of the ER. The details of the molecular processes responsible for the induction of this signalling are incompletely understood, but may involve interactions with BiP. Regulated assembly of VAPB might provide an additional pathway to monitor the levels of membrane-associated components on the cytoplasmic surface of the ER that are unable to interact with the known unfolded protein sensing structures within the lumen. This might explain the seemingly contradictory observations that increased the levels of VAPB induce BiP expression and XBP1 mRNA activation, but that some ER-regulated transcription is reduced (Kanekura et al. 2006; Gkogkas et al. 2008). As an ER-membrane protein, VAPB might be expected to induce an UPR when over expressed. If, however, VAPB also had a regulatory role in ER homeostasis, over expression may disrupt transcriptional regulation of ER stress responses. VAP proteins do not have any structural features that might suggest that they act as transcription factors although they are present in the nuclear membrane. However, one of the yeast VAP homologues, SCS2, was originally identified by its ability to suppress defects in inositol-regulated transcription (Nikawa et al. 1995). In yeast, the UPR and transcriptional regulation of inositol metabolism share regulatory components (Chang et al. 2004). Brickner and Walter have shown that SCS2 is required at the nuclear membrane for the efficient relief of Opi1-mediated transcriptional inhibition (Brickner and Walter 2004). It would be remarkable if a similar mechanism acted in mammals to regulate ATF6α activity. ATF6α is not homologous to Opi1, and no other mammalian proteins are considered Opi1 homologues. However, the mammalian UPR is a striking example of a situation where mammalian cells share a regulatory mechanism with yeast, but employ distinct non-homologous proteins in the process. In yeast, the activation of IRE1 by ER stress or inositol depletion activates the transcription factor Hac1 by a non-conventional mRNA splicing event to remove an intron from the 3′UTR (Bernales et al. 2006). In mammals, the activation of IRE1 induces XBP1 by a similar mRNA processing event. However, in mammals the non-conventional intron is within the coding sequence of the mRNA, and XBP1 shares no significant homology with Hac1. Thus, ER homeostasis in yeast and mammals is controlled by a similar signalling pathway, but via the regulation of distinct proteins. Perhaps Opi1 and ATF6α might be similarly regulated by SCS2 and VAPB respectively, despite sharing no significant structural homology.

Other than the formation of aggregates, the affect of the P56S mutation on VAPB activity remains unclear. Two reports have described potential ALS8 models in *Drosophila* (Chai et al. 2008; Ratnaparkhi et al. 2008). The study of Ratnaparkhi et al. is consistent with a dominant negative effect of the P56S mutation. However, in the report of Chai et al., the mutant protein can rescue the synaptic and motor phenotypes produced by the deletion of the endogenous wild type gene, indicating that it retains biological activity similar to the wild type protein. The reason for this apparent discrepancy is not clear, but may be the result of differing transgene expression levels. Perhaps $VAPB^{P56S}$ is both a gain and loss of function mutation. Thus, if VAPB oligomerisation reduces its activity and P56S increases oligomerisation – as suggested by our cross-linking experiments – then $VAPB^{P56S}$ would be a gain of function mutant with reduced activity. The gain of function would result in increased oligomerisation, which would in turn lead to a reduction in protein activity. However, if the mutation also disrupted a property of the MSP domain, this might also have a dominant negative affect on the endogenous wild type protein. In such a situation, modelling ALS8 in experimental animals will require very careful control of wild type and mutant protein expression levels. However, molecular genetic methods exist to achieve this level of control, and it is hoped that in future more elegant experimental models may elucidate the full properties of VAPB, and in so doing produce valuable information on the potential mechanisms responsible for motor neuron degeneration.

References

Abe, K., M. Aoki, et al. (1996). "Clinical characteristics of familial amyotrophic lateral sclerosis with Cu/Zn superoxide dismutase gene mutations." *J Neurol Sci* **136**(1–2): 108–16.

Amarilio, R., S. Ramachandran, et al. (2005). "Differential regulation of endoplasmic reticulum structure through VAP-Nir protein interaction." *J Biol Chem* **280**(7): 5934–44.

Andersen, P. M. (2006). "Amyotrophic lateral sclerosis associated with mutations in the CuZn superoxide dismutase gene." *Curr Neurol Neurosci Rep* **6**(1): 37–46.

Arrasate, M., S. Mitra, et al. (2004). "Inclusion body formation reduces levels of mutant huntingtin and the risk of neuronal death." *Nature* **431**(7010): 805–10.

Atkin, J. D., M. A. Farg, et al. (2006). "Induction of the unfolded protein response in familial amyotrophic lateral sclerosis and association of protein disulfide isomerase with superoxide dismutase 1." *J Biol Chem* **281**: 30152–65.

Bernales, S., F. R. Papa, et al. (2006). "Intracellular signaling by the unfolded protein response." *Annu Rev Cell Dev Biol* **22**(1): 487–508.

Brickner, J. H. and P. Walter (2004). "Gene recruitment of the activated INO1 locus to the nuclear membrane." *PLoS Biol* **2**(11): e342.

Chai, A., J. Withers, et al. (2008). "hVAPB, the causative gene of a heterogeneous group of motor neuron diseases in humans, is functionally interchangeable with its *Drosophila* homologue DVAP-33A at the neuromuscular junction." *Hum Mol Genet* **17**(2): 266–80.

Chang, H. J., S. A. Jesch, et al. (2004). "Role of the unfolded protein response pathway in secretory stress and regulation of INO1 expression in *Saccharomyces cerevisiae*." *Genetics* **168**(4): 1899–913.

Dominic M., D. J. S. Walsh (2007). "Aβ Oligomers – a decade of discovery." *J Neurochem* **101**(5): 1172–84.

Edwards, D. R., M. M. Handsley, et al. (2008). "The ADAM metalloproteinases." *Mol Aspects Med* **29**(5): 258–89.

Foster, L., M. Weir, et al. (2000). "A functional role for VAP-33 in insulin-stimulated GLUT4 traffic." *Traffic* **1**(6): 512–21.

Gao, L., H. Aizaki, et al. (2004). "Interactions between viral nonstructural proteins and host protein hVAP-33." *J Virol* **78**(7): 3480–8.

Gkogkas, C., S. Middleton, et al. (2008). "VAPB interacts with and modulates the activity of ATF6." *Hum Mol Genet* **17**: 1517–26.

Haass, C. and D. J. Selkoe (2007). "Soluble protein oligomers in neurodegeneration: lessons from the Alzheimer's amyloid beta-peptide." *Nat Rev Mol Cell Biol* **8**(2): 101–12.

Hamamoto, I., Y. Nishimura, et al. (2005). "Human VAP-B is involved in hepatitis C virus replication through interaction with NS5A and NS5B." *J Virol* **79**(21): 13473–82.

Hand, C. and G. Rouleau (2002). "Familial amyotrophic lateral sclerosis." *Muscle Nerve* **25**(2): 135–59.

Ito, K., T. Uchiyama, et al. (2002). "[Different clinical phenotypes of siblings with familial amyotrophic lateral sclerosis showing Cys146Arg point mutation of superoxide dismutase 1 gene]." *Rinsho Shinkeigaku* **42**(2): 175–7.

Kaiser, S. E., J. H. Brickner, et al. (2005). "Structural basis of FFAT motif-mediated ER targeting." *Structure* **13**(7): 1035–45.

Kanekura, K., I. Nishimoto, et al. (2006). "Characterization of amyotrophic lateral sclerosis-linked P56S mutation of vesicle-associated membrane protein-associated protein B (VAPB/ALS8)." *J Biol Chem* **281**(40): 30223–33.

Katayama, T., K. Imaizumi, et al. (2001). "Disturbed activation of endoplasmic reticulum stress transducers by familial Alzheimer's disease-linked Presenilin-1 mutations." *J Biol Chem* **276**(46): 43446–54.

Kawano, M., K. Kumagai, et al. (2006). "Efficient trafficking of ceramide from the endoplasmic reticulum to the Golgi apparatus requires a VAMP-associated protein-interacting FFAT motif of CERT." *J Biol Chem* **281**(40): 30279–88.

King, K. L., M. Stewart, et al. (1992). "Structure and macromolecular assembly of two isoforms of the major sperm protein (MSP) from the amoeboid sperm of the nematode, Ascaris suum." *J Cell Sci* **101**(4): 847–857.

Klein, R. (2009). "Bidirectional modulation of synaptic functions by Eph/ephrin signaling." *Nat Neurosci* **12**(1): 15–20.

Lapierre, L., P. Tuma, et al. (1999). "VAP-33 localizes to both an intracellular vesicle population and with." *J Cell Sci* **112**(Pt 21): 3723–32.

Loewen, C., A. Roy, et al. (2003). "A conserved ER targeting motif in three families of lipid binding proteins." *EMBO J* **22**(9): 2025–35.

Loewen, C. J. and T. P. Levine (2005). "A highly conserved binding site in vesicle-associated membrane protein-associated protein (VAP) for the FFAT motif of lipid-binding proteins." *J Biol Chem* **280**(14): 14097–104.

Marciniak, S. J. and D. Ron (2006). "Endoplasmic reticulum stress signaling in disease." *Physiol Rev* **86**(4): 1133–49.

Marques, V. D., A. A. Barreira, et al. (2006). "Expanding the phenotypes of the Pro56Ser VAPB mutation: proximal SMA with dysautonomia." *Muscle Nerve* **34**(6): 731–9.

Menzies, F., A. Grierson, et al. (2002). "Selective loss of neurofilament expression in Cu/Zn superoxide dismutase." *J Neurochem* **82**(5): 1118–28.

Nikawa, J.-I., A. Murakami, et al. (1995). "Cloning and sequence of the SCS2 gene, which can suppress the defect of IN01 expression in an inositol auxotrophic mutant of *Saccharomyces cerevisiae*." *J Biochem* **118**(1): 39–45.

Nishimura, A., M. Mitne-Neto, et al. (2004a). "A novel locus for late onset amyotrophic lateral sclerosis/motor neurone." *J Med Genet* **41**(4): 315–20.

Nishimura, A. L., M. Mitne-Neto, et al. (2004b). "A mutation in the vesicle-trafficking protein VAPB causes late-onset spinal muscular atrophy and amyotrophic lateral sclerosis." *Am J Hum Genet* **75**(5): 822–31.

Nishimura, Y., M. Hayashi, et al. (1999). "Molecular cloning and characterization of mammalian homologues of vesicle-associated membrane protein-associated (VAMP-associated) proteins." *Biochem Biophys Res Commun* **254**(1): 21–26.

Pennetta, G., P. Hiesinger, et al. (2002). "*Drosophila* VAP-33A directs bouton formation at neuromuscular junctions in a dosage dependent manner." *Neuron* **35**(2): 291–306.

Prosser, D. C., D. Tran, et al. (2008). "FFAT rescues VAPA-mediated inhibition of ER-to-Golgi transport and VAPB-mediated ER aggregation." *J Cell Sci* **121**(18): 3052–61.

Ratnaparkhi, A., G. M. Lawless, et al. (2008). "A *Drosophila* model of ALS: human ALS-associated mutation in VAP33A suggests a dominant negative mechanism." *PLoS ONE* **3**(6): e2334.

Rosen, D. R., T. Siddique, et al. (1993). "Mutations in Cu/Zn superoxide dismutase gene are associated with familial amyotrophic lateral sclerosis." *Nature* **362**: 59–62.

Schroder, M. and R. J. Kaufman (2005). "The mammalian unfolded protein response." *Annu Rev Biochem* **74**: 739–89.

Schymick, J. C., K. Talbot, et al. (2007). "Genetics of sporadic amyotrophic lateral sclerosis." *Hum Mol Genet* **16**(R2): R233–242.

Sepsenwol, S., H. Ris, et al. (1989). "A unique cytoskeleton associated with crawling in the amoeboid sperm of the nematode, *Ascaris suum*." *J Cell Biol* **108**(1): 55–66.

Shen, J., X. Chen, et al. (2002). "ER stress regulation of ATF6 localization by dissociation of BiP/GRP78 binding and unmasking of Golgi localization signals." *Dev Cell* **3**(1): 99–111.

Simon, A. M., R. L. de Maturana, et al. (2009). "Early changes in hippocampal Eph receptors precede the onset of memory decline in mouse models of Alzheimer's disease." *J Alzheimers Dis* **17**: 773–86.

Skehel, P., R. Fabian-Fine, et al. (2000). "Mouse VAP33 is associated with the endoplasmic reticulum and microtubules." *Proc Natl Acad Sci U S A* **97**(3): 1101–6.

Skehel, P., K. Martin, et al. (1995). "A VAMP-binding protein from Aplysia required for neurotransmitter release." *Science* **269**(5230): 1580–3.

Snapp, E. L., R. S. Hegde, et al. (2003). "Formation of stacked ER cisternae by low affinity protein interactions." *J Cell Biol* **163**(2): 257–69.

Soto, C. (2003). "Unfolding the role of protein misfolding in neurodegenerative diseases." *Nat Rev Neurosci* **4**(1): 49–60.

Soussan, L., D. Burakov, et al. (1999). "ERG30, a VAP-33-related protein, functions in protein transport mediated." *J Cell Biol* **146**(2): 301–11.

Talbot, K. (2002). "Motor neurone disease." *Postgrad Med J* **78**(923): 513–519.

Tardif, K. D., K. Mori, et al. (2004). "Hepatitis C virus suppresses the IRE1-XBP1 pathway of the unfolded protein response." *J Biol Chem* **279**(17): 17158–64.

Teuling, E., S. Ahmed, et al. (2007). "Motor neuron disease-associated mutant vesicle-associated membrane protein-associated protein (VAP) B recruits wild-type VAPs into endoplasmic reticulum-derived tubular aggregates." *J Neurosci* **27**(36): 9801–15.

Tomita, T., S. Tanaka, et al. (2006). "Presenilin-dependent intramembrane cleavage of ephrin-B1." *Mol Neurodegener* **1**: 2.

Tompkins, M. M. and W. D. Hill (1997). "Contribution of somal Lewy bodies to neuronal death." *Brain Res* **775**(1–2): 24–9.

Tsuda, H., S. M. Han, et al. (2008). "The amyotrophic lateral sclerosis 8 protein VAPB is cleaved, secreted, and acts as a ligand for Eph receptors." *Cell* **133**(6): 963–977.

Tu, H., L. Gao, et al. (1999). "Hepatitis C virus RNA polymerase and NS5A complex with a SNARE-like." *Virology* **263**(1): 30–41.

Wang, Y., J. Shen, et al. (2000). "Activation of ATF6 and an ATF6 DNA binding site by the endoplasmic reticulum stress response." *J Biol Chem* **275**(35): 27013–20.

Weir, M., A. Klip, et al. (1998). "Identification of a human homologue of the vesicle-associated membrane." *Biochem J* **333**(Pt 2): 247–51.

Welsem, M. v., J. Hogenhuis, et al. (2002). "The relationship between Bunina bodies, skein-like inclusions and neuronal." *Acta Neuropathol* **103**(6): 583–9.

Williamson, T., L. Corson, et al. (2000). "Toxicity of ALS-linked SOD1 mutants." *Science* **288**(5465): 399.

Wyles, J. P., C. R. McMaster, et al. (2002). "Vesicle-associated membrane protein-associated protein-A (VAP-A) interacts with the oxysterol-binding protein to modify export from the endoplasmic reticulum." *J. Biol. Chem.* **277**(33): 29908–18.

Ye, J., R. B. Rawson, et al. (2000). "ER stress induces cleavage of membrane-bound ATF6 by the same proteases that process SREBPs." *Mol Cell* **6**(6): 1355–64.

Yoshida, H., K. Haze, et al. (1998). "Identification of the cis-acting endoplasmic reticulum stress response element responsible for transcriptional induction of mammalian glucose-regulated proteins. Involvement of basic leucine zipper transcription factors." *J Biol Chem* **273**(50): 33741–9.

Zhang, K. and R. J. Kaufman (2005). "The unfolded protein response. A stress signaling pathway critical for health and disease." *Neurology* 2006 **66**(2 Suppl 1): S102–9.

Zheng, Y., B. Gao, et al. (2005). "Hepatitis C virus non-structural protein NS4B can modulate an unfolded protein response." *J Microbiol* **43**(6): 529–36.

Chapter 12
Synaptic Dysfunction in Huntington's Disease

Dervila Glynn and A. Jennifer Morton

Abstract Huntington's disease (HD) is a progressive, inherited, neurodegenerative disorder characterised by movement abnormalities, cognitive impairments and emotional disturbance (Bates et al. 2002). The genetic mutation for HD is an unstable CAG repeat expansion in the *HD* gene (Huntington's Disease Collaborative Research Group 1993). The *HD* gene codes for a protein named huntingtin, and the CAG repeat is translated to an expanded polyglutamine repeat in the disease protein. However, even though the genetic mutation was identified more than 15 years ago, it is still not known how it causes HD. Until recently, the prevailing hypothesis was that the clinical manifestations of HD were due to selective neuronal degeneration in the striatum and cortex. Nevertheless, there is a growing body of work supporting the idea that some of the earliest changes apparent in HD, in particular changes in personality, mood and cognitive performance, may arise as a consequence of synaptic dysfunction. Here, we discuss the idea that synaptic dysfunction (rather than frank cell loss) may underlie early symptoms in HD.

12.1 Introduction

HD is a complex disorder involving motor, cognitive and psychiatric symptoms that progress relentlessly until death occurs, typically 20–25 years after the onset of symptoms. The clinical diagnosis of adult-onset HD is made from the onset of involuntary movement abnormalities although only one half to one third of patients present with chorea (Snell et al. 1993). Throughout the course of the disease, motor symptoms progress to include severe voluntary movement limitations, including severe bradykinesia, rigidity, dystonia, abnormalities in motor speed, fine motor control and gait and speech abnormalities which include dysarthria and dysphagia (Shoulson and Fahn 1979; Folstein et al. 1983; Young et al. 1986). As well, and often in parallel, HD patients suffer progressive cognitive decline. Executive functioning (which includes the

A.J. Morton (✉)
Department of Pharmacology, University of Cambridge, Tennis Court Road, Cambridge CB2 1PD, UK
e-mail: ajm41@cam.ac.uk

A. Wyttenbach and V. O'Connor (eds.), *Folding for the Synapse*,
DOI 10.1007/978-1-4419-7061-9_12,

ability to plan and organise, monitor behaviour, exhibit mental flexibility and to be able to switch from one way of reacting to another) is particularly affected (Brandt and Butters 1986; Lawrence et al. 1999). Procedural memory and psychomotor skills are impaired, with cognitive deficits eventually leading to a global dementia (Lawrence et al. 1996). Another disturbing hallmark of the disease is the progression of emotional changes, such as depression, apathy, irritability, impulsiveness, antisocial and sometimes suicidal behaviour (for references, see Bates et al. 2002). These emotional changes, especially depression and apathy, are often present long before the more obvious visible motor abnormalities become apparent (Jason et al. 1988; Strauss and Brandt 1990).

12.2 Neuropathology of HD

The major pathology of HD is the progressive degeneration of the inhibitory medium spiny γ-amino butyric acid (GABA)-ergic neurons of the striatum, in both the caudate and putamen (Vonsattel et al. 1985; Bates et al. 2002; Gil and Rego 2008). However, the pathological process is not confined to the striatum. Neuronal loss and astrogliosis of the cerebral cortex (particularly in layers III, V and VI) is pronounced (Hedreen et al. 1991), and later degeneration has been reported in other regions, such as the hypothalamus (Kremer et al. 1990), amygdala, cerebellum, thalamus, brain stem nuclei and hippocampus (Vonsattel and DiFiglia 1998; Ross et al. 1999; Davies and Ramsden 2001; for other references, see Bates et al. 2002). Although it is very difficult to study in post-mortem brain, it is clear that HD also causes changes in dendritic morphology. Both proliferative (recurving of distal dendrites, increases in spine density, size and branching) and degenerative (lowered spine density and branching, focal swellings on dendrites) changes in dendrites of striatal medium spiny neurons have been reported in human post-mortem tissue (Graveland et al. 1985; Ferrante et al. 1991). Interestingly, pre-degenerative changes in dendritic morphology have also been observed in animal models of HD (see below). Thus, although HD is unquestionably a neurodegenerative disease during which neuronal loss occurs, post-mortem studies of human brains suggest that the first cognitive and motor symptoms appear in the absence of overt neuronal cell loss (Vonsattel et al. 1985; Myers et al. 1988; Mizuno et al. 2000; Tippett et al. 2007). In light of these findings, it has been proposed that altered neurotransmission might underlie neuronal dysfunction in early HD (Reddy et al. 1999) and that neuronal dysfunction (as opposed to neurodegeneration) may underlie early HD symptomology (Morton et al. 2001; Morton and Edwardson 2001; Li et al. 2003; Freeman and Morton 2004a; Smith et al. 2005).

12.3 Neurotransmitter and Receptor Abnormalities in HD

The earliest studies of HD neurochemistry were made using human post-mortem brain samples. These showed changes in neurotransmitter levels of GABA (Perry et al. 1973), acetylcholine (Bird and Iversen 1974), dopamine (Bird and Iversen 1974), serotonin (Reynolds and Pearson 1987) and noradrenaline (Spokes 1980) in the HD brain. Changes in receptor levels of virtually all transmitter systems investigated have been identified in the striatum of HD patients, including decreases in glutamate (London et al. 1981; Young et al. 1988; Dure et al. 1991), dopamine (Reisine et al. 1977; Lawrence et al. 1998; Glass et al. 2000; Pavese et al. 2003), GABA (Faull et al. 1993; Glass et al. 2000), muscarinic (Hiley and Bird 1974), adenosine (Martinez-Mir et al. 1991), serotonin (Waeber and Palacios 1989; Steward et al. 1993; Wong et al. 1996; Castro et al. 1998), noradrenergic (Waeber et al. 1991), opioid (Weeks et al. 1997) and cannabinoid (Glass et al. 1993; Richfield and Herkenham 1994; Glass et al. 2000) receptors. Interestingly, recent evidence suggests that although striatal cholinergic interneurons do not degenerate in HD, the levels of both vesicular acetylcholine transporter and choline acetyltransferase are significantly decreased indicating abnormal striatal cholinergic neurotransmission (Smith et al. 2006).

While post-mortem studies suggested major deterioration of neurotransmission in HD, advances in functional imaging, such as positron emission tomography (PET), have allowed investigators to study HD patients throughout the course of their disease (Bates et al. 2002; Montoya et al. 2006). PET studies showed reduced levels of dopamine D2 receptors in the caudate and putamen of presymptomatic patients coincident with reduced glucose metabolism (Antonini et al. 1996; Andrews et al. 1999), and more recently the loss of D2 receptors may be from the hypothalamus has been shown (Politis et al. 2008). Reductions in D2 receptor may be some of the earliest changes in HD, since changes are found in presymptomatic gene carriers (van Oostrom et al. 2005). PET studies have also shown altered opioid and benzodiazepine receptor binding in early HD (Holthoff et al. 1993; Weeks et al. 1997).

Although historically the loss of neurotransmitter receptors was attributed to the loss of neurons, PET studies, in particular, suggest that this is probably not the case, since the changes in receptor density in presymptomatic and early symptomatic stages of HD are unlikely to be due directly to the loss of neurons. Early changes in receptor density may not only be the consequence of synaptic abnormalities, but may also contribute to the progression of the disease.

12.4 Evidence from the Mouse Models of HD

Probably, the most compelling evidence of progressive neurological decline related to synaptic dysfunction rather than neurodegeneration comes from mouse models of HD. More than ten different mouse models of HD have been generated (for reviews, see Bates et al. 2002; Menalled and Chesselet 2002; Menalled 2005). The R6/2 mouse is the most commonly used model because it shows early and predictable progressive symptoms. But this mouse is remarkable for the fact that the progressive neurological phenoytype appears in the absence of any obvious neurodegeneration (Davies et al. 1997). Even where degeneration has been reported (Stack et al. 2005), it appears much later than the onset of measurable motor or cognitive symptoms.

12.5 Behavioural Abnormalities in the R6/2 Mouse

The R6/2 mouse expresses the N-terminal portion of human huntingtin, containing a highly expanded polyglutamine repeat (Mangiarini et al. 1996). At weaning, R6/2 mice with CAG repeat lengths of 150–200 are indistinguishable from their normal littermates, and they do not display an overt phenotype until around 8–9 weeks (Mangiarini et al. 1996). Initial studies showed that R6/2 mice develop a number of key features associated with HD, with the overt movement disorder, including an irregular gait, stereotypic grooming activity and a progressive resting tremour (Mangiarini et al. 1996). However, Carter et al. (1999) showed that there were measurable functional deficits present well before the onset of overt symptoms. For example, by 5–6 weeks, R6/2 mice displayed progressive abnormalities in their swimming movements, in beam walking and in maintaining balance on the rotarod. Furthermore, progressive cognitive deficits were detected even earlier (from 3.5 weeks of age; Lione et al. 1999). Notably, the earliest deficits seen are in synaptic physiology (Murphy et al. 2000; Cepeda et al. 2003; Gibson et al. 2005; see also below for more details).

12.6 Neurochemistry of the R6/2 Mouse

Investigations into the pathology and gene expression in the R6/2 mouse have helped substantiate the idea that altered neurotransmission underlies neuronal dysfunction in HD. R6/2 brains display an early decreased expression of a number of neurotransmitter receptors, an abnormality that cannot be explained by neuronal loss. Studies by Cha et al. (1998, 1999) have revealed significant decreases in the expression of the mGluR1, mGluR2 and mGluR3 metabotropic glutamate receptors, D1 and D2 dopamine receptors and adenosine receptors by 8 weeks of age. These changes in protein expression are underpinned by a decrease of mRNA, suggesting that the motor abnormalities observed in R6/2 mice are partly attributable to altered neurotransmission caused by altered transcription. The endocannabinoid system is also impaired, with both the level of CB1 receptor expression and endocannabinoid tissue levels reduced (Lastres-Becker et al. 2001, 2002; Glass 2001; McCaw et al. 2004; Maccarrone et al. 2007).

Luthi-Carter et al. (2000) used microarray studies in the R6/2 mouse to examine changes in gene expression in 6,000 striatal genes. A small number of genes were decreased at 6 weeks of age, prior to the onset of apparent motor abnormalities. Remarkably, these were restricted to genes encoding neurotransmitters, calcium and retinoid signalling pathway components. These results suggest that the dysregulation of mRNAs encoding neurotransmitter receptors and related second messenger systems is an early component of the pathological process.

The changes in neurochemistry offer a classic therapeutic target that is of synaptic transmission. This was exploited in a study aimed at enhancing synaptic function that used a combination "triple" drug treatment of R6/2 mice (Morton et al. 2005). Along with creatine, tacrine (an acetycholine esterase inhibitor that prevents the breakdown of acetylcholine) and moclobemide (a reversible monoamine oxidase inhibitor that prevents the breakdown of noradrenaline and 5-hydroxytryptamine) significantly improved cognitive function in R6/2 mice. In addition, dysregulated gene expression (including genes involved in synaptic transmission) was partially reversed by the triple treatment, and this reversal probably underlies the observed improvements in cognitive function (Morton et al. 2005). This study showed that not only do pre-degenerative changes in synaptic function contribute to cognitive decline in R6/2 mice, but also they are reversible, since they are amenable to drug treatment.

12.7 Changes in Synaptic Proteins in HD

The process of neurotransmitter release or exocytosis is triggered by transient increases in intracellular Ca^{2+} and requires members of the soluble *N*-ethylmaleimide-sensitive factor attachment protein receptor (SNARE) protein family for fusion of the vesicle and target membranes (Fig. 12.1). In the mammalian brain, the SNARE proteins involved in exocytosis fall into two groups; vesicular SNAREs (v-SNAREs) and target membrane SNAREs (t-SNAREs). Synaptic vesicles contain the v-SNARE, synaptobrevin II (also known as vesicle-associated membrane protein (VAMP)), that interacts with the t-SNAREs, SNAP-25 (synaptosomal-associated protein of 25 kDa) and syntaxin 1, to form a ternary SNARE complex (Sollner et al. 1993; Südhof 1995; Fiebig et al. 1999; Lin and Scheller 2000) that forces the vesicle and plasma membranes into close proximity and allows fusion (Jahn and Scheller 2006). After fusion, *N*-ethylmaleimide-sensitive fusion protein (NSF) and α-soluble NSF-attachment protein (α-SNAP) disassemble the SNARE complex (Sollner et al. 1993). The vesicle is then endocytosed and the SNARE proteins are recycled for the next fusion event.

Exocytosis is modulated by numerous regulatory proteins, including the synaptic vesicle proteins, synaptotagmin and rab3A, together with the cytosolic proteins α-SNAP and complexin (Cplx) I and II (Fig.12.1; for reviews, see Südhof 1995; Jahn and Südhof 1999; Lin and Scheller 2000). Interestingly, selective alterations in SNARE-associated proteins have been detected in the R6/2 mouse model of HD (Morton and Edwardson 2001; Freeman and Morton 2004b) and in human HD brains (Morton et al. 2001; DiProspero et al. 2004; Smith et al. 2007). Two of these in particular, Cplx I and II (McMahon et al. 1995), have been implicated in HD as well as in a number of other different neurological diseases (see below). Complexins are small, soluble, regulatory brain proteins (18–19 kDa) (McMahon et al. 1995) involved in the modulation of the SNARE proteins and the process of exocytosis (for reviews, see Lin and Scheller 2000; Rosenmund et al. 2003; Südhof 2004). Several studies have reported roles for complexins in exocytosis (McMahon et al. 1995; Ono et al. 1998; Reim et al. 2001; Tokumaru et al. 2001; Pabst et al. 2002).

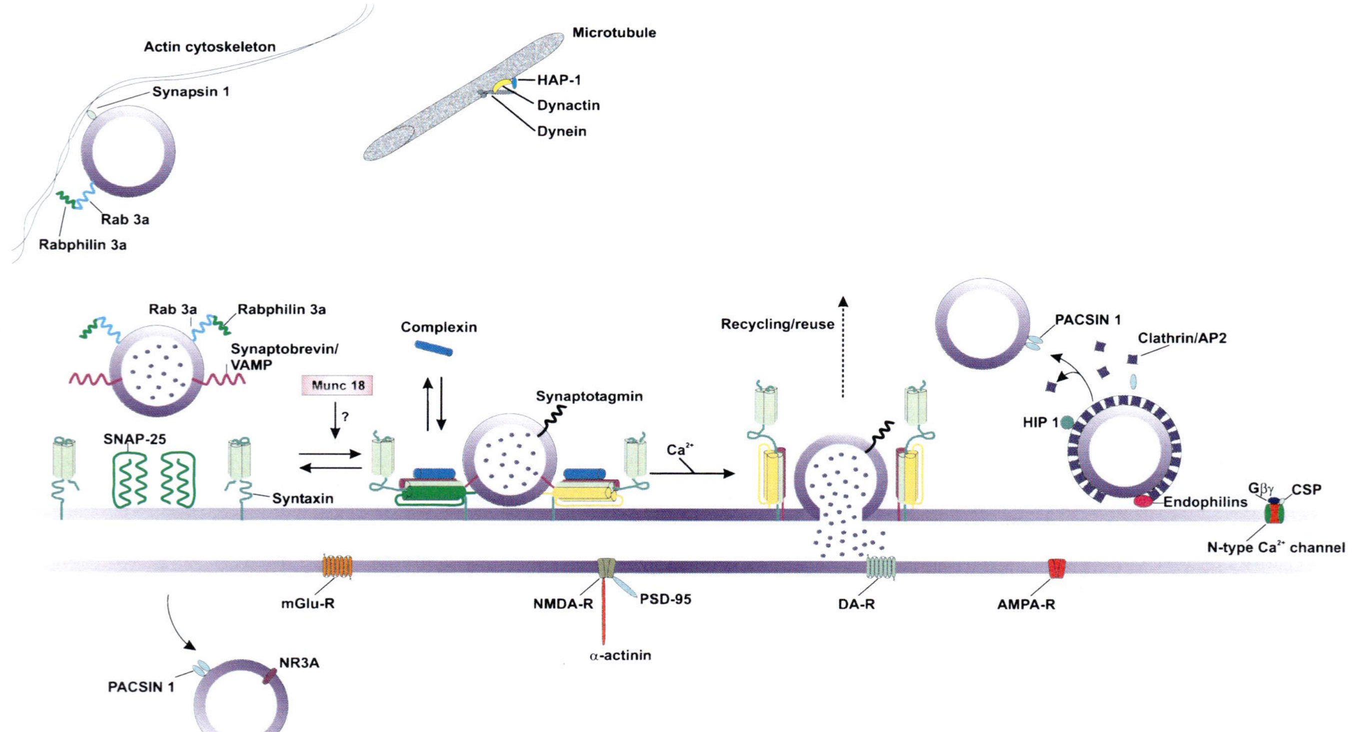

Fig. 12.1 Proteins involved in exocytosis, endocytosis and signalling at the synapse. The vesicle SNARE synaptobrevin II (also known as VAMP) interacts with the target membrane SNAREs, SNAP-25 and syntaxin 1, to form a ternary SNARE (soluble *N*-ethylmaleimide-sensitive factor attachment protein receptor) complex. *AP2* adaptor protein 2, *AMPA-R* α-amino-3-hydroxy-5-methyl-4-isoxazolepropionic acid receptor, *CSP* cysteine string protein, *DA-R* dopamine receptor, *HAP-1* huntingtin-associated protein 1, *mGlu-R* metabotropic glutamate receptor, *NMDA-R* *N*-methyl-D-aspartate receptor, *PACSIN 1* PKC and CK2 substrate in neurons 1, *PSD-95* postsynaptic density 95, *SNAP-25* synaptosomal-associated protein of 25 kDa, *VAMP* vesicle-associated membrane protein

Recent data suggest that complexins operate at a postpriming step stabilising SNARE complexes in a highly fusogenic state prior to exocytosis while simultaneously acting as a clamp that arrests the SNARE complex (Chen et al. 2002; Rizo and Südhof 2002; Giraudo et al. 2006; Schaub et al. 2006; Huntwork and Littleton 2007; Melia 2007; Xue et al. 2007). In addition, Cplx II plays a positive role in Ca^{2+}-triggered exocytosis by facilitating vesicle priming (Cai et al. 2008).

Expression levels and distribution of proteins involved in neurotransmitter release were examined in the striatum and hippocampus of post-mortem human HD brains (Morton et al. 2001), revealing marked depletion of CPLX II, synaptobrevin 2 and rab3A in the striatum. The most distinct change seen in the HD studies was a selective loss of CPLX II, which was found to decline by more than 40% in the striatum of HD brains, including grade 0 patients (patients with no gross striatal atrophy, Vonsattel et al. 1985). The loss of synaptobrevin 2 and CPLX II seen in grade 0 cases is consistent with the idea that a functional alteration in neurotransmission precedes neuronal loss. DiProspero et al. (2004) also reported a reduction in CPLX II in the frontal cortex of HD brains that was marked in some grade 1 HD cases and significantly reduced in all late stage cases. They also reported a significant loss of the presynaptic proteins dynamin and PACSIN 1 beginning in the early stage HD tissue (DiProspero et al. 2004). In addition to changes in complexin expression in HD, Smith et al. (2007) have reported a loss of SNAP 25 (grades I-IV) and rabphilin 3a (grades III-IV) with no change in levels of synaptobrevin II, syntaxin I, rab3a or synaptophysin in prefrontal cortex samples from HD brains.

In addition to studies on human HD brains, a progressive depletion of Cplx II expression levels (mRNA and protein) has been shown in the R6/2 mouse model of HD (Morton and Edwardson 2001; Freeman and Morton 2004b). The earlier study investigated the levels of SNARE proteins and other SNARE-associated proteins (synaptotagmin, synaptophysin, rab3 and 5 and α-SNAP Cplx I and Cplx II). While SNARE proteins were not significantly altered, Cplx II was progressively depleted, until by 16 weeks of age, levels of Cplx II protein expression were only 50% of the level seen at 5 weeks. The SNARE-associated protein, α-SNAP, was also depleted in R6/2 mice from 14 weeks. Subsequent *in situ* hybridisation studies on SNARE and SNARE-associated proteins in the brains of R6/2 mice (Freeman and Morton 2004a) revealed a marked, progressive depletion of Cplx II mRNA expression from 3 weeks of age, particularly in the cerebral cortex, striatum, hippocampus and parafasicular nucleus of the hypothalamus. Small changes in α-SNAP and Cplx I mRNA expression levels were also observed, but only at 15 weeks of age. These studies suggested that the loss of Cplx II protein in R6/2 mouse brains shown previously (Morton and Edwardson 2001) is due to decreased expression levels rather than increased protein degradation.

The selective decrease in complexin levels in HD brains, coupled with the fact that knockout mice show relevant behavioural abnormalities (see below), suggests that the loss of complexins in HD is not simply a consequence of synaptic dysfunction, but that changes in these proteins may actually contribute to the behavioural abnormalities in HD. Although there is no obvious link between huntingtin and complexin function, there is clearly an interaction, since expression of and N-terminal fragment of mutant huntingtin in PC12 cells caused a decrease in exocytosis that was paralleled by a decrease in complexin expression (Edwardson et al. 2003). The decline in exocytosis was rescued by overexpression of Cplx II in the continued expression of mutant huntingtin.

It is now known that changes in Cplx I and Cplx II expression are seen in a number of psychiatric and neurodegenerative disorders (reviewed by Brose 2008), including schizophrenia (Harrison and Eastwood 1998, 2000; Eastwood and Harrison 2000, 2005; Eastwood et al. 2001; Sawada et al. 2002, 2005), bipolar and unipolar disorder (Harrison and Eastwood 2000; Eastwood and Harrison 2000, 2001; Sawada et al. 2002), Parkinson's disease (Basso et al. 2004), Alzheimer's disease (Tannenberg et al. 2006) as well as in animal or cell culture models of depression (Zink et al. 2007), Parkinson's disease (Patel et al. 2007; Zabel et al. 2006), Wernicke's encephalopathy (Hazell and Wang 2005), foetal alcohol syndrome (Barr et al. 2005), alcoholism (Witzmann et al. 2003) and traumatic brain injury (Yi et al. 2006). In a recent review, Brose (2008) discussed whether the altered complexin expression levels reported in these diseases occur secondary to other changes (including changes as a consequence of altered network function, abnormal mRNA or protein stability or transcriptional dysregulation). He suggested that changes in complexin levels are unlikely to have a causal role in disease progression, but they probably contribute to the corresponding symptomology. However, due to the significant changes in complexin expression levels in HD brains (but not of several other synaptic proteins), Brose concluded that complexin changes may be implicated in HD pathogenesis. Our contention is that the profound changes in complexin expression seen in HD are likely to contribute to functional pathogenesis. While, at present, there is no direct evidence to support this hypothesis, it could be tested. For example, a "rescue" experiment (conducted by crossing R6/2 mice with a mouse that over-expresses complexin) could resolve this issue.

The idea that complexins contribute to symptoms in neurological disorders is well supported by evidence from Cplx knockout animals. Both major brain isoforms (Cplx I and Cplx II) play an essential role in the brain, since double knockout mice die at birth (Reim et al. 2001). However, when only one or other isoform of Cplx is knocked out, the mice survive to maturity (Reim et al. 2001; Glynn et al. 2003, 2005), suggesting that they each play important, but complementary roles. Cplx I and Cplx II are differentially distributed in the brain (Freeman and Morton 2004b). As might be expected, the phenotypes of the single knockouts are very different (Glynn et al. 2003, 2005; Glynn and Morton 2006; Drew et al. 2007; Glynn et al. 2007a).

Cplx 1 knockout (*Cplx1*$^{-/-}$) mice exhibit severe neurological symptoms characterized by ataxia from postnatal day 7, deficits in exploratory behaviour, motor behaviour and in tasks reflecting emotional reactivity (Glynn et al. 2005, 2007a). In addition to their pronounced motor phenotype, *Cplx1*$^{-/-}$ mice have pronounced deficits in many aspects of social behaviour but appear to have normal cognitive function (Drew et al. 2007). Comprehensive behavioural testing has shown that although Cplx 2 knockout (*Cplx2*$^{-/-}$) mice appear outwardly normal, they exhibit progressive deficits in both motor and cognitive behaviours (Glynn et al. 2003). Further, *Cplx2*$^{-/-}$ mice have abnormalities in information processing and social behaviour (Glynn and Morton 2006). The abnormal behavioural phenotype of *Cplx2*$^{-/-}$ mice suggests that Cplx II is essential for normal neurological function and is particularly important in higher cognitive function and emotional behaviour.

It is notable that some of the neurological deficits seen in *Cplx2*$^{-/-}$ mice (Glynn et al. 2003) are similar to those seen in the R6/2 mouse model of HD (Carter et al. 1999; Lione et al. 1999), indicating a common mechanism underlying deficits in both lines. In particular, both lines of mice show "reversal" deficits in the two-choice swim tank and the Morris water maze. Recent evidence has strengthened the potential for a functional role of Cplx II in HD symptoms. For example, a similar selective impairment in mossy fibre long-term potentiation (LTP) has been shown in the CA3 region of the hippocampus of both R6/2 mice and *Cplx2*$^{-/-}$ mice (Gibson et al. 2005). The deficit is present in both lines of mice by 3 weeks after birth, long before R6/2 mice show motor behaviour abnormalities. This supports the suggestion that the loss of Cplx II in the R6/2 mouse contributes to the learning deficits observed in this mouse even before the onset of the overt R6/2 phenotype (Gibson et al. 2005). Reduced levels of Cplx II, accompanied by a reduction in neurotransmitter release, were found in PC12 cells expressing the HD mutation (Edwardson et al. 2003). In that study, overexpression of Cplx II rescued the deficit in neurotransmitter release caused by the HD mutation. However, overexpression of Cplx II inhibited neurotransmitter release from normal PC12 cells. This implies that there is an optimal level of Cplx II for normal neurotransmitter release (Edwardson et al. 2003; Glynn et al. 2007b). Further evidence to support this idea comes from expression studies showing that optimal levels of Cplx II are required for epithelial sodium channel (ENaC) activity, since both increased and decreased complexin expression inhibited ENaC currents (Butterworth et al. 2005). Together, these data indicate that there is an optimal intracellular level of Cplx II required for its normal functional role.

12.8 A Direct Role for Mutant Huntingtin in Synaptic Dysfunction

There is strong evidence to suggest that normal huntingtin, the protein product of the *HD* gene, is involved in a host of intracellular functions such as protein trafficking, vesicle transport, clathrin-mediated endocytosis, postsynaptic signalling, transcriptional regulation and antiapoptotic function (for references, see Gil and Rego 2008).

Evidence that huntingtin plays a role in protein trafficking suggests that expansion in the polyglutamine repeat may deleteriously affect neurotransmission.

Huntingtin is found loosely associated with synaptic vesicles (DiFiglia et al. 1995), recycling endosomes, endoplasmic reticulum, Golgi complex and clathrin-coated vesicles (Hilditch-Maguire et al. 2000) and has been shown to interact with some proteins (including huntingtin associated protein 1 (HAP1), Sp1, Src homology 3 domain-containing protein 2c (SH3d2c), huntingtin interacting protein 1 (HIP1), HIP2 HIP14 and glyceraldehyde 3-phosphate dehydrogenase (GADPH)) involved in vesicle trafficking, transcriptional regulation and cytoskeleton organisation (Li et al. 1995; Velier et al. 1998; Zechner et al. 1998; Li et al. 2003; Smith et al. 2005). It has, therefore, been proposed that an expansion in the polyglutamine repeat in huntingtin may lead to abnormal protein interactions that interfere with axonal transport, alter presynaptic vesicle trafficking and modulate neurotransmitter release to ultimately cause neuronal dysfunction (DiFiglia et al. 1995; Velier et al. 1998; Li et al. 2001, 2003; Gunawardena et al. 2003; Szebenyi et al. 2003; Trushina et al. 2004; Sinadinos et al. 2009; see also Chap. 5 on axonal transport).

12.9 Protein Aggregation and the Ubiquitin-Proteasome System

The exact mechanism by which mutant huntingtin causes dysfunction and the resulting neurodegeneration remains unknown. One possibility is that the propensity of mutant huntingtin to form aggregates causes it to gain a toxic function. Mutant huntingtin is recruited into abnormal aggregates of protein in the nucleus and dystrophic neurites of neurons, both in HD brains (Davies et al. 1997; DiFiglia et al. 1997) and in mouse models of HD (Davies et al. 1997). Although the significance of these neuronal intranuclear inclusions is debated, there is no doubt that aggregates can form in axons, dendrites and synapses (for references, see below). It is not clear how or why aggregates form in HD, although the ubiquitin-proteasome system (UPS) has been implicated in this process. The UPS plays an essential role in the degradation of damaged or misfolded proteins (Hershko and Ciechanover 1998), and it has been suggested that mutant huntingtin can contribute to cell toxicity by impairing (blocking) or saturating the UPS, thus causing or contributing to the abnormalities and neurodegeneration seen in HD (Bence et al. 2001; Ciechanover and Brundin 2003; Hartl and Hayer-Hartl 2002). It has been suggested that in a vicious, positive-feedback cycle, the mutant huntingtin inhibits the pathway that destroys it (Bence et al. 2001). Thus, mutant huntingtin accumulates and forms the insoluble intracellular aggregates that, as they continue to build up, result in further impairment of UPS function. Recent evidence has shown that mutant huntingtin decreases synaptic UPS activity in cultured neurons and in the brains of R6/2 mice as discussed in more detail in Chap. 10 (Wang et al. 2008), although evidence from another polyglutamine disease (spinocrebellar ataxia 7) shows that UPS dysfunction is a consequence rather than a cause of neuronal dysfunction (Bowman et al. 2005). The major proteins within aggregates are huntingtin (both mutant and wildtype) and ubiquitin, although aggregates have also been shown to contain many other proteins, including proteasome elements (Wyttenbach et al. 2000; Jana et al. 2001; Wanderer and Morton 2007) and Cplx II (Morton et al. 2001). It has been suggested that the abnormal sequestration of proteins into aggregates depletes their expression elsewhere and thus may impact on normal cellular function (Morton et al. 2001). Specifically, mutant huntingtin may lead to synaptic dysfunction by altering the availability of important synaptic proteins (Li et al. 2003; Smith et al. 2005), thus interfering with its normal cellular functions, including protein trafficking, vesicle transport, endocytosis and anchoring of vesicles and mitochondria to the cytoskeleton.

12.10 Evidence for Disruptions in Axonal Transport in HD

Axonal transport is responsible for the movement of protein and organelle cargoes and is essential to the maintenance of functional synaptic transmission. Recent evidence has shown that the aspects of axonal transport are dysregulated during the early stage HD (Gunawardena and Goldstein 2005). Normal huntingtin protein has been shown to be important in axonal transport (anterograde and retrograde) in the rat sciatic nerve (Block-Galarza et al. 1997) and in Drosophila (Gunawardena et al. 2003). Further, it has been shown that expanded mutant huntingtin expression disrupts axonal transport in larval neurons of *Drosophila melanogaster* (Gunawardena et al. 2003; Lee et al. 2004; Sinadinos et al. 2009), in the squid giant axon system (Szebenyi et al. 2003) and in *Htt*^{Q109} knock in mice (Morfini et al. 2009).

It is possible that normal huntingtin interacts with the axonal transport machinery via its interaction with HAP1. HAP1 interacts with the p150 subunit of dynactin, which in turn interacts with dynein (Engelender et al. 1997; Li et al. 1998). These two proteins need to interact in order for vesicular transport to take place (Waterman-Storer et al. 1997). Binding between HAP1 and huntingtin is enhanced by an expanded polyglutamine repeat (Li et al. 1995). This sequestration of HAP1 by mutant huntingtin reduces its availability for binding to the dynactin/dynein complex and as a result impairs axonal transport. For example, normal huntingtin (in conjunction with HAP1 and dynactin) is required for vesicle trafficking of brain-derived neurotrophic factor (BDNF), a neurotrophin that is important for the survival of striatal neurons and the activity of corticostriatal synapses (Cattaneo et al. 2005). Loss of normal huntingtin function interferes with the anterograde transport of BDNF, thus contributing to its depletion in the striatum (Waterman-Storer et al. 1997).

There has been pathological evidence for axonal transport problems in HD since dystrophic striatal and corticostriatal neurites in HD exhibit many of the characteristics of "blocked" axons. Studies have shown accumulations of vesicles and organelles in swollen axonal projections in association with huntingtin aggregates (DiFiglia et al. 1995; Sapp et al. 1999). Further, huntingtin accumulations have been found in synapses (Morton et al. 2000) and axons of striatal projection neurons in R6/2 and knockin mice as well as in HD patient brains and cultured neurons (Li et al. 2001). Axonal degeneration has been shown to appear in the sciatic nerve of R6/2 mice (Wade et al. 2008) in parallel with synaptic failures at the neuromuscular junction (Ribchester et al. 2004). Finally, recent studies have also strengthened the evidence for abnormal axonal transport in HD since tubulin acetylation is reduced in HD patient brains (Dompierre et al. 2007). Tubulin acetylation promotes the function of kinesin 1, a motor protein that binds to microtubules stimulating axonal transport (Reed et al. 2006). For more insight into the role of axonal transport during HD, the reader is referred to Chap. 5.

12.11 Evidence for Disruptions in Endocytosis in HD

Endocytosis is the process by which many molecules are recycled into cells and as such is important in the recycling of synaptic vesicles and neurotransmission. Several huntingtin interacting and huntingtin-associated proteins are involved in the process of endocytosis (HAP1, HIP14, PACSIN 1, HIP1 and members of the endophilin family) and a growing body of evidence shows that altered interactions with these proteins caused by the mutation in huntingtin further implicate altered pathways of intraneuronal transport and endocytosis in the pathology of HD.

In addition to its major role in axonal transport, evidence suggests that HAP1 is also involved in endocytosis of membrane receptors, such as epidermal growth factor receptor (Li et al. 2002), type 1 inositol (1,4,5)-triphosphate receptor (Tang et al. 2003), GABA type A receptor (Kittler et al. 2004) and nerve growth factor (Rong et al. 2006). HIP14, another huntingtin interacting protein localized predominantly in the brain, has been shown to play a role in intracellular transport and endocytosis (Singaraja et al. 2002). A decreased interaction between mutant huntingtin and HIP14 in HD further implicates a role for altered pathways of intraneuronal transport and endocytosis in the pathology of HD (Singaraja et al. 2002). PACSIN 1/syndapin 1 (a PKC and CK2 substrate in neurons) is a synaptic vesicle protein involved in endocytosis which has been shown to interact with huntingtin. Mutant huntingtin enhances the interaction between the two proteins and although PACSIN 1 levels are normal in brains from HD patients, it is relocalized from the nerve terminals to the cell bodies (Modregger et al. 2002). Endophilin 3 and endophilin B1b are both members of the endophilin family of proteins that bind to huntingtin protein and are known to play roles in endocytosis (Sittler et al. 1998; Modregger et al. 2003). The interaction between endophilin 3 and huntingtin protein is dependent on the length of the CAG expansion and promotes the formation of insoluble polyglutamine-containing aggregates *in vivo* (Sittler et al. 1998). HIP1 has been shown by yeast two-hybrid assays to interact with N-terminal huntingtin (Kalchman et al. 1997; Wanker et al. 1997) although HIP1 has a reduced binding preference to mutated huntingtin. HIP1 and the closely related HIP12 are both orthologous to yeast Sla2p, which is known to be involved in endocytosis (Engqvist-Goldstein et al. 1999). Evidence from Legendre-Guillemin et al. (2002) suggests that HIP1 and HIP12 play related, yet distinct functional roles in clathrin-mediated endocytosis;

HIP1 interacts with clathrin and adaptor protein 2 (AP2), whereas HIP12 interacts with F-actin and to a lesser extent with the clathrin light chain. Further evidence that HIP1 is involved in endocytosis comes from HIP1 knockout ($HIP1^{-/-}$) mice, which show defects in the assembly of clathrin and AP2 protein complexes and α-amino-3-hydroxy-5-methyl-4-isoxazolepropionic acid (AMPA) receptor internalisation (Metzler et al. 2003).

12.12 Changes in Synaptic Plasticity in HD Mice

In addition to changes in receptor levels, abnormalities in glutamatergic (Lievens et al. 2001; Nicniocaill et al. 2001), dopaminergic (Petersen et al. 2002) and cholinergic (Vetter et al. 2003; Smith et al. 2006; Pisani et al. 2007) neurotransmission have been described in R6/1 and R6/2 mice in the absence of any obvious brain pathology. As in the R6/2 mouse (discussed earlier), the R6/1 mouse model expresses exon 1 of the human *HD* gene but with 115 rather than 150 CAG repeats (Mangiarini et al. 1996). The transgene expression in both of these lines of mice is driven by the human huntingtin promoter. The resulting levels of transgene expression are around 31% and 75% of the endogenous huntingtin in the R6/1 and R6/2 models, respectively. The R6/1 line develops its behavioural phenotype much later than R6/2 mice with cognitive deficits evident from 12 to 14 weeks of age and motor impairments seen around 12 weeks (Mazarakis et al. 2005; Milnerwood et al. 2006; Pang et al. 2006; Nithianantharajah et al. 2008). Changes in dendritic morphology have been shown in the R6 mice (Klapstein et al. 2001; Spires et al. 2004). R6/2 mice have decreased spine density, decreased soma size and decreased dendritic area in both striatum and cortex (Klapstein et al. 2001), whereas R6/1 mice show decreased spine density and dendritic spine length in the same brain regions (Spires et al. 2004).

A number of studies have shown that hippocampal LTP is altered in the R6/2 mouse (Murphy et al. 2000; Cepeda et al. 2003, 2007; Gibson et al. 2005). In R6/2 mice alterations in synaptic plasticity occur at the CA1 and dentate granule cell synapses, and are accompanied by impairments in spatial cognitive performance (Murphy et al. 2000; Cummings et al. 2007). In addition, R6/2 mice have selective impairments in mossy fibre LTP, a form of synaptic plasticity widely regarded as a substrate for memory encoding in the CA3 region of the hippocampus, similar to the impairments found in $Cplx2^{-/-}$ mice (see above, Gibson et al. 2005). These changes occur before the onset of an overt motor phenotype, thus supporting the suggestion that altered synaptic plasticity contributes to the presymptomatic changes in cognition found in the R6/2 mouse.

In addition to the R6/2 mouse, hippocampal function has been assessed in transgenic HD YAC46, R6/1 and two different knock in mouse models of HD (Hodgson et al. 1999; Usdin et al. 1999; Milnerwood et al. 2006; Lynch et al. 2007) and in each study, basal neurotransmission at hippocampal synapses appeared normal. However, deficits in hippocampal LTP (Hodgson et al. 1999;

Usdin et al. 1999; Lynch et al. 2007) and long-term depression (LTD; Murphy et al. 2000; Milnerwood et al. 2006) were detected. Striatal electrophysiological changes have also been documented associated with both baseline activity (Levine et al. 1999; Laforet et al. 2001) and synaptic plasticity (Kung et al. 2007). In addition to alterations in *N*-methyl-D-aspartate receptor (NMDAR)-dependent plasticity, there is evidence for NMDAR dysfunction and toxicity in striatal medium spiny neurons in several HD mouse models (Levine et al. 1999; Cepeda et al. 2001; Hansson et al. 2001; Laforet et al. 2001; Zeron et al. 2001, 2002, 2004; Li et al. 2004; Starling et al. 2005; Andre et al. 2006; Shehadeh et al. 2006; Fan and Raymond 2007; Milnerwood and Raymond 2007) and altered glutamate release at the corticostriatal synapse in the R6/2 mouse (Cepeda et al. 2003, 2007). Further, recent data has shown that spontaneous excitatory neurotransmission in the form of AMPA receptor-mediated spontaneous excitatory postsynaptic currents is significantly reduced in full-length human mutant huntingtin bacterial artificial chromosome (BAC HD) mice in striatal neurons at 6 months of age (Gray et al. 2008).

Activity-dependent alterations in synaptic efficacy (synaptic plasticity), such as LTP and LTD, are widely believed to underlie information processing and storage in the brain (Bliss and Collingridge 1993). Reduced plasticity is evident weeks before the first signs of an overt phenotype in several different lines of HD mice, thus supporting the suggestion that altered synaptic plasticity contributes to the cognitive dysfunction seen in HD, particularly in the early stages of the disease.

12.13 Conclusion

The idea that disturbance of synaptic transmission is an important pathophysiological mechanism and underlies symptoms in neurological diseases is not a new one. Indeed, it is the prevailing hypothesis to explain symptoms in schizophrenia (Eastwood 2004) and has also been suggested to be particularly important in the early stages of Alzheimer's disease (Selkoe 2002; Walsh et al. 2002). Growing evidence from mouse models and human studies suggests that in HD, functional abnormalities in brain circuitry can cause abnormal behaviour many years before profound neurodegeneration is seen. This offers real hope for the development of novel therapies in HD. Treatment of neurodegeneration has proven to be an elusive and difficult target, whereas many drugs that have been used successfully to treat neurological symptoms are targeted to the synapse such as the "triple" treatment study (discussed earlier) which reversed cognitive decline in the R6/2 mouse by improving synaptic function (Morton et al. 2005).

References

Andre VM, Cepeda C, Venegas A et al (2006) Altered cortical glutamate receptor function in the R6/2 model of Huntington's disease. J Neurophysiol 95:2108–2119

Andrews TC, Weeks RA, Turjanski N et al (1999) Huntington's disease progression. PET and clinical observations. Brain 122:2353–2363

Antonini A, Leenders KL, Spiegel R et al (1996) Striatal glucose metabolism and dopamine D2 receptor binding in asymptomatic gene carriers and patients with Huntington's disease. Brain 119:2085–2095

Barr AM, Hofmann CE, Phillips AG et al (2005) Prenatal ethanol exposure in rats decreases levels of complexin proteins in the frontal cortex. Alcohol Clin Exp Res 29:1915–1920

Basso M, Giraudo S, Corpillo D et al (2004) Proteome analysis of human substantia nigra in Parkinson's disease. Proteomics 4:3943–3952

Bates G, Harper PS, Jones L (2002) Huntington's disease (3rd Edn) Oxford University Press, Oxford

Bence NF, Sampat RM, Kopito RR (2001) Impairment of the ubiquitin-proteasome system by protein aggregation. Science 292:1552–1555

Bird ED, Iversen LL (1974) Huntington's chorea: measurement of glutamic acid decarboxylase, choline acetyltransferase and dopamine in basal ganglia. Brain 97:457–472

Bliss TV, Collingridge GL (1993) A synaptic model of memory: long-term potentiation in the hippocampus. Nature 361:31–39

Block-Galarza J, Chase KO, Sapp E et al (1997) Fast transport and retrograde movement of huntingtin and HAP 1 in axons. Neuroreport 8:2247–2251

Bowman AB, Yoo SY, Dantuma NP et al (2005) Neuronal dysfunction in a polyglutamine disease model occurs in the absence of ubiquitin-proteasome system impairment and inversely correlates with the degree of nuclear inclusion formation. Hum Mol Genet 14:679–691

Brandt J, Butters N (1986) The neuropsychology of Huntington's disease. Trends Neurosci 9:118–120

Brose N (2008) Altered complexin expression in psychiatric and neurological disorders: cause or consequence? Mol Cells 25:7–19

Butterworth MB, Frizzell RA, Johnson JP et al (2005) PKA-dependent ENaC trafficking requires the SNARE-binding protein complexin. Am J Physiol Renal Physiol 289:F969–F977

Cai H, Reim K, Varoqueaux F et al (2008) Complexin II plays a positive role in Ca^{2+}-triggered exocytosis by facilitating vesicle priming. Proc Natl Acad Sci U S A 105:19538–19543

Carter RJ, Lione, LA, Humby T et al (1999) Characterisation of progressive motor deficits in mice transgenic for the Huntington's disease mutation. J Neurosci 19:3248–3257

Castro ME, Pascual J, Romon T et al (1998) 5-HT1B receptor binding in degenerative movement disorders. Brain Res 790:323–328

Cattaneo E, Zuccato C, Tartari M (2005) Normal huntingtin function: an alternative approach to Huntington's disease. Nat Rev Neurosci 6:919–930

Cepeda C, Ariano MA, Calvert CR et al (2001) NMDA receptor function in mouse models of Huntington disease. J Neurosci Res 66:525–539

Cepeda C, Hurst RS, Calvert CR et al (2003) Transient and progressive electrophysiological alterations in the corticostriatal pathway in a mouse model of Huntington's disease. J Neurosci 23:961–969

Cepeda C, Wu N, Andre VM et al (2007) The corticostriatal pathway in Huntington's disease. Prog Neurobiol 81:253–271

Cha JH, Kosinski CM, Kerner JA et al (1998) Altered brain neurotransmitter receptors in transgenic mice expressing a portion of an abnormal human huntington disease gene. Proc Natl Acad Sci U S A 95:6480–6485

Cha JH, Frey AS, Alsdorf SA et al (1999) Altered neurotransmitter receptor expression in transgenic mouse models of Huntington's disease. Philos Trans R Soc Lond A 354:981–989

Chen X, Tomchick DR, Kovrigin E et al (2002) Three-dimensional structure of the complexin/SNARE complex. Neuron 33:397–409

Ciechanover A, Brundin P (2003) The ubiquitin proteasome system in neurodegenerative diseases: sometimes the chicken, sometimes the egg. Neuron 40:427–446

Cummings DM, Milnerwood AJ, Dallérac GM et al (2007) Abnormal cortical synaptic plasticity in a mouse model of Huntington's disease. Brain Res Bull 72:103–107

Davies SW, Turmaine M, Cozens BA et al (1997) Formation of neuronal intranuclear inclusions underlies the neurological dysfunction in mice transgenic for the HD mutation. Cell 90:537–548

Davies S, Ramsden DB (2001) Huntington's disease. Mol Pathol 54:409–413

DiFiglia M, Sapp E, Chase K et al (1995) Huntingtin is a cytoplasmic protein associated with vesicles in human and rat brain neurons. Neuron 14:1075–1081

DiFiglia M, Sapp E, Chase KO et al (1997) Aggregation of huntingtin in neuronal intranuclear inclusions and dystrophic neurites in brain. Science 277:1990–1993

DiProspero NA, Chen EY, Charles V et al (2004) Early changes in Huntington's disease patient brains involve alterations in cytoskeletal and synaptic elements. J Neurocytol 33:517–533

Dompierre JP, Godin JD, Charrin BC et al (2007) Histone deacetylase 6 inhibition compensates for the transport deficit in Huntington's disease by increasing tubulin acetylation. J Neurosci 27:3571–3583

Drew CJ, Kyd RJ, Morton AJ (2007) Complexin 1 knockout mice exhibit marked deficits in social behaviours but appear to be cognitively normal. Hum Mol Genet 16:2288–2305

Dure LS 4th, Young AB, Penney JB (1991) Excitatory amino acid binding sites in the caudate nucleus and frontal cortex of Huntington's disease. Ann Neurol 30:785–793

Eastwood SL, Harrison PJ (2000) Hippocampal synaptic pathology in schizophrenia, bipolar disorder and major depression: a study of complexin mRNAs. Mol Psychiatry 5:425–432

Eastwood, SL, Cotter D, Harrison PJ (2001) Cerebellar synaptic protein expression in schizophrenia. Neuroscience 105:219–229

Eastwood SL, Harrison PJ (2001) Synaptic pathology in the anterior cingulate cortex in schizophrenia and mood disorders. A review and a Western blot study of synaptophysin, GAP-43 and the complexins. Brain Res Bull 55:569–578

Eastwood SL (2004) The synaptic pathology of schizophrenia: is aberrant neurodevelopment and plasticity to blame? Int Rev Neurobiol 59:47–72

Eastwood SL, Harrison PJ (2005) Decreased expression of vesicular glutamate transporter 1 and complexin II mRNAs in schizophrenia: further evidence for a synaptic pathology affecting glutamate neurons. Schizophr Res 73:159–172

Edwardson JM, Wang CT, Gong B et al (2003) Expression of mutant huntingtin blocks exocytosis in PC12 cells by depletion of complexin II. J Biol Chem 278:30849–30853

Engelender S, Sharp AH, Colomer V et al (1997) Huntingtin-associated protein 1 (HAP1) interacts with the p150Glued subunit of dynactin. Hum Mol Genet 6:2205–2212

Engqvist-Goldstein AE, Kessels MM, Chopra VS et al (1999) An actin-binding protein of the Sla2/Huntingtin interacting protein 1 family is a novel component of clathrin-coated pits and vesicles. J Cell Biol 147:1503–1518

Fan MM, Raymond LA (2007) N-methyl-D-aspartate (NMDA) receptor function and excitotoxicity in Huntington's disease. Prog Neurobiol 81:272–293

Faull RL, Waldvogel HJ, Nicholson LF et al (1993) The distribution of GABAA-benzodiazepine receptors in the basal ganglia in Huntington's disease and in the quinolinic acid-lesioned rat. Prog Brain Res 99:105–123

Ferrante RJ, Kowall NW, Richardson EP Jr (1991) Proliferative and degenerative changes in striatal spiny neurons in Huntington's disease: a combined study using the section-Golgi method and calbindin D28k immunocytochemistry. J Neurosci 11:3877–3887

Fiebig KM, Rice LM, Pollack E et al (1999) Folding intermediates of SNARE complex assembly. Nat Struct Biol 6:117–123

Folstein SE, Jensen B, Leigh RJ et al (1983) The measurement of abnormal movement: methods developed for Huntington's disease. Neurobehav Toxicol Teratol 5:605–609

Freeman W, Morton AJ (2004a) Regional and progressive changes in brain expression of complexin II in a mouse transgenic for the Huntington's Disease mutation. Brain Res Bull 63:45–55

Freeman W, Morton AJ (2004b) Differential messenger RNA expression of complexins in mouse brain. Brain Res Bull 63:33–44

Gibson HE, Reim K, Brose N et al (2005) A similar impairment in CA3 mossy fibre LTP in the R6/2 mouse model of Huntington's disease and in the complexin II knockout mouse. Eur J Neurosci 22:1701–1712

Gil JM, Rego AC (2008) Mechanisms of neurodegeneration in Huntington's disease. Eur J Neurosci 27:2803–2820

Giraudo CG, Eng WS, Melia TJ et al (2006) A clamping mechanism involved in SNARE-dependent exocytosis. Science 283:21211–21219

Glass M, Faull RL, Dragunow M (1993) Loss of cannabinoid receptors in the substantia nigra in Huntington's disease. Neuroscience 56:523–527

Glass M, Dragunow M, Faull RL (2000) The pattern of neurodegeneration in Huntington's disease: a comparative study of cannabinoid, dopamine, adenosine and GABA(A) receptor alterations in the human basal ganglia in Huntington's disease. Neuroscience 97:505–519

Glass M (2001) The role of cannabinoids in neurodegenerative diseases. Prog Neuropsychopharmacol Biol Psychiatry 25:743–765

Glynn D, Bortnick RA, Morton AJ (2003) Complexin II is essential for normal neurological function in mice. Hum Mol Genet 12:2431–3448

Glynn D, Drew CJ, Reim K et al (2005) Profound ataxia in Complexin 1 knockout mice masks a complex phenotype that includes exploratory and habituation deficits. Hum Mol Genet 14:2369–2385

Glynn D, Morton AJ (2006) Deficits in information processing and social behaviours in the Complexin 2 knockout mouse. FENS Abstract 3:A458

Glynn D, Sizemore RJ, Morton AJ (2007a) Early motor development is abnormal in complexin 1 knockout mice. Neurobiol Dis 25:483–495

Glynn D, Reim K, Brose N et al (2007b) Depletion of Complexin II does not affect disease progression in a mouse model of Huntington's disease (HD); support for role for complexin II in behavioural pathology in a mouse model of HD. Brain Res Bull 72:108–120

Graveland GA, Williams RS, DiFiglia M (1985) Evidence for degenerative and regenerative changes in neostriatal spiny neurons in Huntington's disease. Science 227:770–773

Gray M, Shirasaki DI, Cepeda C et al (2008) Full-length human mutant huntingtin with a stable polyglutamine repeat can elicit progressive and selective neuropathogenesis in BACHD mice. J Neurosci 28:6182–6195

Gunawardena S, Her LS, Brusch RG et al (2003) Disruption of axonal transport by loss of huntingtin or expression of pathogenic polyQ proteins in Drosophila. Neuron 40:25–40

Gunawardena S, Goldstein LS (2005) Polyglutamine diseases and transport problems: deadly traffic jams on neuronal highways. Arch Neurol 62:46–51

Hansson O, Guatteo E, Mercuri NB et al (2001) Resistance to NMDA toxicity correlates with appearance of nuclear inclusions, behavioural deficits and changes in calcium homeostasis in mice transgenic for exon 1 of the huntington gene. Eur J Neurosci 14:1492–1504

Harrison PJ, Eastwood SL (1998) Preferential involvement of excitatory neurons in medial temporal lobe in schizophrenia. Lancet 352:1669–1673

Harrison PJ, Eastwood SL (2000) Hippocampal synaptic pathology in schizophrenia, bipolar disorder and major depression: a study of complexin mRNAs. Mol Psychiatry 5:425–532

Hartl FU, Hayer-Hartl M (2002) Molecular chaperones in the cytosol: from nascent chain to folded protein. Science 295:1852–1858

Hazell AS, Wang C (2005) Downregulation of complexin I and complexin II in the medial thalamus is blocked by N-acetylcysteine in experimental Wernicke's encephalopathy. J Neurosci Res 79:200–207

Hedreen JC, Peyser CE, Folstein SE et al (1991) Neuronal loss in layers V and VI of cerebral cortex in Huntington's disease. Neurosci Lett 133:257–261

Hershko A, Ciechanover A (1998) The ubiquitin system. Annu Rev Biochem 67:425–479

Hilditch-Maguire P, Trettel F, Passani LA et al (2000) Huntingtin: an iron-regulated protein essential for normal nuclear and perinuclear organelles. Hum Mol Genet 9:2789–2797

Hiley CR, Bird ED (1974) Decreased muscarinic receptor concentration in post-mortem brain in Huntington's chorea. Brain Res 80:355–358

Hodgson JG, Agopyan N, Gutekunst CA et al (1999) A YAC mouse model for Huntington's disease with full-length mutant huntingtin, cytoplasmic toxicity, and selective striatal neurodegeneration. Neuron 23:181–192

Holthoff VA, Koeppe RA, Frey KA et al (1993) Positron emission tomography measures of benzodiazepine receptors in Huntington's disease. Ann Neurol 34:76–81

The Huntington's Disease Collaborative Research Group (1993) A novel gene containing a trinucleotide repeat that is expanded and unstable on Huntington's disease chromosomes. Cell 72:971–983

Huntwork S, Littleton JT (2007) A complexin fusion clamp regulates spontaneous neurotransmitter release and synaptic growth. Nat Neurosci 10:1235–1237

Jahn R, Südhof TC (1999) Membrane fusion and exocytosis. Annu Rev Biochem 68:863–911

Jahn R, Scheller RH (2006) SNAREs-engines for membrane fusion. Nat Rev Mol Cell Biol 7:631–643

Jana NR, Zemskov EA, Wang GH et al (2001) Altered proteasomal function due to the expression of polyglutamine-expanded truncated N-terminal huntingtin induces apoptosis by caspase activation through mitochondrial cytochrome c release. Hum Mol Genet 10:1049–1059

Jason GW, Pajurkova EM, Suchowersky OEA (1988) Presymptomatic neuropsychological impairment in Huntington's disease. Arch Neurol 45:769–773

Kalchman MA, Koide HB, McCutcheon K et al (1997) HIP1, a human homologue of S. cerevisiae Sla2p, interacts with membrane-associated huntingtin in the brain. Nat Genet 16:44–53

Kittler JT, Thomas P, Tretter V et al (2004) Huntingtin-associated protein 1 regulates inhibitory synaptic transmission by modulating gamma-aminobutyric acid type A receptor membrane trafficking. Proc Natl Acad Sci U S A 101:12736–12741

Klapstein GJ, Fisher RS, Zanjani H et al (2001) Electrophysiological and morphological changes in striatal spiny neurons in R6/2 Huntington's disease transgenic mice. J Neurophysiol 86:2667–2677

Kremer HP, Roos RA, Dingjan G et al (1990) Atrophy of the hypothalamic lateral tuberal nucleus in Huntington's disease. J Neuropathol Exp Neurol 49:371–382

Kung VW, Hassam R, Morton AJ, Jones S (2007) Neuroscience 146:1571–1580

Laforet GA, Sapp E, Chase K et al (2001) Changes in cortical and striatal neurons predict behavioral and electrophysiological abnormalities in a transgenic murine model of Huntington's disease. J Neurosci 21:9112–9123

Lastres-Becker I, Fezza F, Cebeira M et al (2001) Changes in endocannabinoid transmission in the basal ganglia in a rat model of Huntington's disease. Neuroreport 12:2125–2129

Lastres-Becker I, Gomez M, De Miguel R et al (2002) Loss of cannabinoid CB(1) receptors in the basal ganglia in the late akinetic phase of rats with experimental Huntington's disease. Neurotox Res 4:601–608

Lawrence AD, Sahakian BJ, Hodges JR et al (1996) Executive and mnemonic functions in early Huntington's disease. Brain 119:1633–1645

Lawrence AD, Weeks RA, Brooks DJ et al (1998) The relationship between striatal dopamine receptor binding and cognitive performance in Huntington's disease. Brain 121:1343–1355

Lawrence AD, Sahakian BJ, Rogers RD et al (1999) Discrimination, reversal, and shift learning in Huntington's disease mechanisms of impaired response selection. Neuropsychologia 37:1359–1374

Lee WC, Yoshihara M, Littleton JT (2004) Cytoplasmic aggregates trap polyglutamine-containing proteins and block axonal transport in a Drosophila model of Huntington's disease. Proc Natl Acad Sci U S A 101:3224–3229

Legendre-Guillemin V, Metzler M, Charbonneau M et al (2002) HIP1 and HIP12 display differential binding to F-actin, AP2, and clathrin. Identification of a novel interaction with clathrin light chain. J Biol Chem 277:19897–19904

Levine MS, Klapstein GJ, Koppel A et al (1999) Enhanced sensitivity to N-methyl-D-aspartate receptor activation in transgenic and knockin mouse models of Huntington's disease. J Neurosci Res 58:515–532

Li XJ, Li SH, Sharp AH et al (1995) A huntingtin-associated protein enriched in brain with implications for pathology. Nature 378:398–402

Li SH, Gutekunst CA, Hersch SM et al (1998) Interaction of huntingtin-associated protein with dynactin P150Glued. J Neurosci 18:1261–1269

Li H, Li SH, Yu ZX et al (2001) Huntingtin aggregate-associated axonal degeneration is an early pathological event in Huntington's disease mice. J Neurosci 21:8473–8481

Li Y, Chin LS, Levey AI et al (2002) Huntingtin-associated protein 1 interacts with hepatocyte growth factor-regulated tyrosine kinase substrate and functions in endosomal trafficking. J Biol Chem 277:28212–28221

Li JY, Plomann M, Brundin P (2003) Huntington's disease: a synaptopathy? Trends Mol Med 9:414–420

Li L, Murphy TH, Hayden MR et al (2004) Enhanced striatal NR2B-containing N-methyl-D-aspartate receptor-mediated synaptic currents in a mouse model of Huntington disease. J Neurophysiol 92:2738–2746

Lievens JC, Woodman B, Mahal A et al (2001) Impaired glutamate uptake in the R6 Huntington's disease transgenic mice. Neurobiol Dis 8:807–821

Lin RC, Scheller RH (2000) Mechanisms of synaptic vesicle exocytosis. Annu Rev Cell Dev Biol 16:1–49

Lione LA, Carter RJ, Hunt MJ et al (1999) Selective discrimination learning impairments in mice expressing the human Huntington's disease mutation. J Neurosci 19:10428–10437

London ED, Yamamura HI, Bird ED et al (1981) Decreased receptor-binding sites for kainic acid in brains of patients with Huntington's disease. Biol Psychiatry 16:155–162

Luthi-Carter R, Strand A, Peters NL et al (2000) Decreased expression of striatal signaling genes in a mouse model of Huntington's disease. Hum Mol Genet 9:1259–1271

Lynch G, Kramar EA, Rex CS et al (2007) Brain-derived neurotrophic factor restores synaptic plasticity in a knock-in mouse model of Huntington's disease. J Neurosci 27:4424–4434

Maccarrone M, Battista N, Centonze D (2007) The endocannabinoid pathway in Huntington's disease: a comparison with other neurodegenerative diseases. Prog Neurobiol 81:349–379

Mangiarini L, Sathasivam K, Seller M et al (1996) Exon 1 of the HD gene with an expanded CAG repeat is sufficient to cause a progressive neurological phenotype in transgenic mice. Cell 87:493–506

Martinez-Mir MI, Probst A, Palacios JM (1991) Adenosine A2 receptors: selective localization in the human basal ganglia and alterations with disease. Neuroscience 3:697–706

Mazarakis NK, Cybulska-Klosowicz A, Grote H et al (2005) Deficits in experience-dependent cortical plasticity and sensory-discrimination learning in presymptomatic Huntington's disease mice. J Neurosci 25:3059–3066

McCaw EA, Hu H, Gomez GT et al (2004) Structure, expression and regulation of the cannabinoid receptor gene (CB1) in Huntington's disease transgenic mice. Eur J Biochem 271:4909–4920

McMahon HJ, Missler M, Li C et al (1995) Complexins: cytosolic proteins that regulate SNAP receptor function. Cell 83:111–119

Melia TJ Jr (2007) Putting the clamps on membrane fusion: how complexin sets the stage for calcium-mediated exocytosis. FEBS Lett 581:2131–2139

Menalled LB, Chesselet MF (2002) Mouse models of Huntington's disease. Trends Pharmacol Sci 23:32–39

Menalled LB (2005) Knock-in mouse models of Huntington's disease. NeuroRx 2:465–470

Metzler M, Li B, Gan L et al (2003) Disruption of the endocytic protein HIP1 results in neurological deficits and decreased AMPA receptor trafficking. EMBO J 22:3254–3266

Milnerwood AJ, Cummings DM, Dallerac GM et al (2006) Early development of aberrant synaptic plasticity in a mouse model of Huntington's disease. Hum Mol Genet 15:1690–1703

Milnerwood AJ, Raymond LA (2007) Corticostriatal synaptic function in mouse models of Huntington's disease: early effects of huntingtin repeat length and protein load. J Physiol 585:817–831

Mizuno H, Shibayama H, Tanaka F et al (2000) An autopsy case with clinically and molecular genetically diagnosed Huntington's disease with only minimal non-specific neuropathological findings. Clin Neuropathol 19:94–103

Modregger J, DiProspero NA, Charles V et al (2002) PACSIN 1 interacts with huntingtin and is absent from synaptic varicosities in presymptomatic Huntington's disease brains. Hum Mol Genet 11:2547–2558

Modregger J, Schmidt AA, Ritter B et al (2003) Characterization of Endophilin B1b, a brain-specific membrane-associated lysophosphatidic acid acyl transferase with properties distinct from endophilin A1. J Biol Chem 278:4160–4167

Montoya A, Price BH, Menear M et al (2006) Brain imaging and cognitive dysfunctions in Huntington's disease. J Psychiatry Neurosci 31:21–29

Morfini GA, You YM, Pollema SL et al (2009) Pathogenic huntingtin inhibits fast axonal transport by activating JNK3 and phosphorylating kinesin. Nat Neurosci 12:864–871

Morton AJ, Lagan MA, Skepper JN et al (2000) Progressive formation of inclusions in the striatum and hippocampus of mice transgenic for the human Huntington's disease mutation. J Neurocytol 29:679–702

Morton AJ, Edwardson JM (2001) Progressive depletion of CPLXII in a transgenic mouse model of Huntington's disease. J Neurochem 76:166–172

Morton AJ, Faull RLM, Edwardson JM (2001) Abnormalities in the synaptic vesicle fusion machinery in Huntington's disease. Brain Res Bull 56:111–117

Morton AJ, Hunt MJ, Hodges AK et al (2005) A combination drug therapy improves cognition and reverses gene expression changes in a mouse model of Huntington's disease. Eur J Neurosci 21:855–870

Murphy KP, Carter RJ, Lione LA et al (2000) Abnormal synaptic plasticity and impaired spatial cognition in mice transgenic for exon 1 of the human Huntington's disease mutation. J Neurosci 20:5115–5123

Myers RH, Vonsattel JP, Stevens TJ et al (1988) Clinical and neuropathologic assessment of severity in Huntington's disease. Neurology 38:341–347

Nicniocaill B, Haraldsson B, Hansson O et al (2001) Altered striatal amino acid neurotransmitter release monitored using microdialysis in R6/1 Huntington transgenic mice. Eur J Neurosci 13:206–210

Nithianantharajah, J, Barkus, C, Murphy, M et al (2008) Gene–environment interactions modulating cognitive function and molecular correlates of synaptic plasticity in Huntington's disease transgenic mice. Neurobiol Dis 29:490–504

Ono S, Baux G, Sekiguchi M et al (1998) Regulatory roles of complexins in neurotransmitter release from mature presynaptic nerve terminals. Eur J Neurosci 10:2143–2152

Pabst S, Margittai M, Vainius D et al (2002) Rapid and selective binding to the synaptic SNARE complex suggests a modulatory role of complexins in neuroexocytosis. J Biol Chem 277:7838–7848

Pang, TY, Stam, NC, Nithianantharajah, J et al (2006) Differential effects of voluntary physical exercise on behavioural and brain-derived neurotrophic factor expression deficits in Huntington's disease transgenic mice. Neuroscience 141:569–584

Patel S, Sinha A, Singh MP (2007) Identification of differentially expressed proteins in striatum of maneb-and paraquat-induced Parkinson's disease phenotype in mouse. Neurotoxicol Teratol 29:578–585

Pavese N, Andrews TC, Brooks DJ et al (2003) Progressive striatal and cortical dopamine receptor dysfunction in Huntington's disease: a PET study. Brain 126:1127–1135

Perry TL, Hansen S, Kloster M (1973) Huntington's chorea: deficiency of gamma aminobutyric acid in brain. N Engl J Med 288:337–342

Petersen A, Puschban Z, Lotharius J et al (2002) Evidence for dysfunction of the nigrostriatal pathway in the R6/1 line of transgenic Huntington's disease mice. Neurobiol Dis 11:134–146

Pisani A, Bernardi G, Ding J et al (2007) Re-emergence of striatal cholinergic interneurons in movement disorders. Trends Neurosci 30:545–553

Politis M, Pavese N, Tai YF et al (2008) Hypothalamic involvement in Huntington's disease: an in vivo PET study. Brain 131:2860–2869

Reddy PH, Williams M, Tagle DA (1999) Recent advances in understanding the pathogenesis of Huntington's disease. Trends Neurosci 22:248–255

Reed NA, Cai D, Blasius TL et al (2006) Microtubule acetylation promotes kinesin-1 binding and transport. Curr Biol 16:2166–2172

Reim K, Mansour M, Varoqueaux F et al (2001) Complexins regulate a late step in Ca^{2+}-dependent neurotransmitter release Cell 104:71–81

Reisine TD, Fields JZ, Stern LZ et al (1977) Alterations in dopaminergic receptors in Huntington's disease. Life Sci 21:1123–1128

Reynolds GP, Pearson SJ (1987) Decreased glutamic acid and increased 5-hydroxytryptamine in Huntington's disease brain. Neurosci Lett 78:233–238

Ribchester RR, Thomson D, Wood NI et al (2004) Progressive abnormalities in skeletal muscle and neuromuscular junctions of transgenic mice expressing the Huntington's disease mutation. Eur J Neurosci 20:3092–3114

Richfield EK, Herkenham M (1994) Selective vulnerability in Huntington's disease: preferential loss of cannabinoid receptors in lateral globus pallidus. Ann Neurol 36:577–584

Rizo J, Südhof TC (2002) SNAREs and munc-18 in synaptic vesicle fusion. Nat Rev Neurosci 3:641–653

Rong J, McGuire JR, Fang ZH et al (2006) Regulation of intracellular trafficking of huntingtin-associated protein-1 is critical for TrkA protein levels and neurite outgrowth. J Neurosci 26:6019–6030

Rosenmund C, Rettig J, Brose N (2003) Molecular mechanisms of active zone function. Curr Opin Neurobiol 13:509–519

Ross CA, Wood JD, Schilling G et al (1999) Polyglutamine pathogenesis. Philos Trans R Soc Lond B Biol Sci 354:1005–1011

Sapp E, Penney J, Young A et al (1999) Axonal transport of N-terminal huntingtin suggests early pathology of corticostriatal projections in Huntington disease. J Neuropathol Exp Neurol 58:165–173

Sawada K, Young CE, Barr AM et al (2002) Altered immunoreactivity of complexin protein in prefrontal cortex in severe mental illness. Mol Psychiatry 7:484–492

Sawada K, Barr AM, Nakamura M et al (2005) Hippocampal complexin proteins and cognitive dysfunction in schizophrenia. Arch Gen Psychiatry 62:263–272

Schaub JR, Lu X, Doneske B et al (2006) Hemifusion arrest by complexin is relieved by Ca^{2+}-synaptotagmin I. Nat Struct Mol Biol 13:748–750

Selkoe DJ (2002) Alzheimer's disease is a synaptic failure. Science 298:789–791

Shehadeh J, Fernandes HB, Zeron Mullins MM et al (2006) Striatal neuronal apoptosis is preferentially enhanced by NMDA receptor activation in YAC transgenic mouse model of Huntington disease. Neurobiol Dis 21:392–403

Shoulson I, Fahn S (1979) Huntington's disease: clinical care and evaluation. Neurology 29:1–3

Sinadinos C, Burbidge-King T, Soh D et al (2009) Live axonal transport disruption by mutant huntingtin fragments in Drosophila motor neuron axons. Neurobiol Dis 34:389–395

Singaraja RR, Hadano S, Metzler M et al (2002) HIP14, a novel ankyrin domain-containing protein, links huntingtin to intracellular trafficking and endocytosis. Hum Mol Genet 11:2815–2828

Sittler A, Walter S, Wedemeyer N et al (1998) SH3GL3 associates with the Huntingtin exon 1 protein and promotes the formation of polygln-containing protein aggregates. Mol Cell 2:427–436

Smith R, Brundin P, Li JY (2005) Synaptic dysfunction in Huntington's disease: a new perspective. Cell Mol Life Sci 62:1901–1912

Smith R, Chung H, Rundquist S et al (2006) Cholinergic neuronal defect without cell loss in Huntington's disease. Hum Mol Genet 15:3119–3131

Smith R, Klein P, Koc-Schmitz Y et al (2007) Loss of SNAP-25 and rabphilin 3a in sensory-motor cortex in Huntington's disease. J Neurochem 103:115–123

Snell RG, MacMillan JC, Cheadle JP et al (1993) Relationship between trinucleotide repeat expansion and phenotypic variation in Huntington's disease. Nat Genet 4:393–397

Sollner T, Bennett MK, Whiteheart SW et al (1993) A protein assembly-dissambly pathway in vitro that may correspond to sequential steps of synaptic vesicle docking, activation and fusion. Cell 75:2213–2217

Spires TL, Grote HE, Garry S et al (2004) Dendritic spine pathology and deficits in experience-dependent dendritic plasticity in R6/1 Huntington's disease transgenic mice. Eur J Neurosci 19:2799–2807

Spokes EG (1980) Neurochemical alterations in Huntington's chorea: a study of post-mortem brain tissue. Brain 103:179–210

Stack EC, Kubilus JK, Smith KM et al (2005) Chronology of behavioral symptoms and neuropathological sequela in R6/2 Huntington's disease transgenic mice. J Comp Neurol 490:354–370

Starling AJ, Andre VM, Cepeda C et al (2005) Alterations in N-methyl-D-aspartate receptor sensitivity and magnesium blockade occur early in development in the R6/2 mouse model of Huntington's disease. J Neurosci Res 82:377–386

Steward LJ, Bufton KE, Hopkins PC et al (1993) Reduced levels of 5-HT3 receptor recognition sites in the putamen of patients with Huntington's disease. Eur J Pharmacol 242:137–143

Strauss ME, Brandt J (1990) Are there neuropsychologic manifestations of the gene for Huntington's disease in asymptomatic, at-risk individuals? Arch Neurol 47:905–908

Südhof T (1995) The synaptic vesicle cycle: a cascade of protein-protein interactions. Nature 375:645–653

Südhof TC (2004) The synaptic vesicle cycle. Annu Rev Neurosci 27:509–547

Szebenyi G, Morfini GA, Babcock A et al (2003) Neuropathogenic forms of huntingtin and androgen receptor inhibit fast axonal transport. Neuron 40:41–52

Tang TS, Tu H, Chan EY et al (2003) Huntingtin and huntingtin-associated protein 1 influence neuronal calcium signaling mediated by inositol-(1,4,5) triphosphate receptor type 1. Neuron 39:227–239

Tannenberg RK, Scott HL, Tannenberg AE et al (2006) Selective loss of synaptic proteins in Alzheimer's disease: Evidence for an increased severity with APOE varepsilon4. Neurochem Int 49:631–639

Tippett LJ, Waldvogel HJ, Thomas SJ et al (2007) Striosomes and mood dysfunction in Huntington's disease. Brain 130:206–221

Tokumaru H, Umayahara K, Pellegrini LL et al (2001) SNARE Complex oligomerization by synaphin/complexin is essential for synaptic vesicle exocytosis. Cell 104:421–432

Trushina E, Dyer RB, Badger JD, 2nd et al (2004) Mutant huntingtin impairs axonal trafficking in mammalian neurons in vivo and in vitro. Mol Cell Biol 24:8195–8209

Usdin MT, Shelbourne PF, Myers RM et al (1999) Impaired synaptic plasticity in mice carrying the Huntington's disease mutation. Hum Mol Genet 8:839–846

van Oostrom JC, Maguire RP, Verschuuren-Bemelmans CC et al (2005) Striatal dopamine D2 receptors, metabolism, and volume in preclinical Huntington disease. Neurology 65:941–943

Velier J, Kim MT, Schwarz C et al (1998) Wild-type and mutant huntingtins function in vesicle trafficking in the secretory and endocytic pathways. Exp Neurol 152:34–40

Vetter JM, Jehle T, Heinemeyer J et al (2003) Mice transgenic for exon 1 of Huntington's disease: properties of cholinergic and dopaminergic pre-synaptic function in the striatum. J Neurochem 85:1054–1063

Vonsattel JP, Myers RH, Stevens TJ et al (1985) Neuropathological classification of Huntington's disease. J Neuropathol Exp Neurol 44:559–577

Vonsattel JPG, DiFiglia M (1998) Huntington disease. J Neuropathol Exp Neurol 57:369–384

Wade A, Jacobs P, Morton AJ (2008) Atrophy and degeneration in sciatic nerve of presymptomatic mice carrying the Huntington's disease mutation. Brain Res 1188:61–68

Waeber C, Palacios JM (1989) Serotonin-1 receptor binding sites in the human basal ganglia are decreased in Huntington's chorea but not in Parkinson's disease: a quantitative in vitro autoradiography study. Neuroscience 32:337–347

Waeber C, Rigo M, Chinaglia G et al (1991) Beta-adrenergic receptor subtypes in the basal ganglia of patients with Huntington's chorea and Parkinson's disease. Synapse 8:270–280

Walsh DM, Klyubin I, Fadeeva JV et al (2002) Naturally secreted oligomers of amyloid beta protein potently inhibit hippocampal long-term potentiation in vivo. Nature 416:535–539

Wanderer J, Morton AJ (2007) Differential morphology and composition of inclusions in the R6/2 mouse and PC12 cell models of Huntington's disease. Histochem Cell Biol 127:473–484

Wang CE, Zhou H, McGuire JR et al (2008) Suppression of neuropil aggregates and neurological symptoms by an intracellular antibody implicates the cytoplasmic toxicity of mutant huntingtin. J Cell Biol 181:803–816

Wanker EE, Rovira C, Scherzinger E et al (1997) HIP-I: a huntingtin interacting protein isolated by the yeast two-hybrid system. Hum Mol Genet 6:487–495

Waterman-Storer CM, Karki SB, Kuznetsov SA et al (1997) The interaction between cytoplasmic dynein and dynactin is required for fast axonal transport. Proc Natl Acad Sci U S A 94:12180–12185

Weeks RA, Cunningham VJ, Piccini P et al (1997) 11C-diprenorphine binding in Huntington's disease: a comparison of region of interest analysis with statistical parametric mapping. J Cereb Blood Flow Metab 17:943–949

Witzmann FA, Li J, Strother WN et al (2003) Innate differences in protein expression in the nucleus accumbens and hippocampus of inbred alcohol-preferring and -nonpreferring rats. Proteomics 3:1335–1344

Wong EH, Reynolds GP, Bonhaus DW et al (1996) Characterization of [3H]GR 113808 binding to 5-HT4 receptors in brain tissues from patients with neurodegenerative disorders. Behav Brain Res 73:249–252

Wyttenbach A, Carmichael J, Swartz J et al (2000) Effects of heat shock, heat shock protein 40 (HDJ-2), and proteasome inhibition on protein aggregation in cellular models of Huntington's disease. Proc Natl Acad Sci U S A 97:2898–2903

Xue M, Reim K, Chen X et al (2007) Distinct domains of complexin I differentially regulate neurotransmitter release. Nat Struct Mol Biol 14:949–958

Yi JH, Hoover R, McIntosh TK et al (2006) Early, transient increase in complexin I and complexin II in the cerebral cortex following traumatic brain injury is attenuated by N-acetylcysteine. J Neurotrauma 23:86–96

Young AB, Shoulson I, Penney JB (1986) Huntington's disease in Venezuela: neurologic feature and functional decline. Neurology 36:244–249

Young AB, Greenamyre JT, Hollingsworth Z et al (1988) NMDA receptor losses in putamen from patients with Huntington's disease. Science 241:981–983

Zabel C, Sagi D, Kaindl AM et al (2006) Comparative proteomics in neurodegenerative and non-neurodegenerative diseases suggest nodal point proteins in regulatory networking. J Proteome Res 5:1948–1958

Zechner U, Scheel S, Hemberger M et al (1998) Characterization of the mouse Src homology 3 domain gene Sh3d2c on Chr 7 demonstrates coexpression with huntingtin in the brain and identifies the processed pseudogene Sh3d2c-ps1 on Chr 2. Genomics 54:505–510

Zeron MM, Chen N, Moshaver A et al (2001) Mutant huntingtin enhances excitotoxic cell death. Mol Cell Neurosci 17:41–53

Zeron MM, Hansson O, Chen N (2002) Increased sensitivity to N-methyl-D-aspartate receptor-mediated excitotoxicity in a mouse model of Huntington's disease. Neuron 33:849–860

Zeron MM, Fernandes HB, Krebs C et al (2004) Potentiation of NMDA receptor-mediated excitotoxicity linked with intrinsic apoptotic pathway in YAC transgenic mouse model of Huntington's disease. Mol Cell Neurosci 25:469–479

Zink M, Vollmayr B, Gebicke-Haerter PJ et al (2007) Reduced expression of complexins I and II in rats bred for learned helplessness. Brain Res 1144:202–208

Chapter 13
Synaptic Dysfunction in Parkinson's Disease: From Protein Misfolding to Functional Alterations

Tiago Fleming Outeiro and Luísa Vaqueiro Lopes

Abstract Neurodegenerative disorders are devastating human diseases that include Parkinson's, Huntington's and Alzheimer's disease, amyotrophic lateral sclerosis, and the frontal temporal dementias. Although the clinical manifestations of these disorders have been known for some time, our understanding of the molecular underpinnings is only starting to emerge. Protein misfolding and aggregation is a common hallmark among these diseases, and these events are likely to produce a number of cellular and functional alterations. In Parkinson's disease (PD), alpha-synuclein misfolds and forms intracellular inclusions known as Lewy bodies (LBs). Accumulating evidence suggests that these events lead to the disruption of intracellular trafficking which will, undoubtedly, lead to defects in synaptic transmission. Dysfunction in the basal ganglia circuitry is strongly associated with PD, with loss of dopaminergic neurons in the substantia nigra. Here, we present an overview of how misfolding and aggregation of alpha-synuclein might be related to synaptic dysfunction in PD. A deeper understanding of the molecular basis of PD will enable us to devise novel strategies for therapeutic intervention which may be applied to several neurodegenerative disorders that seem to have similar roots.

13.1 Protein Misfolding Diseases

Neurodegenerative disorders are devastating human diseases that include Parkinson's, Huntington's and Alzheimer's disease, amyotrophic lateral sclerosis, and the frontal temporal dementias (Forman et al. 2004). Invariably, these disorders are correlated with aging, which is the major risk factor known to date. While very different in their pathophysiology, they are collectively referred to as protein-misfolding disorders because a common hallmark is the presence of misfolded and aggregated forms of

---

T.F. Outeiro (✉)
Cell and Molecular Neuroscience Unit, Instituto de Medicina Molecular,
Av. Prof. Egas Moniz, 1649-028 Lisboa, Portugal
e-mail: touteiro@gmail.com

A. Wyttenbach and V. O'Connor (eds.), *Folding for the Synapse*,
DOI 10.1007/978-1-4419-7061-9_13,

various proteins in the brains of affected individuals (Caughey and Lansbury 2003; Dobson 2001; Thomas et al. 1995).

Despite the well-known connection between protein misfolding, aggregation, and disease, the manner by which misfolding results in disease is not clearly understood. In some cases, it seems that the deposition of protein aggregates may physically disrupt the functioning of specific cell groups and the respective tissues and organs where those cells are located. In other cases, it seems that the lack of functional protein, due to its recruitment into the aggregates, results in the failure of crucial cellular processes (Muchowski 2002; Soto et al. 2006; Thomas et al. 1995). However, for neurodegenerative diseases, such as Alzheimer's, Parkinson's or the Prion diseases, it appears that the symptoms arise from the destruction of cells by a "gain of toxic function" that results from the aggregation process (via oligomers, protofibrils, amyloid fibrils or other intermediates) or by a combination of both this gain of toxic function and a loss of normal function of the protein (Fig. 13.1) (Caughey and Lansbury 2003; Ross and Poirier 2004).

The "amyloid hypothesis" (developed originally for Alzheimer's disease) posits that the aggregation of proteins into an ordered fibrillar structure is causally related to aberrant protein interactions that culminate in neuronal dysfunction and ultimately neurodegeneration (Chiti and Dobson 2006; Hardy and Selkoe 2002). Proteins, as the main effectors in the cell, play underpinning roles in all biological processes. Thus, it is not surprising that the list of these diseases is continuously expanding, as new proteins are identified and their functions understood. Although different proteins are associated with different disorders, their ability to misfold and form amyloid fibrils seems to be a shared property. Nevertheless, this property is not sufficient to explain the differences in pathology observed in the disorders (Dobson 2003).

13.2 Parkinson's Disease

Parkinson's disease (PD) is the second most common neurodegenerative disorder, exceeded only by Alzheimer's disease (Dawson and Dawson 2003; Forman et al. 2005; Forno 1996; Gasser 2001; Lynch et al. 1997; Moore et al. 2005; Nussbaum and Polymeropoulos 1997; Vila and Przedborski 2004). PD affects about 2% of people over 65 years old and 4–5% of people over 85. Clinical manifestations of PD include severe motor defects characterized by resting muscle tremor ("pill-rolling"), muscle rigidity, bradykinesia, and postural instability (Goedert 2001).

LBs, the pathological hallmark of PD, are concentric hyaline cytoplasmic inclusions that can be visualized by histological analysis. LBs contain the protein alpha-synuclein (aSyn), as well as proteasomal subunits, lysosomal enzymes and molecular chaperones (Irizarry et al. 1998; Spillantini et al. 1997). It is still unclear whether the formation of LBs represents a mechanism developed by neurons to preclude the accumulation of the pathogenic intermediates, which have been proposed to be aSyn oligomers (Ding et al. 2002).

The majority of PD cases arise sporadically and only approximately 10% are familial. These have been genetically linked to the genes encoding aSyn, parkin,

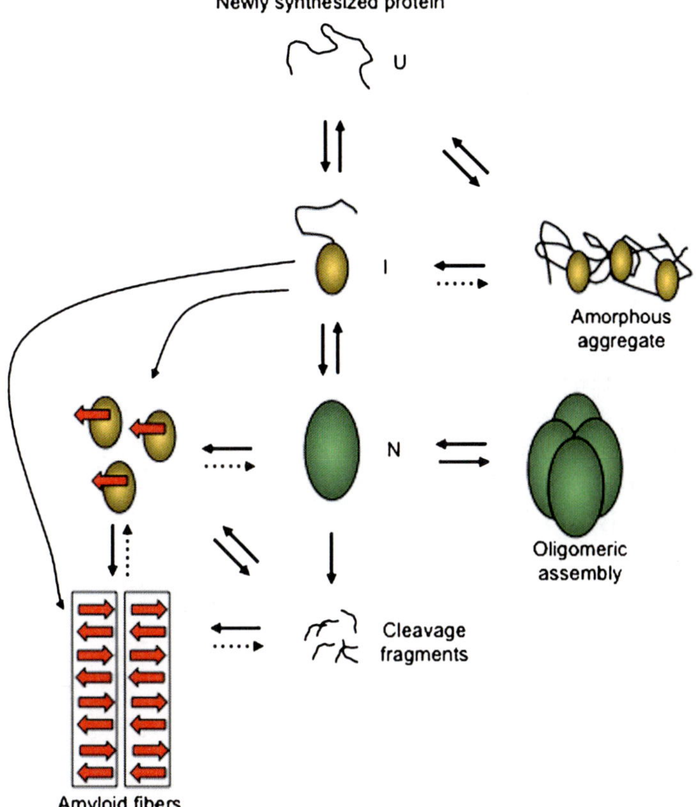

Fig. 13.1 Schematic representation of the folding/unfolding/misfolding reactions. Once proteins are synthesized, they face a wide variety of challenges in the crowded environment of the cell. Throughout their lives, proteins need to change their conformations, allowing unwanted species to accumulate, which may lead to cytotoxicity (gain of toxic function). *U* unfolded, *I* intermediate, *N* native

UCH-L1, DJ-1, PINK1, LRRK2, and ATP13A2 (Gasser 2007). While these cases are rare, it is hoped that understanding the molecular mechanisms underpinning the genetic forms of the disease will provide insight into the pathogenesis of the sporadic forms as well. Therefore, intense research is focused on investigating the genes and proteins involved in PD, both in terms of their normal physiological role as well as their contribution to disease.

13.3 The Basal Ganglia Circuitry

The basal ganglia are a group of related subcortical nuclei and include the neostriatum (caudate nucleus and caudate putamen), the external and internal pallidal segments (GPe, GPi), the subthalamic nucleus (STN), and the substantia nigra with its pars reticulate (SNr) and pars compacta (SNc) (Alexander et al. 1986). They form a complex network of parallel loops that involve specific thalamic and cortical areas. These parallel circuits are divided into "motor," "associative" and limbic loops depending on the function of the cortical area involved (for a review see (Galvan and Wichmann 2008)).

Both the striatum and STN receive glutamatergic afferents from specific areas of the cerebral cortex or thalamus, and transfer the information to the basal ganglia output nuclei, GPi and SNr. The projections between the striatum and GPi/SNr are divided into two separate pathways, a "direct" (monosynaptic) connection and an "indirect" projection, via the intercalated GPe and STN (Fig. 13.2, left). Neurons in the "direct" pathway express dopamine D1 receptors and coexpress the

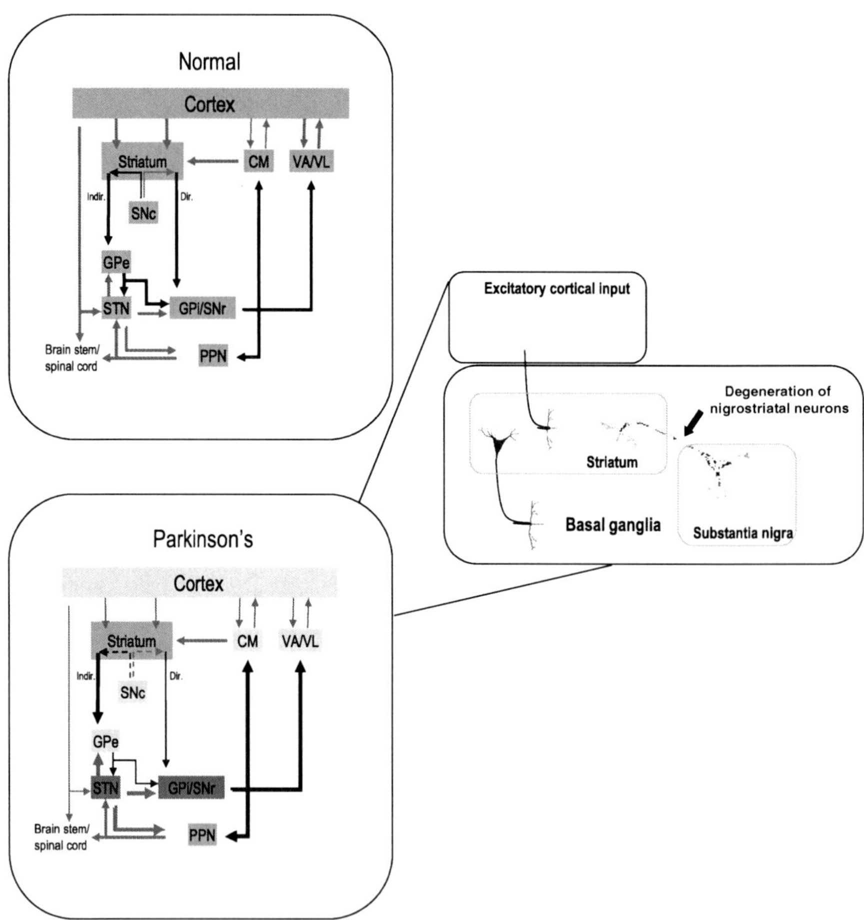

Fig. 13.2 Simplified diagram of the motor "loop" in the basal ganglia in normal (*left panel*) and in Parkinson's condition (*right panel*). *Black arrows* indicate inhibitory connections and *gray arrows* indicate excitatory connections. The thickness of the arrows is an indication of their rate of activity. Dopamine is thought to inhibit neuronal activity in the "indirect" pathway and to excite neurons in the "direct" pathway. In the parkinsonian state, dopamine depletion leads to disinhibition of dopamine D2 receptor striatal neurons in the "indirect" pathway leading ultimately to over-inhibition of

peptides substance P and dynorphin, whereas those in the "indirect" pathway contain dopamine D2 receptors and the peptide enkephalin (Gerfen et al., 1990). Movement-related output is directed from GPi and SNr to the ventral nucleus of the thalamus which in turn projects back to the primary motor cortex. GPi and SNr also project to noncholinergic neurons in the pedunculolantine nucleus (PPN) in the brainstem and to CM/Pf (Galvan and Wichmann 2008).

Activation of the direct pathway neurons in the striatum will inhibit basal ganglia output neurons, which, in turn, will disinhibit the connected thalamocortical neurons and facilitate movement. In contrast, the activation of indirect pathway neurons in the striatum will lead to increased basal ganglia output and, presumably, to suppression of movement. Because most GPi neurons increase their firing rate with movement, the main role of the basal ganglia motor circuit is to inhibit and to stabilize the activity of the thalamocortical neurons. Thus, the direct and indirect pathways have opposing effects on the output function of the basal ganglia (Alexander et al. 1986; DeLong 1990). Dopamine modulates these glutamatergic effects on corticostriatal inputs by exerting a dual effect on striatal neurons: excitatory action by activating D1 receptors in the "direct" pathway and inhibitory action through D2 receptors in the "indirect" pathway (Cepeda et al. 1993). Thus, dopamine's net action may be to reduce GPi/SNr activity, thereby increasing the activity in thalamocortical projection neurons and, through activation of the cerebral cortex, facilitating movement (DeLong and Wichmann 2007) (Fig. 13.2, left).

13.4 Functional Alterations in PD

The prominent motor abnormalities in PD arise in large part from degeneration of neurons in the SNc with the resulting loss of dopamine in the basal ganglia. Abnormal activity in the "motor" loop of the basal ganglia is strongly implicated in the development of parkinsonism (Albin et al. 1989; DeLong 1990; Obeso et al. 2000). Recordings of firing rates in the basal ganglia in animal models of parkinsonism show that spontaneous neuronal discharge in STN, GPi and SNr is increased as compared with normal controls, neuronal discharge in GPe is decreased (Fig. 13.2, right panel). Striatal dopamine loss results in reduced activity in the direct pathway, resulting in disinhibition of the basal ganglia output nuclei. This dopamine loss also leads to greater inhibition of GPe, and consequently, in disinhibition of STN and GPi. The net effect of these

Fig. 13.2 (continued) the thalamo-cortical and brainstem motor centers. The nigrostriatal neurons affected are highlighted in the *top right magnification*. Abbreviations: *CM* centromedian nucleus of thalamus, *CMA* cingulated motor area, *GPe* external segment of the globus pallidus, *GPi* internal segment of the globus pallidus, *PPN* pedunculopontine nucleus, *SNc* substantia nigra pars compacta, *SNr* substantia nigra pars reticulate, *STN* subthalamic nucleus, *VA* ventral anterior nucleus of thalamus, *VL* ventrolateral nucleus of thalalamus. Adapted from (Bender et al. 2006)

changes is an increase in basal ganglia output to brainstem and thalamus which has been postulated to result in over-inhibition of thalamocortical neurons and reduced responsiveness of brainstem and cortical mechanisms involved in motor control (Galvan and Wichmann 2008).

Imaging studies have also suggested that the striatal dopamine concentrations steeply decline preceding the onset of clinical parkinsonism, which is only seen when more than 70% of striatal dopamine is lost. In early phases of PD, dopamine loss affects primarily the posterior putamen (the striatal motor area), but later spreads to involve other nigrostriatal regions. In later stages, more widespread dopamine loss and neuronal degeneration in non-dopaminergic systems, such as the locus coeruleus and the raphe nuclei, may account for some of the non-motor aspects of PD (Galvan and Wichmann 2008).

13.5 aSyn and Parkinson's Disease

Accumulating evidence points to a causative role for the presynaptic protein aSyn in the pathogenesis of PD. aSyn, is a member of the synuclein family of proteins, which includes β- and γ-synuclein. It was initially identified in the electric lobe of *Torpedo californica* for reacting with an antiserum raised against purified cholinergic vesicles (Maroteaux et al. 1988), but its function remains unclear. aSyn is a 14.5 kDa protein, abundant in the brain, which has been involved in a variety of cellular processes (Fig. 13.3) (Cookson 2006; Lucking and Brice 2000).

After an initial report of a non-Aβ component of AD amyloid (NAC), consisting of a 35 amino acid polypeptide generated by cleavage of aSyn, the protein has been associated to several neurodegenerative disorders, as described below.

The *PARK1* locus, which encodes for aSyn, became associated with PD, when a point mutation was found in an Italian kindred afflicted by autosomal dominant PD. The mutation causes a threonine for alanine substitution at position 53 (A53T) (Polymeropoulos et al. 1997). This discovery was then followed by a report identifying aSyn in LBs (Irizarry et al. 1998; Spillantini et al. 1997). Afterwards, another familial form of PD was linked to a mutation in aSyn causing a proline for alanine substitution at position 30 (A30P) (Kruger et al. 1998). More recently, a third mutation consisting of a lysine for glutamate substitution at position 46 (E46K) was discovered to be associated with familial PD (Zarranz et al. 2004). Additionally, duplications and triplications of the *PARK1* locus have also been linked to familial PD (Singleton et al. 2003). The dominant nature of the inherited mutants is thought to reflect a gain rather than a loss of function in the aSyn proteins.

After the initial discovery of aSyn in LBs in PD, the protein was detected in cellular inclusions in several other neurodegenerative diseases, including dementia with Lewy bodies (DLB), multiple system atrophy (MSA), and Hallervorden–Spatz syndrome, now called neurodegeneration with brain iron accumulation type 1 (NBIA). The neurodegenerative diseases that share aSyn pathology as a primary

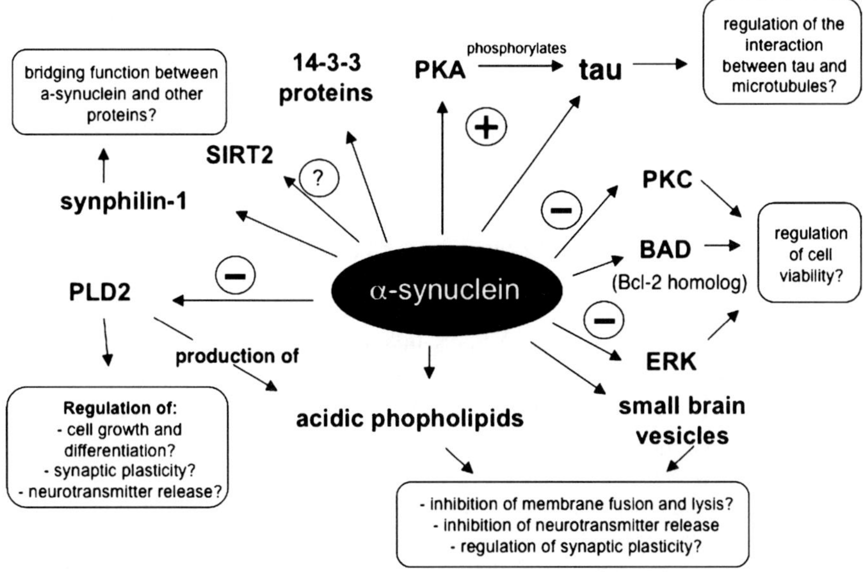

Fig. 13.3 Putative functions and interacting partners of aSyn. The binding partners of aSyn are indicated by *arrows*. "–" and "+" indicate enzyme inhibition or activation by aSyn, respectively. *Boxes* describe potential functions of aSyn interacting with the respective partner. *SIRT2* sirtuin 2, *PLD2* phospholipase D2, *PKC* protein kinase C, *PKA* protein kinase A, *ERK* extracellular-regulated kinase. Adapted from (Lucking and Brice 2000)

feature are collectively known as synucleinopathies (Spillantini and Goedert 2000). It is intriguing that the same protein is associated in different disorders, but the exact molecular mechanisms involved are still unknown.

13.5.1 aSyn: Gain or Loss of Function?

Several genetically tractable model organisms have been employed to study the role of aSyn in PD (Feany and Bender 2000; Masliah et al. 2000; Outeiro and Lindquist 2003). Over-expression studies of aSyn, or its PD-linked mutants, in mouse, rat, fly, nematode, and yeast have all converged on the conclusion that increased levels of aSyn leads to neurotoxicity, possibly due to a gain of toxic function associated with protein aggregation (Auluck et al. 2002; Feany and Bender 2000; Lakso et al. 2003; Lo Bianco et al. 2002; Masliah et al. 2000; Outeiro and Lindquist 2003). However, the presence of LBs in different disorders raises the possibility that sequestration of aSyn into Lewy bodies may decrease effective aSyn levels in nerve terminals and may contribute to neurodegeneration, also due to a loss of normal function (Chandra et al. 2005; Cookson and van der Brug 2007). Nevertheless, despite intense study of aSyn, frustratingly little is known about its normal cellular function and how it contributes to the various diseases (Fig. 13.3) (Gitler and Shorter 2007; Lucking and Brice 2000).

13.5.2 aSyn, Intracellular Trafficking and Synaptic Dysfunction

Through the use of a variety of cellular and animal models, we are gaining insight into the normal and abnormal roles of aSyn in neurons. Although the function of aSyn is still poorly understood, it has been extensively implicated in the regulation of presynaptic secretion. Indeed, aSyn is present in the cytosol of the presynaptic terminal in close proximity to synaptic vesicles (Iwai et al. 1995). Studies focusing specifically on the dopamine system showed that mice lacking aSyn exhibit an increased release of dopamine from nigrostriatal terminals that is mimicked by elevated Ca^{2+} and the reduction of striatal dopamine (Abeliovich et al. 2000). In these mice, however, there are no deficits or abnormalities in neuronal number, morphology or dopaminergic inervation pattern (Abeliovich et al. 2000). This is probably due to the fact that aSyn decreases the dopamine vesicles refilling rate, maintaining a stable dopamine pool (Yavich et al. 2004). Corroborating this, mice with aSyn null mutation or over-expressing A30P aSyn show decreased dopamine release in the striatum in response to prolonged stimulation, a situation in which the releasable pool of dopamine is more rapidly depleted (Yavich et al. 2004, 2005). More recently, it was also shown that A30P mutants display reduced activity of the vesicular monoamine transporter 2 (VMAT2) (Yavich et al. 2005) which is known to induce progressive nigrostriatal neurodegeneration (Goetz et al. 2005).

Studies using the budding yeast, *Saccharomyces cerevisiae*, as a model organism have shown that an increase in the levels of accumulation of aSyn, similar to that observed in patients with multiplications of the *PARK1* gene, leads to intracellular trafficking defects, especially between the ER and Golgi (Cooper et al. 2006; Gitler et al. 2008; Outeiro and Lindquist 2003). Subsequent studies in *Caenorhabditis elegans* models of PD where aSyn was expressed pan-neuronally, confirmed the deleterious effects of aSyn on trafficking and endocytosis (Cooper et al. 2006; Lakso et al. 2003). In addition, a genetic study using mice demonstrated that aSyn ameliorates the inhibition of SNARE complex assembly, the latter being important in exocytosis of neurotransmitter (Chandra et al. 2005; Chua and Tang 2006).

13.6 Concluding Remarks

Endocytosis, an essential process for the recycling of cell surface molecules, has been implicated in the etiopathogenesis not only of PD, but also of AD. On the other hand, tight control of exocytosis is crucial in neurotransmitter release at the synapse. These cellular trafficking pathways are closely linked and under similar regulatory mechanisms. Therefore, aSyn-mediated disruption of essential trafficking events will, undoubtedly, have important consequences on synaptic dysfunction.

One outstanding issue is whether the increased vulnerability of dopaminergic neurons in the substantia nigra is due to the direct impact of aSyn mainly on monoamine pathways, either in vesicle recycling or exocytosis or, alternatively, in presynaptic secretory function in general, irrespective of the neurotransmitter involved. In the latter case, aSyn accumulation in dopaminergic neurons might be secondary and subsequent to neuronal degeneration.

In conclusion, the understanding of the role played by aSyn in synaptic dysfunction promises to open novel avenues for therapeutic intervention in PD and other synucleinopathies.

Acknowledgements TFO is supported by the Michael J. Fox Foundation, Calouste Gulbenkian Foundation, Fundação para a Ciência e Tecnologia and by a Marie Curie International Reintegration Grant from the European Commission. LVL is supported by Fundação para a Ciência e Tecnologia.

References

Abeliovich, A., Schmitz, Y., Farinas, I., Choi-Lundberg, D., Ho, W.H., Castillo, P.E., Shinsky, N., Verdugo, J.M., Armanini, M., Ryan, A., Hynes, M., Phillips, H., Sulzer, D. and Rosenthal, A. (2000) Mice lacking alpha-synuclein display functional deficits in the nigrostriatal dopamine system, *Neuron*, **25**, 239–252.

Albin, R.L., Aldridge, J.W., Young, A.B. and Gilman, S. (1989) Feline subthalamic nucleus neurons contain glutamate-like but not GABA-like or glycine-like immunoreactivity, *Brain Res*, **491**, 185–188.

Alexander, G.E., DeLong, M.R. and Strick, P.L. (1986) Parallel organization of functionally segregated circuits linking basal ganglia and cortex, *Annu Rev Neurosci*, **9**, 357–381.

Auluck, P.K., Chan, H.Y., Trojanowski, J.Q., Lee, V.M. and Bonini, N.M. (2002) Chaperone suppression of alpha-synuclein toxicity in a Drosophila model for Parkinson's disease, *Science*, **295**, 865–868.

Bender, A., Koch, W., Elstner, M., Schombacher, Y., Bender, J., Moeschl, M., Gekeler, F., Muller-Myhsok, B., Gasser, T., Tatsch, K. and Klopstock, T. (2006) Creatine supplementation in Parkinson disease: a placebo-controlled randomized pilot trial, *Neurology*, **67**, 1262–1264.

Caughey, B. and Lansbury, P.T. (2003) Protofibrils, pores, fibrils, and neurodegeneration: separating the responsible protein aggregates from the innocent bystanders, *Annu Rev Neurosci*, **26**, 267–298.

Cepeda, C., Buchwald, N.A. and Levine, M.S. (1993) Neuromodulatory actions of dopamine in the neostriatum are dependent upon the excitatory amino acid receptor subtypes activated, *Proc Natl Acad Sci U S A*, **90**, 9576–9580.

Chandra, S., Gallardo, G., Fernandez-Chacon, R., Schluter, O.M. and Sudhof, T.C. (2005) Alpha-synuclein cooperates with CSPalpha in preventing neurodegeneration, *Cell*, **123**, 383–396.

Chiti, F. and Dobson, C.M. (2006) Protein misfolding, functional amyloid, and human disease, *Annu Rev Biochem*, **75**, 333–366.

Chua, C.E. and Tang, B.L. (2006) alpha-synuclein and Parkinson's disease: the first roadblock, *J Cell Mol Med*, **10**, 837–846.

Cookson, M.R. (2006) Hero versus antihero: the multiple roles of alpha-synuclein in neurodegeneration, *Exp Neurol*, **199**, 238–242.

Cookson, M.R. and van der Brug, M. (2007) Cell systems and the toxic mechanism(s) of alpha-synuclein, *Exp Neurol*, **209**, 5–11.

Cooper, A.A., Gitler, A.D., Cashikar, A., Haynes, C.M., Hill, K.J., Bhullar, B., Liu, K., Xu, K., Strathearn, K.E., Liu, F., Cao, S., Caldwell, K.A., Caldwell, G.A., Marsischky, G., Kolodner, R.D., Labaer, J., Rochet, J.C., Bonini, N.M. and Lindquist, S. (2006) {alpha}-Synuclein blocks ER-golgi traffic and Rab1 rescues neuron loss in Parkinson's models, *Science*, **313**, 3234–328.

Dawson, T.M. and Dawson, V.L. (2003) Molecular pathways of neurodegeneration in Parkinson's disease, *Science*, **302**, 819–822.

DeLong, M.R. (1990) Primate models of movement disorders of basal ganglia origin, *Trends Neurosci*, **13**, 281–285.

DeLong, M.R. and Wichmann, T. (2007) Circuits and circuit disorders of the basal ganglia, *Arch Neurol*, **64**, 20–24.

Ding, T.T., Lee, S.J., Rochet, J.C. and Lansbury, P.T., Jr. (2002) Annular alpha-synuclein protofibrils are produced when spherical protofibrils are incubated in solution or bound to brain-derived membranes, *Biochemistry*, **41**, 10209–10217.

Dobson, C.M. (2001) The structural basis of protein folding and its links with human disease, *Philos Trans R Soc Lond B Biol Sci*, **356**, 133–145.

Dobson, C.M. (2003) Protein folding and misfolding, *Nature*, **426**, 884–890.

Feany, M.B. and Bender, W.W. (2000) A Drosophila model of Parkinson's disease, *Nature*, **404**, 394–398.

Forman, M.S., Lee, V.M. and Trojanowski, J.Q. (2005) Nosology of Parkinson's disease: looking for the way out of a quagmire, *Neuron*, **47**, 479–482.

Forman, M.S., Trojanowski, J.Q. and Lee, V.M. (2004) Neurodegenerative diseases: a decade of discoveries paves the way for therapeutic breakthroughs, *Nat Med*, **10**, 1055–1063.

Forno, L.S. (1996) Neuropathology of Parkinson's disease, *J Neuropathol Exp Neurol*, **55**, 259–272.

Galvan, A. and Wichmann, T. (2008) Pathophysiology of parkinsonism, *Clin Neurophysiol*, **119**, 1459–1474.

Gasser, T. (2001) Genetics of Parkinson's disease, *J Neurol*, **248**, 833–840.

Gasser, T. (2007) Update on the genetics of Parkinson's disease, *Mov Disord*, 22 Suppl 17, S343–S350.

Gerfen, C.R., Engber, T.M., Mahan, L.C., Susel, Z., Chase, T.N., Monsma, F.J., Jr. and Sibley, D.R. (1990) D1 and D2 dopamine receptor-regulated gene expression of striatonigral and striatopallidal neurons, *Science*, **250**, 1429–1432.

Gitler, A.D., Bevis, B.J., Shorter, J., Strathearn, K.E., Hamamichi, S., Su, L.J., Caldwell, K.A., Caldwell, G.A., Rochet, J.C., McCaffery, J.M., Barlowe, C. and Lindquist, S. (2008) The Parkinson's disease protein alpha-synuclein disrupts cellular Rab homeostasis, *Proc Natl Acad Sci U S A*, **105**, 145–150.

Gitler, A.D. and Shorter, J. (2007) Prime time for alpha-synuclein, *J Neurosci*, **27**, 2433–2434.

Goedert, M. (2001) Alpha-synuclein and neurodegenerative diseases, *Nat Rev Neurosci*, **2**, 492–501.

Goetz, C.G., Poewe, W., Rascol, O. and Sampaio, C. (2005) Evidence-based medical review update: pharmacological and surgical treatments of Parkinson's disease: 2001 to 2004, *Mov Disord*, **20**, 523–539.

Hardy, J. and Selkoe, D.J. (2002) The amyloid hypothesis of Alzheimer's disease: progress and problems on the road to therapeutics, *Science*, **297**, 353–356.

Irizarry, M.C., Growdon, W., Gomez-Isla, T., Newell, K., George, J.M., Clayton, D.F. and Hyman, B.T. (1998) Nigral and cortical Lewy bodies and dystrophic nigral neurites in Parkinson's disease and cortical Lewy body disease contain alpha-synuclein immunoreactivity, *J Neuropathol Exp Neurol*, **57**, 334–337.

Iwai, A., Masliah, E., Yoshimoto, M., Ge, N., Flanagan, L., de Silva, H.A., Kittel, A. and Saitoh, T. (1995) The precursor protein of non-A beta component of Alzheimer's disease amyloid is a presynaptic protein of the central nervous system, *Neuron*, **14**, 467–475.

Kruger, R., Kuhn, W., Muller, T., Woitalla, D., Graeber, M., Kosel, S., Przuntek, H., Epplen, J.T., Schols, L. and Riess, O. (1998) Ala30Pro mutation in the gene encoding alpha-synuclein in Parkinson's disease, *Nat Genet*, **18**, 106–108.

Lakso, M., Vartiainen, S., Moilanen, A.M., Sirvio, J., Thomas, J.H., Nass, R., Blakely, R.D. and Wong, G. (2003) Dopaminergic neuronal loss and motor deficits in *Caenorhabditis elegans* overexpressing human alpha-synuclein, *J Neurochem*, **86**, 165–172.

Lo Bianco, C., Ridet, J.L., Schneider, B.L., Deglon, N. and Aebischer, P. (2002) alpha -Synucleinopathy and selective dopaminergic neuron loss in a rat lentiviral-based model of Parkinson's disease, *Proc Natl Acad Sci U S A*, **99**, 10813–10818.

Lucking, C.B. and Brice, A. (2000) Alpha-synuclein and Parkinson's disease, *Cell Mol Life Sci*, **57**, 1894–1908.

Lynch, T., Farrer, M., Hutton, M. and Hardy, J. (1997) Genetics of Parkinson's disease, *Science*, **278**, 1212–1213.

Maroteaux, L., Campanelli, J.T. and Scheller, R.H. (1988) Synuclein: a neuron-specific protein localized to the nucleus and presynaptic nerve terminal, *J Neurosci*, **8**, 2804–2815.

Masliah, E., Rockenstein, E., Veinbergs, I., Mallory, M., Hashimoto, M., Takeda, A., Sagara, Y., Sisk, A. and Mucke, L. (2000) Dopaminergic loss and inclusion body formation in alpha-synuclein mice: implications for neurodegenerative disorders, *Science*, **287**, 1265–1269.

Moore, D.J., West, A.B., Dawson, V.L. and Dawson, T.M. (2005) Molecular pathophysiology of Parkinson's disease, *Annu Rev Neurosci*, **28**, 57–87.

Muchowski, P.J. (2002) Protein misfolding, amyloid formation, and neurodegeneration: a critical role for molecular chaperones? *Neuron*, **35**, 9–12.

Nussbaum, R.L. and Polymeropoulos, M.H. (1997) Genetics of Parkinson's disease, *Hum Mol Genet*, **6**, 1687–1691.

Obeso, J.A., Rodriguez-Oroz, M.C., Rodriguez, M., Lanciego, J.L., Artieda, J., Gonzalo, N. and Olanow, C.W. (2000) Pathophysiology of the basal ganglia in Parkinson's disease, *Trends Neurosci*, **23**, S8–19.

Outeiro, T.F. and Lindquist, S. (2003) Yeast cells provide insight into alpha-synuclein biology and pathobiology, *Science*, **302**, 1772–1775.

Polymeropoulos, M.H., Lavedan, C., Leroy, E., Ide, S.E., Dehejia, A., Dutra, A., Pike, B., Root, H., Rubenstein, J., Boyer, R., Stenroos, E.S., Chandrasekharappa, S., Athanassiadou, A., Papapetropoulos, T., Johnson, W.G., Lazzarini, A.M., Duvoisin, R.C., Di Iorio, G., Golbe, L.I. and Nussbaum, R.L. (1997) Mutation in the alpha-synuclein gene identified in families with Parkinson's disease, *Science*, **276**, 2045–2047.

Ross, C.A. and Poirier, M.A. (2004) Protein aggregation and neurodegenerative disease, *Nat Med*, **10 Suppl**, S10–S17.

Singleton, A.B., Farrer, M., Johnson, J., Singleton, A., Hague, S., Kachergus, J., Hulihan, M., Peuralinna, T., Dutra, A., Nussbaum, R., Lincoln, S., Crawley, A., Hanson, M., Maraganore, D., Adler, C., Cookson, M.R., Muenter, M., Baptista, M., Miller, D., Blancato, J., Hardy, J. and Gwinn-Hardy, K. (2003) alpha-Synuclein locus triplication causes Parkinson's disease, *Science*, **302**, 841.

Soto, C., Estrada, L. and Castilla, J. (2006) Amyloids, prions and the inherent infectious nature of misfolded protein aggregates, *Trends Biochem Sci*, **31**, 150–155.

Spillantini, M.G. and Goedert, M. (2000) The alpha-synucleinopathies: Parkinson's disease, dementia with Lewy bodies, and multiple system atrophy, *Ann N Y Acad Sci*, **920**, 16–27.

Spillantini, M.G., Schmidt, M.L., Lee, V.M., Trojanowski, J.Q., Jakes, R. and Goedert, M. (1997) Alpha-synuclein in Lewy bodies, *Nature*, **388**, 839–840.

Thomas, P.J., Qu, B.H. and Pedersen, P.L. (1995) Defective protein folding as a basis of human disease, *Trends Biochem Sci*, **20**, 456–459.

Vila, M. and Przedborski, S. (2004) Genetic clues to the pathogenesis of Parkinson's disease, *Nat Med*, **10 Suppl**, S58–S62.

Yavich, L., Oksman, M., Tanila, H., Kerokoski, P., Hiltunen, M., van Groen, T., Puolivali, J., Mannisto, P.T., Garcia-Horsman, A., MacDonald, E., Beyreuther, K., Hartmann, T. and Jakala, P. (2005) Locomotor activity and evoked dopamine release are reduced in mice overexpressing A30P-mutated human alpha-synuclein, *Neurobiol Dis*, **20**, 303–313.

Yavich, L., Tanila, H., Vepsalainen, S. and Jakala, P. (2004) Role of alpha-synuclein in presynaptic dopamine recruitment, *J Neurosci*, **24**, 11165–11170.

Zarranz, J.J., Alegre, J., Gomez-Esteban, J.C., Lezcano, E., Ros, R., Ampuero, I., Vidal, L., Hoenicka, J., Rodriguez, O., Atares, B., Llorens, V., Tortosa, E.G., Del Ser, T., Munoz, D.G. and De Yebenes, J.G. (2004) The new mutation, E46K, of alpha-synuclein causes parkinson and Lewy body dementia, *Ann Neurol*, **55**, 164–173.

Chapter 14
Synapses and Alzheimers's Disease: Effect of Immunotherapy?

Nathan C. Denham, James A.R. Nicoll, and Delphine Boche

Abstract Alzheimer's disease (AD) was first described more than 100 years ago; however, the mechanisms underlying its pathogenesis are still poorly understood. Current theories suggest a pivotal role for the protein amyloid-β (Aβ) and many of the novel treatments for AD focus on Aβ. In this chapter, we discuss evidence that Aβ underpins the cognitive decline as a result of direct and indirect toxicity of the peptide on synapses in the cerebral cortex and hippocampus. Furthermore, we will follow the promise that Aβ immunisation holds to alter the natural history of AD, from its beginnings in animal models to the current research on humans. The success seen in mice in preventing both synapse loss and reducing functional decline is yet to be matched in humans and serious adverse events in patients stopped the initial vaccination approach. Research, however, is continuing in human AD aiming to provide a greater understanding of the mechanisms underlying the immune response and the potential effects of immunisation on preventing or reversing cognitive impairment.

14.1 Introduction

Dementia is a global disturbance of higher mental functions of sufficient severity to interfere with social or occupational functioning. Dementia commonly shows at least two abnormalities including memory loss in an otherwise alert patient and impairments in cognition such as language, problem solving, judgement, calculation, attention, perception and praxis (Association 2000). Alongside this, patients may also display neuropsychiatric symptoms that include changes in behaviour, delusions, changes in affect and hallucinations. Impairment of short- and long-term memory, abstract thinking, judgement, other higher cortical functions or personality change result in significant social and economic problems for patients and carers.

D. Boche (✉)
Division of Clinical Neurosciences, University of Southampton, Southampton, UK
e-mail: D.Boche@soton.ac.uk

A. Wyttenbach and V. O'Connor (eds.), *Folding for the Synapse*,
DOI 10.1007/978-1-4419-7061-9_14, © Springer Science+Business Media, LLC 2011

There are a number of different causes of dementia and they are mainly defined by their pathological features.

Alzheimer's disease (AD) is a major neurodegenerative disease and the most common form of dementia affecting 5% of people >65 years and 20% of those >80 years in the developed world (Small et al. 1997). AD starts with an insidious memory loss, progressing over time with a decline in cognition (Ritchie and Lovestone 2002). Clinical symptoms typically begin to appear after 65 years, although AD sufferers can be as young as 45 years in familial early onset cases. The major hallmark of AD is a progressive decline in cognitive performance with typical signs of episodic memory impairment and difficulty forming new memories (Albert 1996). Despite relative sparing of the motor and sensory cortex, some patients suffer from dysphasia, dyspraxia and agnosia. As the disease progresses, attention and executive function decline and behavioural and personality disturbance may occur. Death typically occurs within 5–15 years of disease onset.

14.1.1 Neuropathological Hallmarks of Alzheimer's Disease

Pathologically, AD is characterised by the presence of widespread extracellular foci of amyloid-β (Aβ) protein accumulation, known as plaques, in the limbic system and association cortex. The plaques are composed of filamentous deposits of Aβ40 (40 amino acid residues) and Aβ42 (42 amino acid residues). Neuritic plaques are associated with dystrophic neurites, activated microglia and astrocytes and evidence suggests that they develop over a period of months to years in the course of the disease. Diffuse plaques are different from neuritic plaques as they tend to be amorphous, non-fibrillar structures composed mainly of Aβ42. Although direct evidence is lacking, diffuse plaques are commonly believed to be the precursors to neuritic plaques, and they are found in similar anatomical places in the cortex.

Another key feature of AD is intraneuronal accumulation of the protein tau in the form of paired helical filaments (PHFs), the basic component of which is hyperphosphorylated tau protein. Tau pathology in AD is characterised by the presence of: (1) large bundles of PHFs in the perinuclear cytoplasm of affected neurons called neurofibrillary tangles (NFTs), (2) PHFs in dystrophic neurites localised around Aβ plaques and (3) neuropil threads in distal neuronal processes found between the plaques and tangles. Tau pathology occurs widely throughout the brain, including the hippocampus, entorhinal cortex, parahippocampal gyrus, amygdala and the association cortex (Selkoe 2001; Graham and Lantos 2002).

Although the precise roles each of these lesions take in the pathogenesis of AD has been the subject of debate for many years, the amyloid cascade hypothesis (Selkoe 1991; Hardy and Higgins 1992) has become one of the prominent theories in the development of AD and its central feature is the pivotal role played by the Aβ protein.

14.1.2 Pathophysiology of Alzheimer's Disease: Synaptic Failure

Successful cognitive function requires complex interaction between neurons via synapses. As the principal clinical effect of neurodegenerative disease is cognitive decline, it is not unreasonable to suggest that the root of the decline might be at the synaptic level. A gradual loss of synapses is a part of the ageing process, but it is not known whether a numerical loss of synapses can be compensated for by an increase in the contact area per synapse and thus preserve cognitive function (Uylings and de Brabander 2002). However, for many years, research has suggested that synapse loss is the critical reason for the cognitive decline seen in early AD (Davies et al. 1987; Terry et al. 1991; Lassmann 1996; Sze et al. 1997; Selkoe 2002). The grey matter neuropil, in which the synapses are located, is markedly degenerated in AD (Fig. 14.1). Indeed, the degree of clinical impairment and the severity of the typical neuropathological features of AD are poorly correlated (Lassmann et al. 1993). Terry and colleagues (1991) suggested that synapse loss is the major correlate to declining cognition in AD, observing a strong association between synaptic density in the neocortex and cognitive testing. Quantitative morphometric studies in the frontal and temporal cortex of AD patients compared with controls have confirmed that there is a decline in both the total synaptic density (between 25 and 36%) and in the number of synapses per neuron (a loss of 38%) (Davies et al. 1987). Therefore, synaptic loss is the best correlate to the cognitive deficits, unlike plaques and NFTs, which correlate relatively poor (Terry et al. 1991), and is proportional to the cognitive decline (Lassmann 1996). The loss in synapses is greatest in the acetylcholine, noradrenaline and 5-hydroxytryptamine systems (Small et al. 2001), however, with individual variability between patients. Immunohistochemical studies showed that cognitive decline in AD is correlated with significant reductions (up to 55%) in the synaptic protein synaptophysin in the hippocampus (Masliah 1995; Sze et al. 1997). Synaptophysin is located on presynaptic vesicles and a reduction in staining for this protein in the hippocampus suggests a significant loss of hippocampal synapses, which offers an anatomical basis for

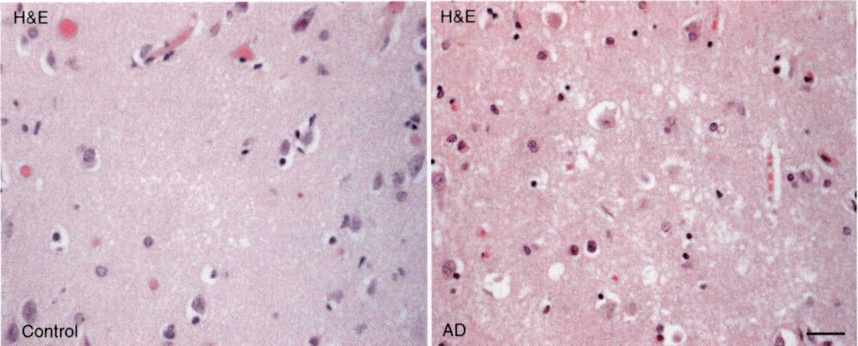

Fig. 14.1 *Cortical grey matter stained with haematoxylin and eosin (H&E)*. In Alzheimer's disease (AD) there is degeneration of the neuropil, compared with the control, reflecting loss of synapses and small neuronal processes. Scale bar = 50 μm

memory loss in AD (Masliah 1995). A decrease of 45% in synaptophysin was also observed in the frontal and inferior parietal cortex when compared with age-matched controls (Masliah et al. 1991). This suggests a pathological mechanism in AD for the loss of these synapses (Lassmann 1996). Transgenic mice that bear mutations in the gene encoding β-amyloid precursor protein (APP), a rare cause of familial AD, have a decrease in synaptophysin immunoreactivity, which appears at the age of 2–3 months, before the accumulation of Aβ protein (Hsia et al. 1999). Morphological, functional and behavioural studies in the Tg2576 transgenic mouse model, also mimicking a familial form of AD, show at 4 months a deficit of synaptic integrity and behaviour before impairment of LTP at 5 months and amyloid deposition at 6 months (Jacobsen et al. 2006). Findings in this animal model support a role for synapse loss as an early feature of the disease. Subtle alterations in the synaptic connections, particularly of the hippocampus, may be responsible for the memory impairment and early cognitive decline seen around the time of diagnosis. A loss of 25% of synaptophysin was observed in patients with mild cognitive function. This synaptic dysfunction may occur before the frank neuronal loss that is observed in autopsies of patients with late-stage AD (Masliah et al. 2001). Since the loss of synaptic structure and function seems so critical to the early neurodegeneration in AD, therapies aimed at preventing this loss are likely to be the most effective.

14.2 Aβ and Its Role in Alzheimer's Disease

14.2.1 Physiology of Aβ

Amyloid β-protein (Aβ) is derived from APP, which is a heterogenous group of proteins expressed ubiquitously throughout the body. Three isoforms of APP are created by alternative splicing of a single gene; however, each one has a basic structure of a large extracellular N terminus, a single transmembrane-spanning region and a shorter cytoplasmic C terminus (Kang et al. 1987). APP has the highest expression in the body on neuronal cell membranes in the brain and can undergo a variety of proteolytic cleavages by a number of secretase enzymes. The essential enzymes required for the creation of Aβ are β- and γ-secretases (Haass and Selkoe 1993). Neither the production of Aβ nor the action of γ-secretase is pathological as Aβ can be detected in the CSF and plasma of normal individuals (Giedraitis et al. 2007). Evidence suggests that most of the Aβ in the brain in AD comes from the cleavage of APP in neurons as observed by *in situ* hybridization (Bahmanyar et al. 1987; Kang et al. 1987).

14.2.2 The Amyloid Cascade Hypothesis

The amyloid cascade hypothesis (Fig. 14.2) is one of the leading theories to explain the pathogenesis of AD and critical to this cascade is Aβ (Hardy and Higgins 1992). Over many years, the pathogenesis of AD has been fiercely debated between those

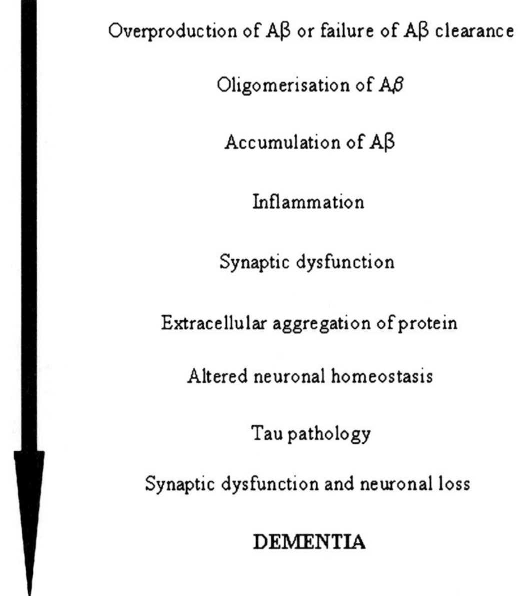

Fig. 14.2 *Amyloid cascade hypothesis.* Schema of the events involved in the pathophysiological cascade leading to Alzheimer's disease, known as the amyloid cascade hypothesis

who believe the primary cause is Aβ (Hardy and Higgins 1992; Hardy and Selkoe 2002) and those who believe the primary cause is tau accumulation (Braak and Braak 1995; Nagy et al. 1995; Braak et al. 1996; Terry 1996; Naslund et al. 2000). Much evidence exists for and against the relevance of both proteins; however, the amyloid cascade hypothesis provides unification as it suggests the first step in the pathogenesis is abnormal Aβ accumulation and that this leads to tau pathology as a consequence. Although the causal links are lacking, the fact that the cytoskeletal abnormalities and dysfunction in axonal transport attributed to tau pathology could be a consequence of the accumulation of Aβ has been of major support for the amyloid cascade hypothesis, as it helps to explain how the accumulation of Aβ and tau can come together and result in AD (Hardy et al. 1998; Small and Duff 2008). Much of the support for the amyloid cascade hypothesis comes from genetic studies (see below), particularly in that familial AD is caused by aberrations in Aβ production, and that risk factors for sporadic AD may be based on the inability to clear Aβ (Wavrant-DeVrieze et al. 1999; Myers et al. 2000; Weller and Nicoll 2003; Weller et al. 2008).

14.2.3 The Genetics of Alzheimer's Disease

14.2.3.1 Aβ Peptide

The major pieces of evidence that link Aβ to AD pathogenesis is the research that has come from patients affected by the Down's syndrome (DS) and familial cases of AD. DS, which is usually due to trisomy 21, i.e. triplication of chromosome 21 on which the *APP* gene is located (St George-Hyslop et al. 1987), is associated with the development of an AD-type pathology (Wisniewski et al. 1985; de la Monte and Hedley-Whyte 1990; Oyama et al. 1994). Familial and sporadic cases of AD are phenotypically similar, both having the same symptoms and the same neuropathology; however, familial cases tend to have a younger age of onset (Selkoe 2001) with specific point mutations in the *APP* gene (St George-Hyslop et al. 1987; Goate et al. 1991). Multiple mutations in the *APP* gene have been found (St George-Hyslop et al. 1987), all leading to an increased production of Aβ, as altering the cleavage sites alters the proteolytic processing of APP. Mutations around codon 717 of the *APP* gene (Goate et al. 1991; Chartier-Harlin et al. 1991) favour processing by the γ-secretase leading to an elevated production and deposition of Aβ42 (Scheuner et al. 1996; Mann et al. 1996). The identification of mutations in the *APP* gene permitted the creation of transgenic mice in which incorporation of the human *APP* gene with a point mutation mimics AD-type neuropathology with mainly Aβ deposition (Games et al. 1995; Hsiao et al. 1996).

The presenilin (*PSEN*) genes were also implicated by positional cloning in genetic forms of AD. Mutations in *PSEN* genes are the most common cause of autosomal dominant AD, with a number of missense mutations being found in both *PSEN1* and *PSEN2* (respectively on chromosomes 14 and 1) (Sherrington et al. 1995; Levy-Lahad et al. 1995). Transgenic mice with mutant *PSEN* genes show an increase in the production of Aβ42 without abnormal pathology (Citron et al. 1997; Duff et al. 1996), and when coupled with an APP mutation, the production of Aβ increases markedly (Borchelt et al. 1997; Holcomb et al. 1998). Research into the presenilins suggests that they are necessary for normal γ-secretase function on APP, either because presenilins are the catalytic site of γ-secretase itself or an essential co-factor necessary for normal action.

The final gene linked to AD, but is not a direct familial cause, is the apolipoprotein E (*APOE)* gene located on chromosome 19 (Saunders et al. 1993). Genetics studies in the 1990s identified the *APOE* gene as the major genetic risk factor for sporadic or late-onset AD (Strittmatter et al. 1993). ApoE proteins are key proteins to facilitate lipid transport (Brown and Goldstein 1986) and metabolism within the nervous system to maintain structural and functional integrity of membranes and synapses (Mahley 1988). ApoE also acts as an Aβ-scavenging receptor molecule that regulates Aβ concentration through internalisation via apoE receptors and the endosomal/lysosomal pathway (Poirier 2000). In addition, recently, it has been revealed that Aβ has an essential physiological role in lipid homeostasis (Grimm et al. 2005). This evidence suggests that clearance of Aβ is likely to be regulated by the apoE–Aβ interactions. Three different isoforms of apoE occur (ApoE E2, E3 and E4), which differ in amino acids at position 112 and 158, according to the genotype of the individual (Mahley 1988). The increased risk of developing AD associated with the possession of apoE E4 may be due to its inability to internalise, and therefore clear, extracellular Aβ into endosomes/lysosomes; AD patients with the ε4 allele show an increased plaque burden in the brain. In addition, the biochemical difference of apoE E4 may induce the promotion of lipid rafts that have a suitable environment for the amyloidogenic processes (Chauhan 2003). Some studies have suggested a more profound loss of pre-synaptic cholinergic markers in AD patients with *APOE* ε4 (Lai et al. 2006). In addition, ε4 was associated with a lower concentration of synaptic proteins in the temporal cortex of normal brains, which might help to explain the epidemiological studies related to *APOE* ε4 and AD (Love et al. 2006). It has been recently observed that mice expressing the human *APOE* ε4 allele have less total apoE than those expressing the two other human variants (Riddell et al. 2008), and therefore that apoE4 might be associated with less efficient synaptic repair.

14.2.3.2 Tau Protein

Mutations in the tau protein are also of interest in AD. Tau mutations have been shown to cause frontotemporal dementia, which is a disease typified by tau NFTs (Goedert et al. 1998; Spillantini and Goedert 1998). Since this form of dementia has little associated Aβ pathology, it is likely that abnormal tau accumulation cannot induce the formation of Aβ plaques (Hardy et al. 1998).

Further research in mice has created transgenic animals with mutant tau and mutant APP (Lewis et al. 2001). Results from these animals show that double mutants undergo increased formation of tau tangles when compared with mice with mutant tau alone, suggesting that altered Aβ production influences tau formation and neurofibrillary pathology. Recent evidence in a triple transgenic animal model of AD, with point mutations in APP, *PSEN1* and tau genes, indicates that Aβ accumulation precedes the tau pathology in a cascade of events that ultimately leads to the cognitive alterations (Oddo et al. 2003a, b).

14.2.4 Synaptotoxicity of Aβ

The specific neurotoxic nature of Aβ is a matter for great debate, particularly as it is critical and central to the amyloid cascade hypothesis. The main evidence used to portray Aβ as a non-toxic molecule is that the number and place of Aβ deposits within the cortex does not correlate well with the degree of cognitive impairment in life (Naslund et al. 2000). Furthermore, Aβ deposits can be found in the normal ageing brains of individuals without AD (Arriagada et al. 1992; Price and Morris 1999) and some researchers have even suggested that Aβ has a protective role in the brain to prevent inflammatory damage (Lee et al. 2007).

Recently, research has shifted the emphasis away from the insoluble fibrillar Aβ found in plaques to soluble Aβ oligomers (i.e. aggregates of a few molecules of Aβ). The degree of dementia has been shown to correlate better with the soluble fraction of Aβ assayed biochemically than with histologically determined plaque counts (Naslund et al. 2000; Lue et al. 1999). A primary role for Aβ oligomers and protofibrils in causing damage before the formation of fibrillar plaques has been supported by evidence showing that protofibrils cause neuronal damage *in vitro* (Hartley et al. 1999) and that neuronal damage precedes fibrillar plaque formation in mice (Hsia et al. 1999). The specific toxicity caused by non-fibrillar Aβ has been invested by many groups including Walsh and colleagues (Walsh et al. 2002) and Wang and co-workers (2002) who looked at the effects of Aβ on the hippocampus. Walsh *et al.* isolated soluble oligomeric Aβ from cell cultures and injected it directly into the hippocampus in rats. They showed that the Aβ oligomers inhibited long term potentiation (LTP). This represents a major step towards understanding AD symptoms as there is evidence that LTP underlies the molecular mechanisms of memory formation (Elgersma and Silva 1999; Roman et al. 1999; Luscher et al. 2000; Martin et al. 2000). They further showed that Aβ oligomers disrupt synaptic plasticity *in vivo* and are directly synaptotoxic at concentrations found in the human cortex (Walsh et al. 2002). The work of Wang *et al.* also supports this hypothesis by showing inhibition of LTP by Aβ oligomers in the dentate gyrus of rat hippocampal slices. They suggested the possible disruption in memory came from Aβ creating an imbalance between LTP and LTD (long-term depression) (Wang et al. 2002). In addition, it was observed that a decrease in synapses is critically dependent on cortical Aβ levels (Mucke et al. 2000).

The seminal link between Aβ, synaptic loss and cognitive decline was made in transgenic mice where the increases in Aβ was associated with the appearance of behavioural defects including deficits in spatial learning tasks and memory (Hsiao et al. 1996). Recently, it has been observed that Aβ dimers extracted from human AD brains where sufficient to inhibit LTP, enhance LTD and reduce dendritic spine density in the normal rodent hippocampus (Shankar et al. 2008). These effects were associated with impairment in the memory of learned behaviour in normal rats. Aβ dimers are the smallest Aβ aggregates in human brain to perturb the synaptic physiology and they were the major toxic species extracted from the human AD brains (Shankar et al. 2008). Aβ monomers do not appear to impair LTP (Klyubin et al. 2008). Together, these observations are consistent with the view that the symptoms of AD might result from the accumulation of toxic Aβ species leading to synaptic loss and cognitive decline.

The exact mechanisms that result in Aβ synaptotoxicity and neurotoxicity are unknown. Many lines of evidence suggest multiple roles for Aβ in altering the local environment in the cortex. *In vitro* studies suggest that Aβ interacts directly with specific neuronal potassium channels to induce reversible increases in action potential spontaneity resulting in subtle changes in synaptic transmission (Ye et al. 2003). *In vitro* and *in vivo* evidence suggests that Aβ has an essential role in oxidative stress partly because of its single methionine residue that allows it to participate in free radical species generation, protein oxidation, DNA damage and lipid peroxidation (Varadarajan et al. 1999, 2001; Yatin et al. 1999;). *In vitro* studies on neuronal culture have also shown that Aβ peptides decrease the number and activity of synaptic N-methyl-D aspartate (NMDA) glutamate receptors, alpha-amino-3-hydroxy-5-methyl-4-isoxazolepropionic acid (AMPA) glutamate receptors (Snyder et al. 2005) and more specifically of the glutamate receptor-1 (GluR1), a subunit of the AMPA glutamate receptor (Almeida et al. 2005). Recently, this finding has been confirmed after the application of oligomeric Aβ isolated from human AD brains to neuronal cultures (Shankar et al. 2007), and among the different Aβ species, dimeric Aβ was sufficient to induce the receptor changes (Shankar et al. 2008).

14.3 Rationale for Immunisation Against Aβ

The amyloid cascade hypothesis states that the age-related accumulation of Aβ leads to direct and indirect neurotoxicity, which depletes cortical synaptic density and results in cognitive decline (Hardy and Higgins 1992). Since Aβ is the critical element in this cascade, it is a potential therapeutic target, and hence the novel idea of Aβ vaccination was developed. The main aims of vaccination are to produce one or all of the following: enhance Aβ clearance from the brain, disrupt Aβ fibrils, prevent fibril formation, reduce oligomeric Aβ and block the toxic effects of Aβ. It is proposed that this will then prevent or reverse the synaptic loss in AD, hopefully altering the natural history of the disease.

14.3.1 Animal Models – Aβ Immunisation in Mice: Effect on the Pathology

The prospects of using immunisation to attenuate or prevent Aβ-associated pathology in AD lead to experiments involving transgenic mice to investigate whether an immune response could alter the course of the disease. In a seminal experiment, transgenic mice that expressed the human *APP* gene with a point mutation (Games et al. 1995) were immunised using synthetic human Aβ42 in order to provoke an immune response against the Aβ plaques which develop with age in this model (Schenk et al. 1999). The vaccinated mice subsequently produced serum antibody levels to Aβ, which were not found in controls. Autopsies on these mice showed that immunisation with Aβ42 significantly reduced the amyloid plaque pathology in the hippocampus when compared with controls. Further experiments on the ability of Aβ immunisation to reduce plaque pathology have confirmed the initial observation (Bard et al. 2000; Games et al. 2000; DeMattos et al. 2001).

To investigate the nature of the removal of amyloid plaques from the brains of immunised mice, further work was carried out in APP mice in which peripheral injections of Aβ antibodies (passive immunisation) were sufficient to reduce the Aβ burden in the brain (Bard et al. 2000). This meant that Aβ antibodies alone were necessary and sufficient to reduce pathology. Different mechanisms of Aβ removal have been suggested. Following immunisation, the antibodies produced cross the blood–brain barrier to attach to amyloid plaques leading to Aβ clearance (Bard et al. 2000). This disaggregation of fibrillar plaques is supported by the *in vitro* work of Solomon and colleagues (Solomon et al. 1997) and the *in vivo* experiments of McLaurin and colleagues (2002) who both suggest that antibodies also prevent new plaque formation (Solomon et al. 1996). Another major mechanism for the clearance of Aβ after immunisation is mediated via phagocytic microglia (Bard et al. 2000, 2003; Wilcock et al. 2003, 2004).

Following the development of triple transgenic mice, in which both Aβ and tau accumulate with age (Oddo et al. 2003a, b), it has been shown that passive immunisation to Aβ in this model not only led to a reduction in extracellular Aβ plaques but also a decrease in intracellular Aβ and early tau pathology (Oddo et al. 2004). Subsequently, when the passive immunisation is halted, the Aβ pathology returns before occurrence of the tau pathology. This finding supports the critical role for Aβ in causing tau pathology and suggests that immunisation may prevent AD progression by stopping both Aβ- and tau-associated toxicity. However, these studies showed that only early tau was removed, not the hyperphosphorylated tau seen in advanced AD (Oddo et al. 2004).

14.3.2 Animal Models – Aβ Immunisation in Mice: Effect on the Progression of the Disease

The prospect of removing Aβ from the cortex using immunisation was promising however, it was important to seek evidence that the removal of plaques had a beneficial effect on brain function.

The TgCRND8 murine model of AD expresses a mutant β*APP* gene and shows both Aβ pathology and spatial learning defects. To explore the behavioural consequence of Aβ immunisation, these mice were actively immunised and their ability to perform in the Morris water maze test was assessed. The immunisation significantly prevented the cognitive decline that was seen in unimmunised transgenic mice; however, the level of cognition in immunised mice was still below that expected of normal, non-transgenic mice (Janus et al. 2000). This provided evidence that immunisation slowed the rate of cognitive decline in mice with an AD-like phenotype that was further confirmed by the work of Morgan who used a working memory task (water maze and radial arm maze combined) to show that immunisation protects mice from developing learning and memory defects (Morgan et al. 2000).

A key test for Aβ vaccination would be whether it could alter the course of AD by preventing the synaptic degeneration responsible for the cognitive decline seen in patients. The previous work in mice had only alluded to changes in amyloid plaques, which has been found to correlate poorly with cognition (Terry et al. 1991). Therefore, Buttini and colleagues decided to measure synaptic degeneration by comparing synaptic densities in the cortex of mice that had been immunised (both active and passive) against controls. They showed that immunisation prevented synaptic loss in the hippocampus and frontal cortex of mice when compared with controls, not only linking Aβ immunisation to the preservation of cognition, but also the potential of a vaccine as an efficient therapy (Buttini et al. 2005).

14.3.3 Immunisation of AD Patients: The Pathology

The success of Aβ immunisation in the mouse models led to clinical trials in humans with AD. The main hypothesis underpinning the human vaccine is that Aβ accumulation in AD is abnormal and leads to the neurodegenerative processes underlying the cognitive decline seen in patients. In 2000, Elan Pharmaceuticals started a phase 1 clinical trial in the United Kingdom involving 80 patients with mild to moderate AD (Bayer et al. 2005). The vaccine, named AN1792, consisted of a synthetic form of Aβ42 and QS-21 (saponin) as an adjuvant. The primary aims of the phase 1 trial were to assess the safety, tolerability and immunogenicity of AN1792. Results showed that the vaccine was well tolerated and led to variable Aβ antibody production in patients (Bayer et al. 2005). A second clinical trial was started in 2001 involving 370 individuals with mild to moderate AD. This double-blind randomised controlled trial assigned 300 patients to AN-1792 and 70 patients to placebo in order to evaluate the clinical impact of immunisation. In January 2002, this trial was suspended after 4 individuals presented with suspected neuroinflammation, and by the end of February, a further 11 individuals were identified with the same clinical deterioration. In total, 18 patients (6%) had symptoms and laboratory findings consistent with meningoencephalitis (Orgogozo et al. 2003). The details of the trial were analysed and all individuals with encephalitis had been assigned AN1792, with no cases among the placebo patients. The majority of those affected had received 2 doses of AN1792 and presented on average 40 days after the second injection (i.e. booster). There was no correlation between antibody titres to Aβ and occurrence or severity of encephalitis, suggesting the antibodies themselves were not likely to be directly responsible. Follow-up studies of these individuals reported that 12 recovered back to normal; however, 6 of the 18 were left with disabling cognitive or neurological complications (Orgogozo et al. 2003). Autopsy results on 2 individuals with meningoencephalitis, who died from other causes, showed focal T cell infiltrates in the meninges, blood vessel walls and cerebral cortex (Nicoll et al. 2003; Ferrer et al. 2004). These patients also had diffuse white matter abnormalities with a decreased density of myelinated fibres and an extensive macrophage infiltrate (Nicoll et al. 2003; Ferrer et al. 2004), features that were not observed in a case without encephalitis (Masliah et al. 2005). The reason for the cases of T cell encephalitis remains unknown, although it has been suggested that the formulation used in the second clinical trial to stabilise the Aβ peptide might be responsible for this side effect (Schenk et al. 2004).

Further investigation of the AN1792 immunotherapy showed that the antibodies produced were selective for Aβ plaques, diffuse Aβ deposits and Aβ in the walls of blood in the brain (Hock et al. 2003). These antibodies were not produced in individuals who received a placebo, and they did not cross react with the APP protein or its derivatives including Aβ oligomers. This suggested that the main target of the antibodies was insoluble forms of Aβ; however, Hock noted that their assays may not have been sensitive enough to exclude the possibility of antibodies reacting with soluble Aβ (Hock et al. 2003). However, the immunostaining showed variability between immunised cases, suggesting variations in the patients' antibody affinity to Aβ. During the initial trial, only 50% of the patients had positive antibody titres at some point (Bayer et al. 2005), implying that for immunisation to occur, multiple injections are necessary to lead to a significant likelihood of an antibody response in AD.

Many reports have now been published describing the effects of immunisation on the pathology of AD patients (Nicoll et al. 2003, 2006; Ferrer et al. 2004; Masliah et al. 2005; Bombois et al. 2007; Boche et al. 2008), which showed evidence of a significant reduction in the load of Aβ plaques within the cerebral cortex. In some cases residual plaques showed an abnormal morphology, surrounded by activated microglia containing Aβ (Nicoll et al. 2006). A significant increase in the vascular load of Aβ (i.e. cerebral amyloid angiopathy (Boche et al. 2008)) suggested immunisation-mediated transport of Aβ from the plaques to the blood vessels. The areas of the cortex from which plaques had been removed showed evidence for removal of tau-containing plaque-associated dystrophic neurites even though tau in the form of NFTs and neuropil threads were still present (Nicoll et al. 2006).

The effects of immunisation-induced plaque removal on synaptic density have not yet been studied in detail in humans, but in Fig. 14.2 an example of synaptophysin immunostaining in immunised AD cortex is shown comparing an area in which plaques have been removed (Fig. 14.3f) and areas where the plaques remain (Fig. 14.3c). There is no clear difference, suggesting that plaque removal may not influence the synaptic density in immunised AD. In a sense this would not be surprising because the synaptic loss in AD is distributed widely throughout the cortex and not localised to the plaques. This can be seen from the figure; the location of the plaques seen in the Aβ-immunostained image (Fig. 14.3b) cannot be determined by examining the corresponding synaptophysin-immunostained image (Fig. 14.3c). The loss of synapses is not where the plaques are located; therefore, the presence of plaques in the cortex is not the direct cause of the synaptic loss. However, this pattern would be consistent with an effect on synapses of soluble forms of Aβ diffusing through the extracellular space, intraneuronal Aβ or non-Aβ-related mechanisms.

14.3.4 Immunisation of AD Patients: The Clinical Course

Limited data are available on the effect of Aβ immunisation in altering the clinical course of AD, but initial reports were encouraging. Bayer reported a significant difference in daily activity (DAD) scores but not change in mini-mental state examination (MMSE) scores (Bayer et al. 2005); whereas Gilman reported a significant difference in neurological test battery (NTB) scores but neither MMSE or DAD scores (Gilman et al. 2005). Further research on cognitive changes in response to immunisation was carried out by Hock (Hock et al. 2003). His results showed that immunised patients when compared with non-immunised controls

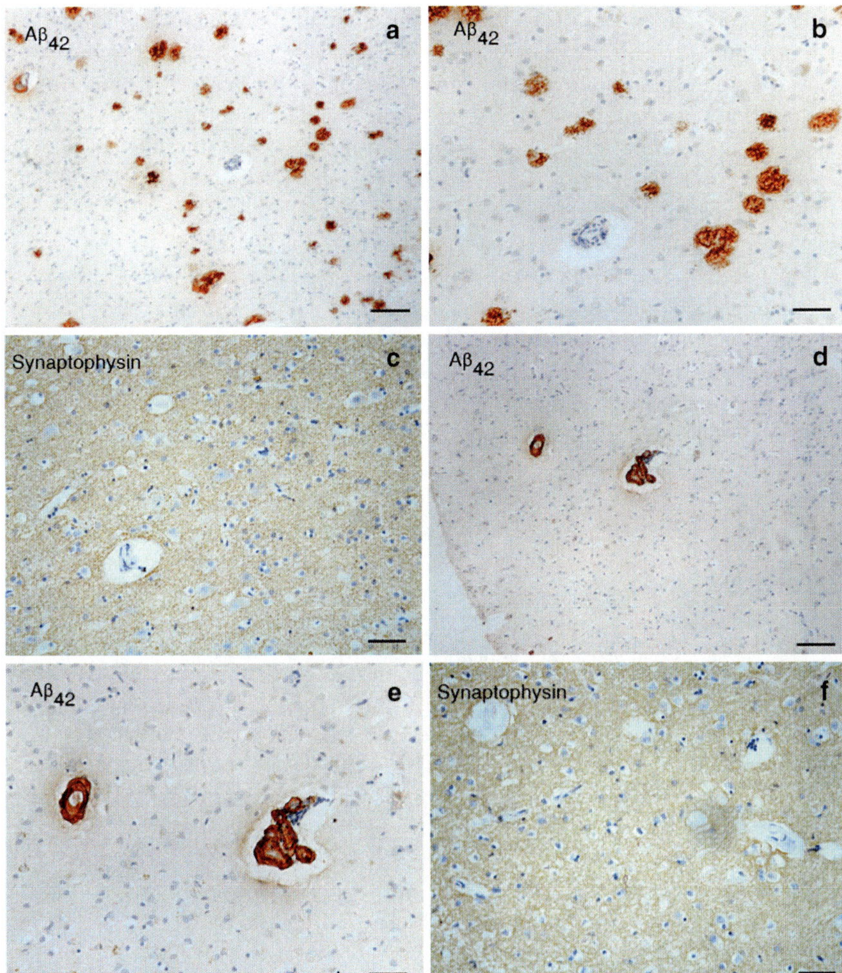

Fig. 14.3 *Synaptophysin staining vs. Aβ42 staining in human immunised AD brain.* Immunostaining for Aβ42 and synaptophysin in the cerebral cortex of an AD patient immunised against Aβ42. In this case, there has been patchy removal of the plaques, with the presence of areas with Aβ42 plaques (**a, b**) and without Aβ42 plaques (**d, e**). The synaptophysin immunostaining of the corresponding areas in the same brain (with or without Aβ42 plaques, respectively **c** and **f**) shows that (1) synaptophysin loss in AD is not localised to the plaques and (2) removal of plaques by immunisation does not result in an obvious increase in synaptophysin staining

showed less cognitive decline based on MMSE and DAD results. His results suggested increasing protection from decline occurred as the titres of serum antibodies to Aβ increased; however, there was no significant difference when he compared decline to affinity of antibodies to Aβ.

Our 6-year clinical and neuropathological study of the AD patients involved in the first clinical trial from Elan Pharmaceuticals confirmed that Aβ plaques are removed from the brain following the immunisation (as predicted by animal studies) but this did not prolong the survival time to death and did not halt the development of dementia. Therefore, it appears that immunisation did not stop the neurodegenerative process (Holmes et al. 2008). The active immunisation produced antibodies mainly directed against the N-terminal part of the Aβ peptide (Lee et al. 2005) and these antibodies have been observed to neutralise the LTP deficit observed in hippocampal mouse slice cultures exposed to extracted Aβ from human AD brains, by precipitating the soluble Aβ (Shankar et al. 2008). These findings are currently challenging the amyloid hypothesis that aggregated Aβ is the underlying cause of Alzheimer's disease (St George-Hyslop and Morris 2008), although data on oligomeric Aβ are currently lacking. The key specific feature or features of the multifaceted neurodegenerative pathology of AD which lead to synaptic dysfunction and dementia remain unknown.

References

Albert M S (1996). Cognitive and neurobiologic markers of early Alzheimer disease. Proc Natl Acad Sci U S A 93:13547–51

Almeida C G, Tampellini D, Takahashi R H, et al. (2005). Beta-amyloid accumulation in APP mutant neurons reduces PSD-95 and GluR1 in synapses. Neurobiol Dis 20:187–98

Arriagada P V, Marzloff K and Hyman B T (1992). Distribution of Alzheimer-type pathologic changes in nondemented elderly individuals matches the pattern in Alzheimer's disease. Neurology 42:1681–8

Association A P (2000). Diagnostic and Statistical Manual of Mental Disorders, Washington, DC

Bahmanyar S, Higgins G A, Goldgaber D, et al. (1987). Localization of amyloid beta protein messenger RNA in brains from patients with Alzheimer's disease. Science 237:77–80

Bard F, Barbour R, Cannon C, et al. (2003). Epitope and isotype specificities of antibodies to beta-amyloid peptide for protection against Alzheimer's disease-like neuropathology. Proc Natl Acad Sci U S A 100:2023–8

Bard F, Cannon C, Barbour R, et al. (2000). Peripherally administered antibodies against amyloid beta-peptide enter the central nervous system and reduce pathology in a mouse model of Alzheimer disease. Nat Med 6:916–9

Bayer A J, Bullock R, Jones R W, et al. (2005). Evaluation of the safety and immunogenicity of synthetic Abeta42 (AN1792) in patients with AD. Neurology 64:94–101

Boche D, Zotova E, Weller R O, et al. (2008). Consequence of A{beta} immunization on the vasculature of human Alzheimer's disease brain. Brain 131:3299–310

Bombois S, Maurage C A, Gompel M, et al. (2007). Absence of beta-amyloid deposits after immunization in Alzheimer disease with Lewy body dementia. Arch Neurol 64:583–7

Borchelt D R, Ratovitski T, van Lare J, et al. (1997). Accelerated amyloid deposition in the brains of transgenic mice coexpressing mutant presenilin 1 and amyloid precursor proteins. Neuron 19:939–45

Braak H and Braak E (1995). Staging of Alzheimer's disease-related neurofibrillary changes. Neurobiol Aging 16:271–8; discussion 278–84

Braak H, Braak E, Bohl J, et al. (1996). Age, neurofibrillary changes, A beta-amyloid and the onset of Alzheimer's disease. Neurosci Lett 210:87–90

Brown M S and Goldstein J L (1986). A receptor-mediated pathway for cholesterol homeostasis. Science 232:34–47

Buttini M, Masliah E, Barbour R, et al. (2005). Beta-amyloid immunotherapy prevents synaptic degeneration in a mouse model of Alzheimer's disease. J Neurosci 25:9096–101

Chartier-Harlin M C, Crawford F, Houlden H, et al. (1991). Early-onset Alzheimer's disease caused by mutations at codon 717 of the beta-amyloid precursor protein gene. Nature 353:844–6

Chauhan N B (2003). Membrane dynamics, cholesterol homeostasis, and Alzheimer's disease. J Lipid Res 44:2019–29

Citron M, Westaway D, Xia W, et al. (1997). Mutant presenilins of Alzheimer's disease increase production of 42-residue amyloid beta-protein in both transfected cells and transgenic mice. Nat Med 3:67–72

Davies C A, Mann D M, Sumpter P Q, et al. (1987). A quantitative morphometric analysis of the neuronal and synaptic content of the frontal and temporal cortex in patients with Alzheimer's disease. J Neurol Sci 78:151–64

de la Monte S M and Hedley-Whyte E T (1990). Small cerebral hemispheres in adults with Down's syndrome: contributions of developmental arrest and lesions of Alzheimer's disease. J Neuropathol Exp Neurol 49:509–20

DeMattos R B, Bales K R, Cummins D J, et al. (2001). Peripheral anti-A beta antibody alters CNS and plasma A beta clearance and decreases brain A beta burden in a mouse model of Alzheimer's disease. Proc Natl Acad Sci U S A 98:8850–5

Duff K, Eckman C, Zehr C, et al. (1996). Increased amyloid-beta42(43) in brains of mice expressing mutant presenilin 1. Nature 383:710–3

Elgersma Y and Silva A J (1999). Molecular mechanisms of synaptic plasticity and memory. Curr Opin Neurobiol 9:209–13

Ferrer I, Boada Rovira M, Sanchez Guerra M L, et al. (2004). Neuropathology and pathogenesis of encephalitis following amyloid-beta immunization in Alzheimer's disease. Brain Pathol 14:11–20

Games D, Adams D, Alessandrini R, et al. (1995). Alzheimer-type neuropathology in transgenic mice overexpressing V717F beta-amyloid precursor protein. Nature 373:523–7

Games D, Bard F, Grajeda H, et al. (2000). Prevention and reduction of AD-type pathology in PDAPP mice immunized with A beta 1–42. Ann N Y Acad Sci 920:274–84

Giedraitis V, Sundelof J, Irizarry M C, et al. (2007). The normal equilibrium between CSF and plasma amyloid beta levels is disrupted in Alzheimer's disease. Neurosci Lett 427:127–31

Gilman S, Koller M, Black R S, et al. (2005). Clinical effects of Abeta immunization (AN1792) in patients with AD in an interrupted trial. Neurology 64:1553–62

Goate A, Chartier-Harlin M C, Mullan M, et al. (1991). Segregation of a missense mutation in the amyloid precursor protein gene with familial Alzheimer's disease. Nature 349:704–6

Goedert M, Crowther R A and Spillantini M G (1998). Tau mutations cause frontotemporal dementias. Neuron 21:955–8

Graham D I and Lantos P L (2002). Greenfield's Neuropathology. London, Arnold

Grimm M O, Grimm H S, Patzold A J, et al. (2005). Regulation of cholesterol and sphingomyelin metabolism by amyloid-beta and presenilin. Nat Cell Biol 7:1118–23

Haass C and Selkoe D J (1993). Cellular processing of beta-amyloid precursor protein and the genesis of amyloid beta-peptide. Cell 75:1039–42

Hardy J, Duff K, Hardy K G, et al. (1998). Genetic dissection of Alzheimer's disease and related dementias: amyloid and its relationship to tau. Nat Neurosci 1:355–8

Hardy J and Selkoe D J (2002). The amyloid hypothesis of Alzheimer's disease: progress and problems on the road to therapeutics. Science 297:353–6

Hardy J A and Higgins G A (1992). Alzheimer's disease: the amyloid cascade hypothesis. Science 256:184–5

Hartley D M, Walsh D M, Ye C P, et al. (1999). Protofibrillar intermediates of amyloid beta-protein induce acute electrophysiological changes and progressive neurotoxicity in cortical neurons. J Neurosci 19:8876–84

Hock C, Konietzko U, Streffer J R, et al. (2003). Antibodies against beta-amyloid slow cognitive decline in Alzheimer's disease. Neuron 38:547–54

Holcomb L, Gordon M N, McGowan E, et al. (1998). Accelerated Alzheimer-type phenotype in transgenic mice carrying both mutant amyloid precursor protein and presenilin 1 transgenes. Nat Med 4:97–100

Holmes C, Boche D, Wilkinson D, et al. (2008). Long term effect of Abeta42 immunization in Alzheimer's disease: follow-up of a randomised, placebo-controlled phase I trial. Lancet 372:216–23

Hsia A Y, Masliah E, McConlogue L, et al. (1999). Plaque-independent disruption of neural circuits in Alzheimer's disease mouse models. Proc Natl Acad Sci U S A 96:3228–33

Hsiao K, Chapman P, Nilsen S, et al. (1996). Correlative memory deficits, Abeta elevation, and amyloid plaques in transgenic mice. Science 274:99–102

Jacobsen J S, Wu C C, Redwine J M, et al. (2006). Early-onset behavioral and synaptic deficits in a mouse model of Alzheimer's disease. Proc Natl Acad Sci U S A 103:5161–6

Janus C, Pearson J, McLaurin J, et al. (2000). Abeta peptide immunization reduces behavioural impairment and plaques in a model of Alzheimer's disease. Nature 408:979–82

Kang J, Lemaire H G, Unterbeck A, et al. (1987). The precursor of Alzheimer's disease amyloid A4 protein resembles a cell-surface receptor. Nature 325:733–6

Klyubin I, Betts V, Welzel A T, et al. (2008). Amyloid beta protein dimer-containing human CSF disrupts synaptic plasticity: prevention by systemic passive immunization. J Neurosci 28:4231–7

Lai, M K, Tsang S W, Garcia-Alloza, M, et al. (2006). Selective effects of the APOE epsilon4 allele on presynaptic cholinergic markers in the neocortex of Alzheimer's disease. Neurobiol Dis 22(3): 555–561

Lassmann H (1996). Patterns of synaptic and nerve cell pathology in Alzheimer's disease. Behav Brain Res 78:9–14

Lassmann H, Fischer P and Jellinger K (1993). Synaptic pathology of Alzheimer's disease. Ann N Y Acad Sci 695:59–64

Lee H G, Zhu X, Castellani R J, et al. (2007). Amyloid-beta in Alzheimer disease: the null versus the alternate hypotheses. J Pharmacol Exp Ther 321:823–9

Lee M, Bard F, Johnson-Wood K, et al. (2005). Abeta42 immunization in Alzheimer's disease generates Abeta N-terminal antibodies. Ann Neurol 58:430–5

Levy-Lahad E, Wasco W, Poorkaj P, et al. (1995). Candidate gene for the chromosome 1 familial Alzheimer's disease locus. Science 269:973–7

Lewis J, Dickson D W, Lin W L, et al. (2001). Enhanced neurofibrillary degeneration in transgenic mice expressing mutant tau and APP. Science 293:1487–91

Love S, Siew L K, Dawbarn D, et al. (2006). Premorbid effects of APOE on synaptic proteins in human temporal neocortex. Neurobiol Aging 27:797–803

Lue L F, Kuo Y M, Roher A E, et al. (1999). Soluble amyloid beta peptide concentration as a predictor of synaptic change in Alzheimer's disease. Am J Pathol 155:853–62

Luscher C, Nicoll R A, Malenka R C, et al. (2000). Synaptic plasticity and dynamic modulation of the postsynaptic membrane. Nat Neurosci 3:545–50

Mahley R W (1988). Apolipoprotein E: cholesterol transport protein with expanding role in cell biology. Science 240:622–30

Mann D M, Iwatsubo T, Ihara Y, et al. (1996). Predominant deposition of amyloid-beta 42(43) in plaques in cases of Alzheimer's disease and hereditary cerebral hemorrhage associated with mutations in the amyloid precursor protein gene. Am J Pathol 148:1257–66

Martin S J, Grimwood P D and Morris R G (2000). Synaptic plasticity and memory: an evaluation of the hypothesis. Annu Rev Neurosci 23:649–711

Masliah E (1995). Mechanisms of synaptic dysfunction in Alzheimer's disease. Histol Histopathol 10:509–19

Masliah E, Hansen L, Adame A, et al. (2005). Abeta vaccination effects on plaque pathology in the absence of encephalitis in Alzheimer disease. Neurology 64:129–31

Masliah E, Mallory M, Alford M, et al. (2001). Altered expression of synaptic proteins occurs early during progression of Alzheimer's disease. Neurology 56:127–9

Masliah E, Terry R D, Alford M, et al. (1991). Cortical and subcortical patterns of synaptophysin-like immunoreactivity in Alzheimer's disease. Am J Pathol 138:235–46

McLaurin J, Cecal R, Kierstead M E, et al. (2002). Therapeutically effective antibodies against amyloid-beta peptide target amyloid-beta residues 4–10 and inhibit cytotoxicity and fibrillogenesis. Nat Med 8:1263–9

Morgan D, Diamond D M, Gottschall P E, et al. (2000). A beta peptide vaccination prevents memory loss in an animal model of Alzheimer's disease. Nature 408:982–5

Mucke L, Masliah E, Yu G Q, et al. (2000). High-level neuronal expression of abeta 1–42 in wild-type human amyloid protein precursor transgenic mice: synaptotoxicity without plaque formation. J Neurosci 20:4050–8

Myers A, Holmans P, Marshall H, et al. (2000). Susceptibility locus for Alzheimer's disease on chromosome 10. Science 290:2304–5

Nagy Z, Esiri M M, Jobst K A, et al. (1995). Relative roles of plaques and tangles in the dementia of Alzheimer's disease: correlations using three sets of neuropathological criteria. Dementia 6:21–31

Naslund J, Haroutunian V, Mohs R, et al. (2000). Correlation between elevated levels of amyloid beta-peptide in the brain and cognitive decline. JAMA 283:1571–7

Nicoll J A, Barton E, Boche D, et al. (2006). Abeta Species Removal After Abeta42 Immunization. J Neuropathol Exp Neurol 65:1040–1048

Nicoll J A, Wilkinson D, Holmes C, et al. (2003). Neuropathology of human Alzheimer disease after immunization with amyloid-beta peptide: a case report. Nat Med 9:448–52

Oddo S, Billings L, Kesslak J P, et al. (2004). Abeta immunotherapy leads to clearance of early, but not late, hyperphosphorylated tau aggregates via the proteasome. Neuron 43:321–32

Oddo S, Caccamo A, Kitazawa M, et al. (2003a). Amyloid deposition precedes tangle formation in a triple transgenic model of Alzheimer's disease. Neurobiol Aging 24:1063–70

Oddo S, Caccamo A, Shepherd J D, et al. (2003b). Triple-transgenic model of Alzheimer's disease with plaques and tangles: intracellular Abeta and synaptic dysfunction. Neuron 39:409–21

Orgogozo J M, Gilman S, Dartigues J F, et al. (2003). Subacute meningoencephalitis in a subset of patients with AD after Abeta42 immunization. Neurology 61:46–54

Oyama F, Cairns N J, Shimada H, et al. (1994). Down's syndrome: up-regulation of beta-amyloid protein precursor and tau mRNAs and their defective coordination. J Neurochem 62:1062–6

Poirier J (2000). Apolipoprotein E and Alzheimer's disease. A role in amyloid catabolism. Ann N Y Acad Sci 924:81–90

Price J L and Morris J C (1999). Tangles and plaques in nondemented aging and "preclinical" Alzheimer's disease. Ann Neurol 45:358–68

Riddell D R, Zhou H, Atchison K, et al. (2008). Impact of apolipoprotein E (ApoE) polymorphism on brain ApoE levels. J Neurosci 28:11445–53

Ritchie K and Lovestone S (2002). The dementias. Lancet 360:1759–66

Roman F S, Truchet B, Marchetti E, et al. (1999). Correlations between electrophysiological observations of synaptic plasticity modifications and behavioral performance in mammals. Prog Neurobiol 58:61–87

Saunders A M, Strittmatter W J, Schmechel D, et al. (1993). Association of apolipoprotein E allele epsilon 4 with late-onset familial and sporadic Alzheimer's disease. Neurology 43:1467–72

Schenk D, Barbour R, Dunn W, et al. (1999). Immunization with amyloid-beta attenuates Alzheimer-disease-like pathology in the PDAPP mouse. Nature 400:173–7

Schenk D, Hagen M and Seubert P (2004). Current progress in beta-amyloid immunotherapy. Curr Opin Immunol 16:599–606

Scheuner D, Eckman C, Jensen M, et al. (1996). Secreted amyloid beta-protein similar to that in the senile plaques of Alzheimer's disease is increased in vivo by the presenilin 1 and 2 and APP mutations linked to familial Alzheimer's disease. Nat Med 2:864–70

Selkoe D J (1991). The molecular pathology of Alzheimer's disease. Neuron 6:487–98

Selkoe D J (2001). Alzheimer's disease: genes, proteins, and therapy. Physiol Rev 81:741–66

Selkoe D J (2002). Alzheimer's disease is a synaptic failure. Science 298:789–91

Shankar G M, Bloodgood B L, Townsend M, et al. (2007). Natural oligomers of the Alzheimer amyloid-beta protein induce reversible synapse loss by modulating an NMDA-type glutamate receptor-dependent signaling pathway. J Neurosci 27:2866–75

Shankar G M, Li S, Mehta T H, et al. (2008). Amyloid-beta protein dimers isolated directly from Alzheimer's brains impair synaptic plasticity and memory. Nat Med 14:837–42

Sherrington R, Rogaev E I, Liang Y, et al. (1995). Cloning of a gene bearing missense mutations in early-onset familial Alzheimer's disease. Nature 375:754–60

Small D H, Mok S S and Bornstein J C (2001). Alzheimer's disease and Abeta toxicity: from top to bottom. Nat Rev Neurosci 2:595–8

Small G W, Rabins P V, Barry P P, et al. (1997). Diagnosis and treatment of Alzheimer disease and related disorders. Consensus statement of the American Association for Geriatric Psychiatry, the Alzheimer's Association, and the American Geriatrics Society. JAMA 278:1363–71

Small S A and Duff K (2008). Linking Abeta and tau in late-onset Alzheimer's disease: a dual pathway hypothesis. Neuron 60:534–42

Snyder E M, Nong Y, Almeida C G, et al. (2005). Regulation of NMDA receptor trafficking by amyloid-beta. Nat Neurosci 8:1051–8

Solomon B, Koppel R, Frankel D, et al. (1997). Disaggregation of Alzheimer beta-amyloid by site-directed mAb. Proc Natl Acad Sci U S A 94:4109–12

Solomon B, Koppel R, Hanan E, et al. (1996). Monoclonal antibodies inhibit in vitro fibrillar aggregation of the Alzheimer beta-amyloid peptide. Proc Natl Acad Sci U S A 93:452–5

Spillantini M G and Goedert M (1998). Tau protein pathology in neurodegenerative diseases. Trends Neurosci 21:428–33

St George-Hyslop P H and Morris J C (2008). Will anti-amyloid therapies work for Alzheimer's disease? Lancet 372:180–2

St George-Hyslop P H, Tanzi R E, Polinsky R J, et al. (1987). The genetic defect causing familial Alzheimer's disease maps on chromosome 21. Science 235:885–90

Strittmatter W J, Saunders A M, Schmechel D, et al. (1993). Apolipoprotein E: high-avidity binding to beta-amyloid and increased frequency of type 4 allele in late-onset familial Alzheimer disease. Proc Natl Acad Sci U S A 90:1977–81

Sze C I, Troncoso J C, Kawas C, et al. (1997). Loss of the presynaptic vesicle protein synaptophysin in hippocampus correlates with cognitive decline in Alzheimer disease. J Neuropathol Exp Neurol 56:933–44

Terry R D (1996). The pathogenesis of Alzheimer disease: an alternative to the amyloid hypothesis. J Neuropathol Exp Neurol 55:1023–5

Terry R D, Masliah E, Salmon D P, et al. (1991). Physical basis of cognitive alterations in Alzheimer's disease: synapse loss is the major correlate of cognitive impairment. Ann Neurol 30:572–80

Uylings H B and de Brabander J M (2002). Neuronal changes in normal human aging and Alzheimer's disease. Brain Cogn 49:268–76

Varadarajan S, Kanski J, Aksenova M, et al. (2001). Different mechanisms of oxidative stress and neurotoxicity for Alzheimer's A beta(1–42) and A beta(25–35). J Am Chem Soc 123:5625–31

Varadarajan S, Yatin S, Kanski J, et al. (1999). Methionine residue 35 is important in amyloid beta-peptide-associated free radical oxidative stress. Brain Res Bull 50:133–41

Walsh D M, Klyubin I, Fadeeva J V, et al. (2002). Naturally secreted oligomers of amyloid beta protein potently inhibit hippocampal long-term potentiation in vivo. Nature 416:535–9

Wang H W, Pasternak J F, Kuo H, et al. (2002). Soluble oligomers of beta amyloid (1–42) inhibit long-term potentiation but not long-term depression in rat dentate gyrus. Brain Res 924:133–40

Wavrant-DeVrieze F, Lambert J C, Stas L, et al. (1999). Association between coding variability in the LRP gene and the risk of late-onset Alzheimer's disease. Hum Genet 104:432–4

Weller R O and Nicoll J A (2003). Cerebral amyloid angiopathy: pathogenesis and effects on the ageing and Alzheimer brain. Neurol Res 25:611–6

Weller R O, Subash M, Preston S D, et al. (2008). Perivascular drainage of amyloid from the brain and its failure in cerebral amyloid angiopathy and Alzheimer's disease. Brain Pathol 18:253–66

Wilcock D M, DiCarlo G, Henderson D, et al. (2003). Intracranially administered anti-Abeta antibodies reduce beta-amyloid deposition by mechanisms both independent of and associated with microglial activation. J Neurosci 23:3745–51

Wilcock D M, Munireddy S K, Rosenthal A, et al. (2004). Microglial activation facilitates Abeta plaque removal following intracranial anti-Abeta antibody administration. Neurobiol Dis 15:11–20

Wisniewski K E, Wisniewski H M and Wen G Y (1985). Occurrence of neuropathological changes and dementia of Alzheimer's disease in Down's syndrome. Ann Neurol 17:278–82

Yatin S M, Varadarajan S, Link C D, et al. (1999). In vitro and in vivo oxidative stress associated with Alzheimer's amyloid beta-peptide (1–42). Neurobiol Aging 20: 325–30; discussion 339–42

Ye C P, Selkoe D J and Hartley D M (2003). Protofibrils of amyloid beta-protein inhibit specific K+ currents in neocortical cultures. Neurobiol Dis 13:177–90

Chapter 15
Prion Protein Misfolding at the Synapse

Zuzana Šišková, V. Hugh Perry, and Ayodeji A. Asuni

Abstract The synapse has emerged as a major target for the misfolding insults that underlie prion disease and many other proteinopathies (e.g., Alzheimer's disease (AD)). This common theme in the pathogenesis of these disorders indicates that analogous degenerative processes could be at play when increasing extracellular and/or intracellular accumulation of misfolded proteins leads to eventual cell loss. Similar therapeutic strategies may thus be effective in various central nervous system amyloidoses. Animal models of prion disease provide good evidence for specific synaptic degeneration within defined anatomical pathways of the hippocampus. Biochemical, histological, and electron microscopy studies have documented disintegrating synaptic structures during the early asymptomatic stage of disease, which has lead to the hypothesis that degenerative pathways are engaged locally at the synapse during an early key stage of neurodegeneration. Mirroring this, synapse loss precedes neuronal loss in early AD, and is more closely correlated with cognitive impairment than are plaques and tangles. As in other protein misfolding neurodegenerative disorders, it is likely that in prion disease, pathological prion protein conformers are present and actively participate in disease pathogenesis at the synapse. Despite this fundamental understanding, there has been little systematic study of the evidence for pathological accumulation of prion protein in either the presynaptic or postsynaptic specializations, or indeed the role of cellular pathways and synaptic proteins associated with these pathologies. This chapter will review key signaling pathways and processes implicated in biochemical changes that misfolded prion protein triggers at the synapse. Knowing what these changes are may well lead to new drug targets that would then enable us to prevent neuronal cell loss.

Z. Šišková (✉)
School of Biological Sciences, Mailpoint 840 LD 80B, South Lab and Path Block, Southampton General Hospital, Southampton, SO16 6YD, UK
e-mail: z.siskova@soton.ac.uk

A. Wyttenbach and V. O'Connor (eds.), *Folding for the Synapse*,
DOI 10.1007/978-1-4419-7061-9_15,

15.1 Prion Diseases

Prion disease or transmissible spongiform encephalopathies (Table 15.1) are characterized by similar clinical and neuropathological changes, generally exhibiting symptoms of cognitive and motor dysfunction. Prion diseases are progressive, invariably fatal, and presently incurable. The hallmark of all these diseases, whether sporadic, inherited, or acquired, is that they involve the aberrant conversion of the host-encoded cellular prion protein (PrP^{C}) into scrapie prion protein (PrP^{Sc}) isoform and has potential for further propagation. In many instances, amyloid plaques containing PrP^{Sc} aggregates are formed (Prusiner 1997). The presence of PrP^{Sc} undeniably accompanies neurodegeneration in prion diseases; however, several lines of evidence strongly indicate that its effect is only neurotoxic if native, host-encoded PrP^{C} is expressed by neuronal cells, further supported by the observation that $PrP^{-/-}$ mice are resistant to the development of the pathology (Bueler et al. 1993).

15.1.1 The Cellular Prion Protein: Genetic Background, Structural and Biochemical Properties

PrP^{C} is the product of a single-copy gene PRNP first identified by Basler et al. (1986). The human PRNP is on chromosome 20 within the prion protein locus that contains two more genes: DOPPEL and PRNT. All of these genes appear to be evolutionarily related, although their low sequence homology implies that they could be functionally distinct. PRNP expression is widespread and developmentally regulated (Miele et al. 2003) with high levels of expression in the central nervous system (CNS) (Aguzzi and Polymenidou 2004). DOPPEL is also widely expressed in fetal tissues, but is only found in testis throughout adulthood and PRNT expression is restricted to adult testis (Makrinou et al. 2002). Structural studies of human recombinant prion protein revealed a structured globular domain (C-terminus) and an N-terminal flexible region. The N-terminal portion of the protein that appears mostly unstructured consists of two well-defined regions. The first octapeptide repeat region contains five repeats of an octameric amino acid sequence (Riek et al. 1997), and in vitro studies identified it as the binding site for several metal ions including copper (Stockel et al. 1998; Brockes 1999) (reviewed in Vassallo and Herms 2003). The second region consists of a highly hydrophobic and conserved segment that was initially named the transmembrane region. A secretory signal peptide is located at the extreme N-terminus of the protein (residues 1–22). PrP^{C} can have either one or two N-linked complex glycosylation sites linked to the C-terminus of the protein, but can exist in a non-glycosylated form (Fig. 15.1). PrP^{C} is a glycosyl phosphatidyl inositol (GPI)-linked extracellular membrane protein, typically enriched in detergent-resistant membranes.

Table 15.1 Overview of the various prion diseases

Disease abbreviation	Natural host	Year and country of first report	Proposed mechanism of pathogenesis
Scrapie	Sheep and goats	1732	Infection in genetically susceptible sheep
Transmissible mink encephalopathy (TME)	Mink	1947	Infection from sheep or cattle
Chronic wasting disease (CWD)	Mule deer and elk	1980	Unknown
Bovine spongiform encephalopathy (BSE)	Cattle	1986	Infection with contaminated meat and bone meal (MBM)
Exotic ungulate encephalopathy (EUE)	Nyala, greater kudu and oryx	1986	Infection with contaminated MBM
Feline spongiform encephalopathy (FSE)	Cats	1986	Infection with contaminated MBM
Sporadic Creutzfeldt-Jakob disease (CJD)	Humans	1920, Germany	Somatic mutations or spontaneous conversion of PrP^C
Familial CJD	Humans	1924, Germany	Germline mutations in PRNP gene
Gerstmann-Sträussler-Scheinker disease (GSS)	Humans	1928, Austria	Germline mutations in PRNP gene
Kuru	Humans (Fore people)	1957, New Guinea	Infection through ritualistic cannibalism
Latrogenic CJD	Humans	1974, USA	Infection from contaminated human growth hormone, dura mater grafts
Fatal familial insomnia (FFI)	Humans	1986, Italy	Germline mutations in PRNP gene (D178N and M129)
Variant CJD (vCJD)	Humans	1996, UK	Infection from bovine prions
Sporadic fatal insomnia	Humans	1999, USA	Unknown, not associated with germline mutations in PRNP gene (D178N and M129)

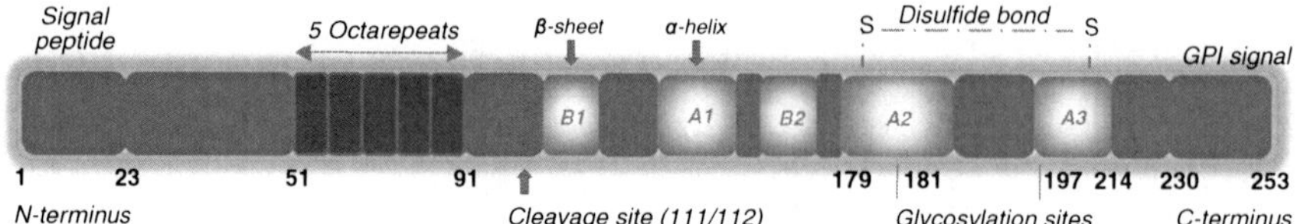

Fig. 15.1 Structure of cellular prion protein. The C-terminal domain of the protein is structured, containing α-helical and β-sheeted regions in contrast to more flexible N-terminal region (from amino acids 23–160). The N-terminal signal peptide (1–23) is cleaved off in the endoplasmic reticulum. GPI anchor attachment on the serine 230 is dependent on the cleavage of the C-terminal hydrophobic signal peptide. The protein has two glycosylation sites (residues 181 and 197) and disulfide bond linking 179 and 214. The five octapeptide repeats region (amino acids 51–91) represents the most conserved region of the protein, normally cleaved at 111/112 throughout its metabolic cycle

15.1.2 The Physiological Functions of PrP^C

PrP^C is highly expressed both in the developing and mature nervous system, mainly in synaptic membranes of neuronal cells and in astrocytes (Moser et al. 1995). Both PrP^C and its mRNA are readily detectable in the cerebellum, cerebral cortex, thalamus, hippocampus, and medulla oblongata (Makrinou et al. 2002; McLennan et al. 2004). In addition to cells of the CNS, cells within the immune system also constitutively express PrP^C. Antigen-presenting cells such as dendritic cells and monocytes express high levels of PrP^C (Burthem et al. 2001), and PrP^C is abundantly present on T lymphocytes, particularly on the CD8$^+$T subset (Durig et al. 2000; Politopoulou et al. 2000). This is also consistent with the involvement of the lymphoreticular system in prion accumulation and development of prion disease (Klein et al. 1997; Montrasio et al. 2000; Prinz et al. 2003). Several physiological functions of PrP^C within the nervous system will be discussed hence.

15.1.2.1 Neuroprotective Functions

Initial studies on different lines of PrP$^{-/-}$ mice yielded contradictory results with regard to the physiological function of PrP^C. Lines generated with modifications only in the open reading frame (Bueler et al. 1993; Bueler et al. 1992; Manson et al. 1994) produced animals that developed no detectable pathologies and were resistant to prion infection. In contrast to these findings, other PrP$^{-/-}$ lines developed ataxia and loss of Purkinje cells (Sakaguchi et al. 1996; Moore et al. 1999). Initially, these findings pointed to a role for PrP^C in the development or survival of Purkinje cells, but this was interpreted differently after the discovery of DOPPEL or PRND localized downstream from PRNP gene. In all the three lines that developed ataxia and neuronal pathology, chimeric transcripts of DOPPEL under control of PRNP promoter were formed as result of genetic manipulation, subsequently leading to overexpression of Doppel protein in the brain that could increase oxidative damage in the absence of PrP^C (Sakaguchi et al. 1996; Moore et al. 1999). Recently, a study of transgenic mice that express a form of murine PrP^{C} that lacks a conserved block of 21 amino acids (residues 105–125) in the unstructured, N-terminal domain of the protein, described development of lethal, severe neurodegenerative disease within one week of birth. Importantly, this phenotype could be reversed by co-expression of wild-type PrP^{C} (Li et al. 2007).

PrP^{C} has been proposed to play an anti-apoptotic role through interacting with BAX, the pro-apoptotic gene playing an important role in death signaling in post-mitotic neurons of the CNS (Yuan and Yankner 2000). Several studies indicated that PrP^{C} could be a potent inhibitor of Bax-mediated release of cytochrome *c* from mitochondria and subsequent cell death (Roucou et al. 2003; Roucou and LeBlanc 2005). A further role has been proposed in the regulation of the activity of an excitatory ionotropic glutamate receptor N-methyl-d-aspartate receptor (NMDA-R) subtype and hence control of excitotoxic lesions (Khosravani et al. 2008). These suggested functions are consistent with the observation that the infarct volume following an ischemic challenge in PrP$^{-/-}$ mice is substantially larger when compared to that in wild-type mice (McLennan et al. 2004) in which PrP^{C} levels normally increase after ischemia (Weise et al. 2004).

15.1.2.2 A Role for PrP^{C} in Protection Against Oxidative Stress?

In vitro studies provided the first hints that PrP^{C} could have an antioxidative effect. For example, neurons lacking PrP^{C} were more vulnerable to hydrogen peroxide toxicity than wild-type cells, and increased susceptibility to peroxide corresponded to a significant lower activity of glutathione peroxidase (GPX) in these neurons (White et al. 1999). Prior studies suggested that PrP^{C} could have a role in the transport of copper and zinc required for antioxidant enzymes and could, therefore, influence their activity (Brown and Besinger 1998). It has been proposed that PrP^{C} itself may have superoxide dismutase activity that could mediate its antioxidative function (Brown et al. 1999) but other studies failed to detect such activity (Hutter et al. 2003; Jones et al. 2005). However, PrP$^{-/-}$ mice exhibited higher levels of oxidative damage to both protein and lipids in the brain when compared with wild-type controls (Klamt et al. 2001). Reduced numbers of mitochondria and alterations in their morphology have been described in mice lacking PrP^{C} (Miele et al. 2002).

15.1.2.3 Involvement of PrP^{C} in Neuronal Development, Differentiation, and Neurite Outgrowth

PrP^{C} is abundantly expressed in the developing brain, its expression levels correlated in a dose-dependent manner with the differentiation of multipotent neural precursors into mature neurons in vitro, and PrP^{C} also increased cellular proliferation in vivo (Steele et al. 2006). These results suggested that PrP^{C} could have a role in proliferation and differentiation of neuronal precursor cells during development and neurogenesis in the adulthood (Steele et al. 2006). Several signal transduction pathways, including the non-receptor Src-related family member p59^{Fyn}, PI3 kinase/Akt, cAMP-dependent protein kinase-A (PKA), and MAP kinase, have been implicated in neurite outgrowth elicited by binding to PrP^C (Chen et al. 2003). Activation of $p59^{Fyn}$ kinase in this context appears to be dependent on interactions between neural cell adhesion molecule (NCAM) and PrP^C in the lipid rafts. When these interactions were perturbed, NCAM/PrP^C-dependent neurite outgrowth was arrested (Santuccione et al. 2005). However, a critical role of PrP^C in neuronal development may be regarded as not essential since $PrP^{-/-}$ mice are relatively, if not almost completely, normal (see above) apart from abnormalities related to the circadian activity rhythms (Tobler et al. 1996).

15.1.2.4 PrP^C (at the Synapse): Functional Implications

PrP^C localizes to presynaptic membranes, synaptic vesicles (Fournier et al. 1995; Gohel et al. 1999; Moya et al. 2000), the postsynaptic membrane (Godsave et al. 2008), and a number of compartments of the secretory pathway (Laine et al. 2001). Although PrP^C was found with the same frequency within the synaptic specialization and peri-synaptically, it was almost completely excluded from synaptic vesicles (Godsave et al. 2008). Unexpectedly, PrP^C was also found in the cytosol in subpopulations of neurons in the hippocampus, neocortex, thalamus, and cerebellum (Ford et al. 2002). Cytosolic PrP^C could have altered susceptibility to aggregation, suggesting that these neurons might play a significant role in the pathogenesis of prion diseases (Mironov et al. 2003).

The view that PrP^C contributes to synaptic signaling is further supported by electrophysiological studies of hippocampal neurons in $PrP^{-/-}$ mice. Several studies have drawn attention to alterations in $GABA_A$ receptor-mediated currents, glutamatergic synaptic transmission, and long-term potentiation (Carleton et al. 2001; Collinge et al. 1994; Mallucci et al. 2002). However, other authors failed to detect such changes (Lledo et al. 1996) or documented increases in glutamatergic synaptic transmission in $PrP^{-/-}$ mice, whereas decreases have previously been reported (Maglio et al. 2004). Another line of evidence supporting a role for PrP^C in synaptic signaling has come from mice with selective deletion of PrP^C that exhibited deficits in hippocampal-dependent spatial learning, together with reduction in paired-pulse facilitation and long-term potentiation in the dentate gyrus (Criado et al. 2005).

The neuromuscular junction is another site where PrP^C is highly expressed (Gohel et al. 1999). In a mouse phrenic-diaphragm preparation, nanomolar concentrations of recombinant PrP^C were able to induce potentiation of the acetylcholine release. The effect was mainly presynaptic with an increase in the amplitude of the miniature end-plate currents, and facilitation of synaptic transmission was also noted (Re et al. 2006). Exposure of rat fetal hippocampal neurons to recombinant PrP^C resulted in increased formation of synaptic-like contacts after 7 days in culture indicating that synapse formation could be regulated by PrP^C (Kanaani et al. 2005).

Thus there are several aspects of synaptic function in which PrP^C could exert its role: (i) via direct interaction with synaptic proteins, (ii) a capacity to participate in cellular signaling, (iii) endocytosis, (iv) cellular adhesion, and (v) PrP^C and copper interactions (Fig. 15.2). These possibilities are considered below.

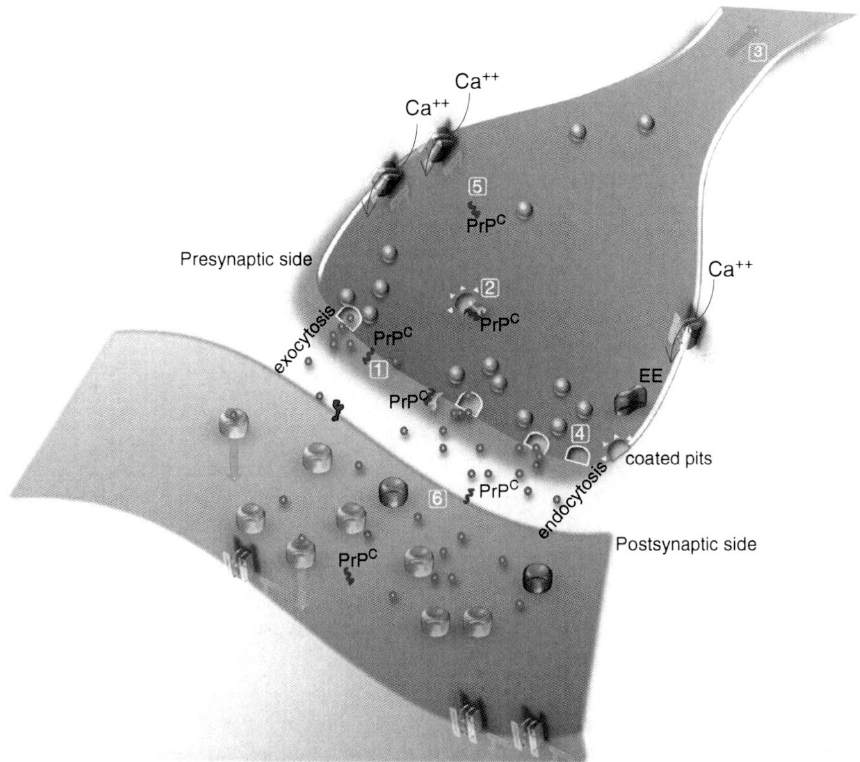

Fig. 15.2 Putative functions of PrP^C. PrP^C functions likely are exerted via distinctly localized PrP^C forms. Following its synthesis within the cell body, PrP^C is subsequently transported via the anterograde transportation machinery operating within axons and anchored into the plasma membrane on the presynaptic side of synaptic junctions (1). Furthermore, PrP^C can be positioned on the opposite, postsynaptic side where is it expressed either constitutively or transported (6). The PrP^C endocytic capacity depends on clathrin coated-pits, recruiting dynamin, rab proteins, and Grb2 (4). Next, PrP^C is internalized after specific binding with the cell membrane or extracellular receptors (2) that further have the capacity to facilitate cellular signaling that could lead to neuroprotection and other putative functions (3). Another function for PrP^C has been proposed in regulating neurotransmission mainly due to its membrane-independent, cytosolic form, potentially interacting with synapsin 1 and synaptophysin, and thus could imply involvement in the process of synaptic vesicular recycling (5)

Interactions of PrP^C with synaptic proteins: A few synaptic proteins have been suggested as candidates for interactions with PrP^C and interestingly, all of them play a role in synaptic vesicle function. Synapsins are a family of highly conserved proteins involved in the formation of synapses and in the regulation of neurotransmitter release (De Camilli et al. 1990). Their putative function is to cross-link synaptic vesicles to each other and to cytoskeletal proteins, thereby providing a control of vesicle release. Phosphorylation by a range of kinases is thought to regulate this process. Synapsin 1b protein is proposed to be a direct interacting partner with PrP^C, as determined by yeast two-hybrid screen and co-immunoprecipitation (Spielhaupter and Schatzl 2001). The C-terminus of synapsin 1b can be phosphorylated by calcium/calmodulin kinase II, also associated with synaptic vesicles (Greengard et al. 1993). The proposed PrP^{C}-interacting region of synapsin 1b contains some of these phosphorylation sites and it was suggested that their interaction could be regulated by phosphorylation, and in vitro recombinant PrP^{C} was shown to be the template for phosphorylation (Negro et al. 2000). This is further supported by tissue co-distribution of both proteins and from cell fractionation experiments where both synapsin 1b and PrP^{C} were detected in Golgi fractions (Spielhaupter and Schatzl 2001), in synaptic vesicles, and in presynaptic membranes (Gohel et al. 1999).

Similarly, synaptophysin, a major presynaptic transmembrane protein localized at synaptic vesicles (Hergersberg et al. 2000; Jahn and Sudhof 1994), and gamma-aminobutyricacid decarboxylase (GAD), an enzyme that synthesizes the neurotransmitter molecule GABA from glutamate, co-immunoprecipitated with PrP^{C} in the same raft fractions (Keshet et al. 2000).

A different approach using a chemical cross-linker, followed by co-immunoprecipitation was applied to identify PrP^{C}-interacting proteins. This approach yielded characterization of two further potential partners – syntaxin-binding protein 1 (STXBP1) and brain abundant signal protein 1 (BASP1) (Petrakis and Sklaviadis 2006). STXBP1 and BASP1 are mainly localized in lipid rafts and participate in signal transduction by regulating neurotransmitter release and synaptic function. STXBP1 is membrane associated in the presence of syntaxin (Swanson et al. 1998; Zakharov et al. 2003).

Cellular signaling: In many cell types, the attachment of a protein to the membrane by a GPI anchor and its localization to lipid rafts are indications of involvement in cellular signaling (Shmerling et al. 1998), but few studies have addressed specific targets of GPI-mediated signaling of PrP^{C}. Infection of GPI-negative PrP^{C} transgenic mice resulted in transmissible amyloid disease without clinical symptoms of scrapie pathology, suggesting that the GPI anchor might contribute to pathogenesis (Chesebro et al. 2005; Rogers et al. 1993). Modulation of PrP^{C} signaling capacity by cross-linking or interaction with another protein leads to changes in several downstream signaling pathways, such as cAMP, Akt, Fyn, and Erk1/2 (Mouillet-Richard et al. 2000; Schneider et al. 2003). Antibody-mediated cross-linking of PrP^{C} resulted in the activation of p59^{Fyn} in neuronal cells, mediated downstream by Erk1/2 phosphorylation (Mouillet-Richard et al. 2000; Schneider et al. 2003). Additionally, PrP^{C} cross-linking was reported to modulate the function of serotonergic receptors in neuronal cells (Mouillet-Richard et al. 2005). Cross-linking of PrP^{C} using three different purified, endotoxin-free, PrP^{C}-specific monoclonal antibodies triggered rapid and extensive apoptosis in hippocampal and cerebellar neurons in vivo. The authors hypothesized that antibody binding could interrupt a PrP^{C}-dependent neuronal survival signaling mediated through an association between PrP^{C} and another molecule (Solforosi et al. 2004). Signals elicited by interactions of PrP^{C} with stress-inducible protein I (STI1) provided protection from apoptosis via cAMP/PKA and Erk signaling (Chiarini et al. 2002; Zanata et al. 2002). Additionally, interaction with STI1 induced neuritogenesis by activation of the MAPK pathway (Lopes et al. 2005). Signaling transduction via cAMP/

PKA and Erk pathways was suggested to co-regulate both neurite outgrowth and neuronal survival (Chen et al. 2003; Santuccione et al. 2005).

Endocytosis: Endocytosis is an important aspect of synaptic activity. A growing body of data supports the idea that PrP^C exerts its functions through endocytosis. Endocytosis plays a crucial role in PrP^C trafficking: here, involvement of coated-pits, dynamin, rab family proteins, and $p59^{Fyn}$ have been implicated (Campana et al. 2005; Fischer von Mollard et al. 1994; Magalhaes et al. 2002). Endocytosis of PrP^C might represent a mechanism for the down-regulation of PrP^C on synaptic membranes or could additionally lead to co-internalization of another interacting partner. This could further modulate the activity of signal transduction pathways at the synapse by degradation of co-internalized molecules. PrP^C is rapidly internalized from the cell membrane, although the exact mechanism remains controversial because both caveolae (Kaneko et al. 1997; Vey et al. 1996) and clathrin-dependent endocytosis have been described (Taylor et al. 2005). PrP^C endocytosis could be mediated by signaling of low-density lipoprotein receptor-related protein 1 (LRP1)–PrP^C complex (Taylor and Hooper 2007). Conversely, recent studies suggested that a conserved macropinocytic mechanism could be involved in PrP^C uptake (Wadia et al. 2008).

Cellular adhesion: A role for PrP^C in cellular adhesion has been suggested by several studies. Tantalizingly, all cell surface molecules that show affinity to PrP^C, such as NCAM (Schmitt-Ulms et al. 2001), laminin (Graner et al. 2000), laminin receptor precursor (Gauczynski et al. 2001), LRP1 (Taylor and Hooper 2007), and glycans (Pan et al. 2002), are known to have a synaptic localization (Ushkaryov et al. 1992; Waites et al. 2005; Zhai and Bellen 2004). It was also proposed that PrP^C could stabilize opposing synaptic membranes through these adhesive interactions (Sales et al. 1998).

Prion protein and copper: Copper is an essential transitional metal that is required for the activity of multiple mammalian enzymes (Macreadie 2008; O'Dell 1976). Copper is present in the synaptic cleft of glutaminergic neurons along with zinc (Ward et al. 1998) at concentrations of 100–250 μM and 300 μM, respectively (Kardos et al. 1989; Watt and Hooper 2003). Copper has been associated with prion disease (Brown 2002), and PrP^C is able to bind copper primarily within the octarepeat region in vivo (Jackson et al. 2001; Stockel et al. 1998), as well as along the C-terminal domain of protein fragments (Cereghetti et al. 2001). The possible synaptic role of PrP^C has been exemplified by electrophysiological findings in $PrP^{-/-}$ animals as outlined above, and it was suggested that PrP^C could regulate the copper content of the synaptic cleft. The latter property of PrP^C may also endow PrP^C with the activity of a complex copper-dependent superoxide dismutase (Daniels and Brown 2002; Lasmezas 2003) and may be involved in the normal function of PrP^C (Brown et al. 1999). These observations have led to the suggestion that the reduced copper binding of PrP^{Sc} and associated reduction of antioxidant activity are a part of the pathogenesis of prion disease (Brown 2001). This is further supported by the finding that copper is reduced up to 50% in the brains of sporadic CJD patients (Wong et al. 2001). Despite these observations, some researchers have failed to demonstrate SOD activity associated with prion (Hutter et al. 2003; Jones et al. 2005). The loss of copper from the CJD brain is not altogether surprising given the extensive neuronal and synaptic loss.

15.1.3 Conformational Changes of Native PrP^C and Propagation of Its Misfolded Isoform

The physiological function of a protein can only be fulfilled once a protein adopts its appropriate conformation. Multiple lines of evidence show that accumulation of misfolded proteins leads to synaptic dysfunction, neuronal apoptosis and brain damage (reviewed in Soto and Estrada 2008). Conversion of native, membrane-anchored cellular prion protein (PrP^C) into its infectious scrapie prion protein isoform mostly named as PrP^{Sc}, or PK-resistant PrP (PrP^{res}) is believed to constitute an essential step in prion diseases (Prusiner 1991). It was suggested by some that this conversion is a posttranslational modification of a folding process without the involvement of any covalent modification (Stahl et al. 1993). PrP^{Sc} has different physicochemical properties, characterized by an increase in β-sheet conformation, insolubility in non-denaturing detergents, and partial resistance against digestion with proteinase K (PK) when compared to PrP^C (Caughey et al. 1991; McKinley et al. 1983; Safar et al. 1993). Molecules that might promote PrP^C conversion to PrP^{Sc} and further propagation of its misfolded isoform have been sought extensively. In contrast to native yeast, in which a role for molecular chaperones has been strongly implicated during prion propagation (Tuite and Cox 2003), in mammalian systems evidence for a PrP chaperone or any binding partner is missing (Telling et al. 1995). Endogenous RNA molecules have the potential to act as cellular cofactors for PrP^{Sc} conversion (Adler et al. 2003) and it has been reported that RNA is able to enhance the stochiometric conversion of PrP^C to PrP^{Sc} in vitro (Deleault et al. 2003).

15.1.4 PrP^{Sc} (at the Synapse): Pathophysiological Implications

The conversion of PrP^C to the pathological conformer, PrP^{Sc}, is commonly associated with disease and neurodegeneration (Aguzzi et al. 2007; Mallucci et al. 2003), leading to the notion that PrP^{Sc} is neurotoxic, but accumulating evidence shows neurodegeneration in a prion model in the absence of PrP^{Sc} (Barron et al. 2007; Lasmezas et al. 1997; Manson et al. 1999; Piccardo et al. 2007). This evidence suggests that PrP^{Sc} is neither necessary nor sufficient for neurodegeneration. Furthermore, although the presence of PrP^{Sc} can often accompany neurodegeneration in prion diseases, several lines of evidence strongly indicate that its effect is only neurotoxic if native, host-encoded PrP^C is expressed on neuronal cells, further supported by observation that PrP null mice are resistant to development of the pathology (Bueler et al. 1993). PrP^{Sc} accumulates in prion-affected individuals in the form of amorphous aggregates, but what is also clear is that in addition to PrP^{Sc} propagation in the brain, other factors such as glycosylation contribute to ensuing neurodegeneration, as evidenced by studies in prion models with altered glycosylation (Rudd et al. 2001).

The central question is then whether this is either a loss of PrP^C function (Unterberger et al. 2005; Wong et al. 2000) (see Sect. 15.1.2.1 on putative neuroprotective functions of PrP^C) or more likely, a toxic gain of function associated with

PrP^{Sc} or other as-yet uncharacterized pathological conformers or misfolded PrP intermediates (PrP^{Int}) (Aguzzi et al. 2001; Prusiner 1998; Soto and Castilla 2004; Soto 1999; Stefani and Dobson 2003) (see below). However, these alternatives may not be mutually exclusive. A hypothetical sequence of events might be

$$PrP^{C} \Rightarrow PrP^{Int} \left\{ \begin{array}{c} \text{Uncharacterized, oligomeric PrP pathological isoforms} \\ \text{with varying degrees of protease sensitivity.} \\ \text{Neurotoxic.} \end{array} \right\} \Leftrightarrow PrP^{Sc}$$

The emergence of synaptic compartments as a target of PrP and other misfolding diseases such as AD highlights the significance of sub-compartment processes in modulating disease progression in both extracellular and intracellular misfolding diseases (Chiesa et al. 2008; Chiesa et al. 2005; Fein et al. 2008; Fuhrmann et al. 2007; Gylys et al. 2004). In particular, recent advances in studies on AD provide evidence for the existence of small, misfolded oligomers of Aβ that target synaptic structure or impair synaptic function (Haass and Selkoe 2007; Shankar et al. 2008) (see Chap. 14). As in other neurodegenerative disorders, such as HD, PD, and tauopathies (Friedhoff et al. 1998; Lashuel et al. 2002; Marchut and Hall 2006), it is possible that pathological conformers similar to PrP^{Int} are present and actively participate in prion pathogenesis at the synapse.

The propensity for protein misfolding and interconversions between different states of a protein at the synapse is likely governed by molecular chaperones working in concert with quality control and degradation mechanisms (Kleizen and Braakman 2004; Nishikawa et al. 2005). Importantly, a number of other proteins that give rise to disease pathways have also been implicated in or act as modulators of degenerative pathways (Yu and Lyubchenko 2009). Equally intriguing is the accumulating evidence for the site where PrP^{Sc} is formed, and its subcellular distribution following anterograde axonal transport to nerve terminal (Campana et al. 2005). Some studies using isolated synaptosomes support a synaptic PrP^{Sc} localization associated with dendritic degeneration (Bouzamondo-Bernstein et al. 2004; Ishikura et al. 2005) and suggest that PrP^{Sc} is localized to vesicles resembling early/recycling endosomes (Godsave et al. 2008). The discussions that follow will reflect the fact that without converging around any one proposed mechanism, the evidence to date supports that prion misfolding and the presence or accumulation of PrP^{Int} or PrP^{Sc} at the synapse implicate multiple signaling pathways required to coordinate cellular homeostasis. Identification, characterization, and understanding the possible interactions of PrP^{Int} or PrP^{Sc} with the protein components of synapses will lead to a better understanding of these pathways, the mechanisms of neurotransmission, and prion pathogenesis.

15.1.4.1 Prion Protein and the Nerve Terminal

Synaptophysin immunocytochemistry has been used in a number of studies to detect changes in synaptic density during prion disease progression (Jeffrey et al. 2000; Toggas et al. 1996), and our studies and others in experimental prion models, where infectious material is injected into the hippocampus, have shown a significant reduction in synaptophysin staining in the stratum radiatum prior to the onset of clinical symptoms (Cunningham et al. 2003; Jeffrey et al. 2000). Several other synaptic proteins, including synaptosomal-associated protein of molecular weight 25 kDa (SNAP-25), syntaxin-1 and synapsin-1, have been shown to be reduced or absent at the late stage in prion disease (Siso et al. 2002; Gray et al. 2009). Within the presynaptic nerve terminal, neurotransmitter-containing vesicles are clustered near the exocytotic sites of the presynaptic plasma membrane, in the active zones, and the precise alignment of presynaptic and postsynaptic specializations are crucial for efficient synaptic transmission at nerve terminals (Sanes and Lichtman 1999). It is feasible that PrP^{Sc} or PrP^{Int} selectively interfere with these highly regulated processes, including synaptic vesicle transport, docking/tethering, priming, fusion, endocytosis, recycling, and neurotransmitter refilling (Sudhof 2004). Biochemical studies in cell culture showed that accumulation of misfolded prion inhibits the formation of the SNARE complexes involving synaptobrevin, syntaxin, and SNAP-25 that play an essential role in neurotransmitter release (Sandberg and Low 2005). However, in vivo accumulation of misfolded PrP^{Sc} in the ME7 prion model, although deleterious to synaptic function in the hippocampus, did not disrupt the formation of SNARE complexes (Asuni et al. 2008) and is, therefore, unlikely to account for synaptic dysfunction.

The postsynaptic density (PSD) is a specialized sub-membranous cytoskeleton containing a wide assortment of proteins that serve to organize the de novo clustering and anchoring of postsynaptic receptors and the juxtaposed presynaptic membrane specialization through intercellular connections anchored in the PSD (Bresler et al. 2004; Garner et al. 2000; Kim and Sheng 2004). At the postsynaptic membrane, endocytic proteins such as clathrin and dynamin form part of the NMDA-R/postsynaptic density-95 (PSD-95) protein complex (Okamoto et al. 2001; Gray et al. 2003; Tu et al. 1998). Biochemically, synaptosomal fractionation methods have been used to show evidence of presynaptic localization of the PrP^C (Herms et al. 1999), and we have utilized synaptosomal preparations from the hippocampus of ME7-animals to illustrate the differential partitioning of PrP into various synaptic compartments. We find that PrP^{Sc} has accumulated in the pH 8 triton X-100 insoluble fraction that is also heavily enriched in the PSD components (Fig. 15.3) (Phillips et al. 2001), which would support a role for PrP^{Sc} modulation of NMDA-R/PSD-95 complex in prion disease. Future studies exploring the impact of PrP^{Sc} at the PSD are thus warranted. Selective vulnerability of ionotropic glutamate receptors has previously been implicated in the pathogenesis of prion disease (Scallet and Ye 1997; Ferrer and Puig 2003), but as with other pathways implicated in degenerative process, several aspects of the molecular mechanism responsible for it remain to be fully elucidated in prion disease. There is also evidence that implicates glutamate receptor activated Ca^{2+} signaling disruptions in the etiology of other neurological diseases (LaFerla 2002; Mattson et al. 2000; Stutzmann 2005).

15.1.4.2 Mitochondria, Apoptosis, and Oxidative Stress

Mitochondrial function is compromised in prion disease (Ferreiro et al. 2008a, b), and mitochondrial changes develop long before symptoms emerge (Martin 2007).

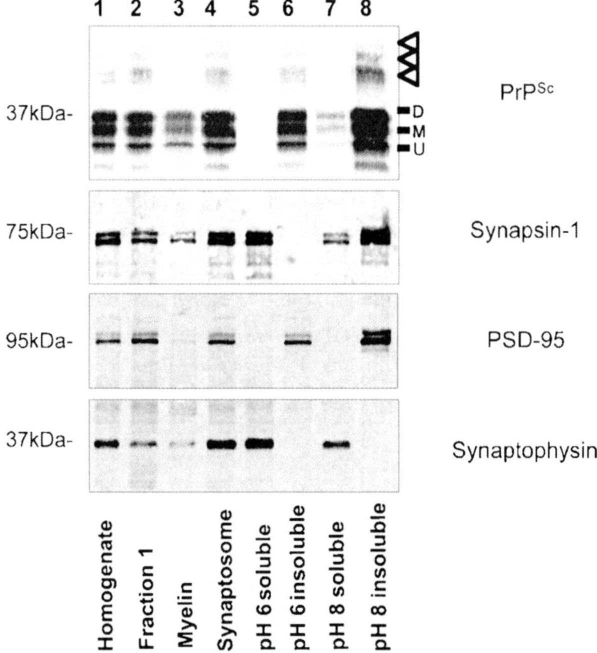

Fig. 15.3 Localization of PrP^{Sc} in an animal model of prion disease (ME7). PrP^{Sc} is localized to the PSD of synaptosomes purified from the hippocampus of ME7 animals with prion-mediated synaptic degeneration. PrP immunoreactivity with 6H4 mAb, as well as markers of the presynaptic compartment (synaptophysin, SY38 mAb, and synapsin 1, 10H5 mAb) and postsynaptic side (PSD-95, PSD-95 mAb) from various subcellular compartments analyzed by means of Western blotting. (1) protein ladder; (2) Homogenate; (3) crude membrane fraction; (4) nuclei-associated membranes; (5) synaptosomes; (6) pH 6 soluble fraction; (7) pH 6 insoluble; (8) pH 8 soluble; (9) pH 8 insoluble

This may imply that increasing accumulation of misfolded proteins can induce an age-related change in mitochondrial activity. This in turn may result in an insufficient mitochondrial capacity to buffer large Ca^{2+} elevations or lead to lowered ATP levels sufficient to disrupt cellular homeostasis. However, this may not be exact, as neurons appear to have significant reserve capacities and can harness alternative sources for ATP generation. They can maintain relatively stable ATP levels even under prolonged fasting conditions. Although apoptotic cell death is associated with neurodegenerative diseases such as AD, PD, and HD, and is likely a key mechanism of neuronal loss (Jesionek-Kupnicka et al. 1997), in prion disease it has not been particularly well documented and thus represents an area of future studies. Both pre- and post-synaptic sites are altered by PrP^{Sc}, which would likely disrupt neuronal circuitry and may initiate apoptotic cell death and associated neurological deficits (Belichenko et al. 2000).

Several in vitro studies have reported that accumulation of PrP^{Sc} was associated with decreased proteasome activity, increased intracellular reactive oxygen species (ROS) levels, and increased expression of caspases, that is triggered by the mitochondrial release of cytochrome *c* and its subsequent complex formation with

Apaf-1 and culminates in the activation of effector caspases (Carimalo et al. 2005; Lee et al. 2007; Watt et al. 2005). It is also likely that both death (caspase-dependent apoptosis) and survival (PI3-kinase/Akt and SLT2 mitogen-activated protein kinase (MAPK)) pathways are initiated as the cellular response to misfolded PrP^{Int} or PrP^{Sc} (Hetz et al. 2003; Lee et al. 2005; Lopes et al. 2005). Evidence for oxidative stress late in prion disease has also been presented through the detection of oxidative damage to important biomolecules, such as DNA, proteins, and lipids, and through altered activities of anti-oxidative enzymes (Choi et al. 1998; Choi et al. 2000; Guentchev et al. 2000; Kim et al. 1999; Kim et al. 2000; Minghetti et al. 2000; Thackray et al. 2002). The generation of ROS-induced oxidative stress also triggers a variety of signal molecules such as MAPKs, which induce cell apoptosis (Duranteau et al. 1998).

Several studies have reported the activation of intracellular cell signaling kinases by amyloid aggregates in different in vitro models of prion or prion peptide treatments (Gavin et al. 2005; Pietri et al. 2006; White et al. 2001). Moreover, Lee et al. reported increased activation of MAPK in hamster brains infected with scrapie (Lee et al. 2005). The results suggest that increased MAPK activation is associated with an increased ROS production in these conditions (Gavin et al. 2005; Pamplona et al. 2008; Pietri et al. 2006). Conversely, ligand-induced ROS species, especially H_2O_2, can act as intracellular messenger, and ROS are necessary for maintaining the functions of signaling molecules (such as Src kinases, Janus kinases, MAPKs, NF-B, AP-1, p53, and c-Myb) that regulate physiological and pathophysiological cellular functions (Rhee 1999; Thannickal and Fanburg 2000).

15.1.4.3 Prion Protein at the Synapse and Under-Explored Mechanisms

The dynamic processes associated with the formation of new synapses or remodeling of existing synapses likely require the targeted delivery of synaptic components to the sites of axodendritic contact, a process that begins with the transport of synaptic carrier vesicles along the secretory pathway (Horton and Ehlers 2004). Components of the translational machinery such as ribosomes, translation factors, numerous RNAs, and components of posttranslational secretory pathways have been identified in synapses and dendrites, where de novo synthesis of proteins may be required to supplement the plasticity and vital transsynaptic activity that occur at the synapse (Wang and Tiedge 2004). There is some in vitro evidence for anterograde and retrograde transport of GFP-PrP^{C}, which were completely inhibited by AMP-PNP/vanadate which stops the kinesin family/dynein family-driven movements, respectively (Borchelt et al. 1994; Hachiya et al. 2004). Presynaptic components, including precursors of synaptic vesicle (SV) and active zone (AZ) compartments, mitochondria, and proteins essential for SV release, are also transported to the synapse by kinesin motors moving along microtubules (MT) (Bradke and Dotti 1998; Goldstein and Philp 1999; Hirokawa 1998; Roos and Kelly 2000). We speculate that as the capacity for growth or shrinkage/degeneration exists at the synapse, essential cargo is required at the postsynaptic spine and the presynaptic nerve terminal. Increasing accumulation of PrP^{Int} or PrP^{Sc} may overwhelm the transport of these cargos. Such transport deficiencies associated with protein aggregation have been described in several neurodegenerative diseases (Gunawardena et al. 2003; Roy et al. 2005; Xia et al. 2003). In prion disease, evidence for this remains to be fully explored, although no change in disease incubation time was observed after peripheral prion infection in mice with a heterozygous mutation of dynein, a motor protein involved in the transport of axonal cargo along the cytoskeleton (Hafezparast et al. 2005).

The microtubule-associated protein tau forms insoluble pathological aggregates in AD, and is implicated in Aβ-induced neurodegenerative processes (Rapoport et al. 2002). Like PrP^{Sc}, both tau and Aβ have been localized to the synapse (Takahashi et al. 2010), where Aβ misfolds and oligomerizes into neurotoxic species and tau is hyperphosphorylated. Hyperphosphorylation of the tau and abrogation of its microtubule-supporting role in maintaining axon structure lead to alteration of cellular transport of essential materials, degeneration of synapses, and ultimately the death of neurons (Leuba et al. 2008; Rauk 2008). The consequence of this altered transport is a progressive degeneration initiated at the distal end of the processes and moving retrogradely toward the cell body (Theiss and Meller 2000). Transgenic mice overexpressing four-repeat tau with impaired axonal transport displayed comparable incubation times to control animals following intraneural infection with scrapie (Kunzi et al. 2002). Studies on BSE-infected bovine prion protein transgenic animals described an increase in hyperphosphorylated tau (Bautista et al. 2006), and it has been shown that there is also an increase in CSF phospho-tau levels associated with sporadic CJD (Kapaki et al. 2001; Sarac et al. 2008). In prion disease, the role of tau-dysfunction in synaptic transport deficits remains to be fully elucidated.

15.1.5 Prion and Glial Cells

Increasing evidence supports the idea that dysregulation of the interaction patterns between glial cells and neurons underlies neuronal degeneration, and is in line with the notion that increasing accumulation of pathologically conformed PrP^{Int} or PrP^{Sc} disrupts communication between these cell types, and may contribute to the synaptic impairment associated with neuronal degeneration. However, other explanations cannot be discounted. The coordinated synaptic exchange of anterograde and retrograde signals between pre- and postsynaptic neurons also involves the surrounding glia. Together they constitute the tripartite synapse, with the astrocyte as the third partner (Araque 2006; Haber et al. 2006). The dynamic structural changes associated with astrocytes can help control the degree of neuron–glial communication at hippocampal synapses (Araque et al. 2001; Hirrlinger et al. 2004; Tiruchinapalli et al. 2003), and recent studies suggest that altered patterns in the glia–neuron interactions constitute early molecular events within the cascade of cellular signals that lead to neurodegeneration (Halassa et al. 2007; Mitterauer 2004). There thus may be a role for PrP^{Sc} in disrupting neuron–glial interactions that could in turn lead to synaptic degeneration.

15.1.5.1 Perspective

The early stages of many neurodegenerative diseases and age-related degeneration are characterized by compromised synaptic functions, neurite damage, and eventual cell death. Additional contributions to neurotoxicity in the hippocampus during prion disease include misfolded proteins, activated glial cells, oxidative stress, apoptosis, and/or complement activation (Gray et al. 1999, reviewed in Caughey and Baron 2006). Extensive studies in a prion disease model with a clearly defined hippocampal circuit (ME7 model) have yielded many interesting observations, and improved our understanding of synaptic degeneration in that structure. For example, consistent observations of disintegrating synaptic structures during a pre-clinical stage of disease support that the pathological cascade is initiated locally at the synapse during an early, key stage of neurodegeneration (Scott et al. 1994; Jeffrey et al. 2000; Cunningham et al. 2003; Gray et al. 2009; Siskova et al. 2009). The early neuropathological, behavioral, biochemical, and electrophysiological changes in the course of the disease have been widely described (Guenther et al. 2001; Jeffrey et al. 1995; Johnston et al. 1998; Kim et al. 1999; Lee et al. 1999); however, further studies will be required to elucidate whether these processes are driven by the increasing deposition of pathological conformers of PrP^{Int} or PrP^{Sc} at the synapse, and if synaptic compartments in other brain regions also exhibit similar vulnerability to that observed in the hippocampus.

References

Adler, V., et al. 2003. Small, highly structured RNAs participate in the conversion of human recombinant PrP(Sen) to PrP(Res) in vitro. J Mol Biol. 332, 47–57

Aguzzi, A., et al. 2001. Spongiform encephalopathies: insights from transgenic models. Adv Virus Res. 56, 313–52

Aguzzi, A., et al. 2007. Insights into prion strains and neurotoxicity. Nat Rev Mol Cell Biol. 8, 552–61

Aguzzi, A., Polymenidou, M. 2004. Mammalian prion biology: one century of evolving concepts. Cell. 116, 313–27

Araque, A., 2006. Astrocyte-neuron signaling in the brain – implications for disease. Curr Opin Investig Drugs. 7, 619–24

Araque, A., et al. 2001. Dynamic signaling between astrocytes and neurons. Annu Rev Physiol. 63, 795–813

Asuni, A. A., et al. 2008. Unaltered SNARE complex formation in an in vivo model of prion disease. Brain Res. 1233, 1–7

Barron, R. M., et al. 2007. High titers of transmissible spongiform encephalopathy infectivity associated with extremely low levels of PrPSc in vivo. J Biol Chem. 282, 35878–86

Basler, K., et al. 1986. Scrapie and cellular PrP isoforms are encoded by the same chromosomal gene. Cell. 46, 417–28

Bautista, M. J., et al. 2006. BSE infection in bovine PrP transgenic mice leads to hyperphosphorylation of tau-protein. Vet Microbiol. 115, 293–301

Belichenko, P. V., et al. 2000. Dendritic and synaptic alterations of hippocampal pyramidal neurones in scrapie-infected mice. Neuropathol Appl Neurobiol. 26, 143–9

Borchelt, D. R., et al. 1994. Rapid anterograde axonal transport of the cellular prion glycoprotein in the peripheral and central nervous systems. J Biol Chem. 269, 14711–4

Bouzamondo-Bernstein, E., et al. 2004. The neurodegeneration sequence in prion diseases: evidence from functional, morphological and ultrastructural studies of the GABAergic system. J Neuropathol Exp Neurol. 63, 882–99

Bradke, F., Dotti, C. G., 1998. Membrane traffic in polarized neurons. Biochim Biophys Acta. 1404, 245–58

Bresler, T., et al. 2004. Postsynaptic density assembly is fundamentally different from presynaptic active zone assembly. J Neurosci. 24, 1507–20

Brockes, J. P., 1999. Topics in prion cell biology. Curr Opin Neurobiol. 9, 571–7

Brown, D. R., 2001. Prion and prejudice: normal protein and the synapse. Trends Neurosci. 24, 85–90

Brown, D. R., 2002. Copper and prion diseases. Biochem Soc Trans. 30, 742–5

Brown, D. R., Besinger, A., 1998. Prion protein expression and superoxide dismutase activity. Biochem J. 334 (Pt 2), 423–9

Brown, D. R., et al. 1999. Normal prion protein has an activity like that of superoxide dismutase. Biochem J. 344 (Pt 1), 1–5

Bueler, H., et al. 1993. Mice devoid of PrP are resistant to scrapie. Cell. 73, 1339

Bueler, H., et al. 1992. Normal development and behaviour of mice lacking the neuronal cell-surface PrP protein. Nature. 356, 577–82

Burthem, J., et al. 2001. The normal cellular prion protein is strongly expressed by myeloid dendritic cells. Blood. 98, 3733–8

Campana, V., et al. 2005. The highways and byways of prion protein trafficking. Trends Cell Biol. 15, 102–11

Carimalo, J., et al. 2005. Activation of the JNK-c-Jun pathway during the early phase of neuronal apoptosis induced by PrP106–126 and prion infection. Eur J Neurosci. 21, 2311–9

Carleton, A., et al. 2001. Dose-dependent, prion protein (PrP)-mediated facilitation of excitatory synaptic transmission in the mouse hippocampus. Pflugers Arch. 442, 223–9

Caughey, B., Baron, G. S., 2006. Prions and their partners in crime. Nature. 443, 803–10

Caughey, B., et al. 1991. N-terminal truncation of the scrapie-associated form of PrP by lysosomal protease(s): implications regarding the site of conversion of PrP to the protease-resistant state. J Virol. 65, 6597–603

Cereghetti, G. M., et al. 2001. Electron paramagnetic resonance evidence for binding of Cu(2+) to the C-terminal domain of the murine prion protein. Biophys J. 81, 516–25

Chen, S., et al. 2003. Prion protein as trans-interacting partner for neurons is involved in neurite outgrowth and neuronal survival. Mol Cell Neurosci. 22, 227–33

Chesebro, B., et al. 2005. Anchorless prion protein results in infectious amyloid disease without clinical scrapie. Science. 308, 1435–9

Chiarini, L. B., et al. 2002. Cellular prion protein transduces neuroprotective signals. EMBO J. 21, 3317–26

Chiesa, R., et al. 2008. Aggregated, wild-type prion protein causes neurological dysfunction and synaptic abnormalities. J Neurosci. 28, 13258–67

Chiesa, R., et al. 2005. Bax deletion prevents neuronal loss but not neurological symptoms in a transgenic model of inherited prion disease. Proc Natl Acad Sci USA. 102, 238–43

Choi, S. I., et al. 1998. Mitochondrial dysfunction induced by oxidative stress in the brains of hamsters infected with the 263 K scrapie agent. Acta Neuropathol. 96, 279–86

Choi, Y. G., et al. 2000. Induction of heme oxygenase-1 in the brains of scrapie-infected mice. Neurosci Lett. 289, 173–6

Collinge, J., et al. 1994. Prion protein is necessary for normal synaptic function. Nature. 370, 295–7

Criado, J. R., et al. 2005. Mice devoid of prion protein have cognitive deficits that are rescued by reconstitution of PrP in neurons. Neurobiol Dis. 19, 255–65

Cunningham, C., et al. 2003. Synaptic changes characterize early behavioural signs in the ME7 model of murine prion disease. Eur J Neurosci. 17, 2147–55

Daniels, M., Brown, D. R., 2002. Purification and preparation of prion protein: synaptic superoxide dismutase. Methods Enzymol. 349, 258–67

De Camilli, P., et al. 1990. The synapsins. Annu Rev Cell Biol. 6, 433–60

Deleault, N. R., et al. 2003. RNA molecules stimulate prion protein conversion. Nature. 425, 717–20
Duranteau, J., et al. 1998. Intracellular signaling by reactive oxygen species during hypoxia in cardiomyocytes. J Biol Chem. 273, 11619–24
Durig, J., et al. 2000. Differential constitutive and activation-dependent expression of prion protein in human peripheral blood leucocytes. Br J Haematol. 108, 488–95
Fein, J. A., et al. 2008. Co-localization of amyloid beta and tau pathology in Alzheimer's disease synaptosomes. Am J Pathol. 172, 1683–92
Ferreiro, E., et al. 2008a. Involvement of mitochondria in endoplasmic reticulum stress-induced apoptotic cell death pathway triggered by the prion peptide PrP(106–126). J Neurochem. 104, 766–76
Ferreiro, E., et al. 2008b. The release of calcium from the endoplasmic reticulum induced by amyloid-beta and prion peptides activates the mitochondrial apoptotic pathway. Neurobiol Dis. 30, 331–42
Ferrer, I., Puig, B., 2003. GluR2/3, NMDAepsilon1 and GABAA receptors in Creutzfeldt-Jakob disease. Acta Neuropathol. 106, 311–8
Fischer von Mollard, G., et al. 1994. Rab proteins in regulated exocytosis. Trends Biochem Sci. 19, 164–8
Ford, M. J., et al. 2002. A marked disparity between the expression of prion protein and its message by neurones of the CNS. Neuroscience. 111, 533–51
Fournier, J. G., et al. 1995. Ultrastructural localization of cellular prion protein (PrPc) in synaptic boutons of normal hamster hippocampus. C R Acad Sci III. 318, 339–44
Friedhoff, P., et al. 1998. A nucleated assembly mechanism of Alzheimer paired helical filaments. Proc Natl Acad Sci USA. 95, 15712–7
Fuhrmann, M., et al. 2007. Dendritic pathology in prion disease starts at the synaptic spine. J Neurosci. 27, 6224–33
Garner, C. C., et al. 2000. Molecular determinants of presynaptic active zones. Curr Opin Neurobiol. 10, 321–7
Gauczynski, S., et al. 2001. The 37-kDa/67-kDa laminin receptor acts as the cell-surface receptor for the cellular prion protein. EMBO J. 20, 5863–75
Gavin, R., et al. 2005. PrP(106–126) activates neuronal intracellular kinases and Egr1 synthesis through activation of NADPH-oxidase independently of PrPc. FEBS Lett. 579, 4099–106
Godsave, S. F., et al. 2008. Cryo-immunogold electron microscopy for prions: toward identification of a conversion site. J Neurosci. 28, 12489–99
Gohel, C., et al. 1999. Ultrastructural localization of cellular prion protein (PrPc) at the neuromuscular junction. J Neurosci Res. 55, 261–7
Goldstein, L. S., Philp, A. V., 1999. The road less traveled: emerging principles of kinesin motor utilization. Annu Rev Cell Dev Biol. 15, 141–83
Graner, E., et al. 2000. Cellular prion protein binds laminin and mediates neuritogenesis. Brain Res Mol Brain Res. 76, 85–92
Gray, B. C., et al. 2009. Selective presynaptic degeneration in the synaptopathy associated with ME7-induced hippocampal pathology. Neurobiol Dis. 35, 63–74
Gray, F., et al. 1999. Neuronal apoptosis in Creutzfeldt-Jakob disease. J Neuropathol Exp Neurol. 58, 321–8
Gray, N. W., et al. 2003. Dynamin 3 is a component of the postsynapse, where it interacts with mGluR5 and Homer. Curr Biol. 13, 510–5
Greengard, P., et al. 1993. Synaptic vesicle phosphoproteins and regulation of synaptic function. Science. 259, 780–5
Guentchev, M., et al. 2000. Evidence for oxidative stress in experimental prion disease. Neurobiol Dis. 7, 270–3
Guenther, K., et al. 2001. Early behavioural changes in scrapie-affected mice and the influence of dapsone. Eur J Neurosci. 14, 401–9
Gunawardena, S., et al. 2003. Disruption of axonal transport by loss of huntingtin or expression of pathogenic polyQ proteins in Drosophila. Neuron. 40, 25–40

Gylys, K. H., et al. 2004. Synaptic changes in Alzheimer's disease: increased amyloid-beta and gliosis in surviving terminals is accompanied by decreased PSD-95 fluorescence. Am J Pathol. 165, 1809–17

Haass, C., Selkoe, D. J., 2007. Soluble protein oligomers in neurodegeneration: lessons from the Alzheimer's amyloid beta-peptide. Nat Rev Mol Cell Biol. 8, 101–12

Haber, M., et al. 2006. Cooperative astrocyte and dendritic spine dynamics at hippocampal excitatory synapses. J Neurosci. 26, 8881–91

Hachiya, N. S., et al. 2004. Anterograde and retrograde intracellular trafficking of fluorescent cellular prion protein. Biochem Biophys Res Commun. 315, 802–7

Hafezparast, M., et al. 2005. Prion disease incubation time is not affected in mice heterozygous for a dynein mutation. Biochem Biophys Res Commun. 326, 18–22

Halassa, M. M., et al. 2007. The tripartite synapse: roles for gliotransmission in health and disease. Trends Mol Med. 13, 54–63

Hergersberg, M., et al. 2000. Deletions in the spinal muscular atrophy gene region in a newborn with neuropathy and extreme generalized muscular weakness. Eur J Paediatr Neurol. 4, 35–8

Herms, J., et al. 1999. Evidence of presynaptic location and function of the prion protein. J Neurosci. 19, 8866–75

Hetz, C., et al. 2003. Caspase-12 and endoplasmic reticulum stress mediate neurotoxicity of pathological prion protein. EMBO J. 22, 5435–45

Hirokawa, N., 1998. Kinesin and dynein superfamily proteins and the mechanism of organelle transport. Science. 279, 519–26

Hirrlinger, J., et al. 2004. Astroglial processes show spontaneous motility at active synaptic terminals in situ. Eur J Neurosci. 20, 2235–9

Horton, A. C., Ehlers, M. D., 2004. Secretory trafficking in neuronal dendrites. Nat Cell Biol. 6, 585–91

Hutter, G., et al. 2003. No superoxide dismutase activity of cellular prion protein in vivo. Biol Chem. 384, 1279–85

Ishikura, N., et al. 2005. Notch-1 activation and dendritic atrophy in prion disease. Proc Natl Acad Sci USA. 102, 886–91

Jackson, G. S., et al. 2001. Location and properties of metal-binding sites on the human prion protein. Proc Natl Acad Sci USA. 98, 8531–5

Jahn, R., Sudhof, T. C., 1994. Synaptic vesicles and exocytosis. Annu Rev Neurosci. 17, 219–46

Jeffrey, M., et al. 1995. Early unsuspected neuron and axon terminal loss in scrapie-infected mice revealed by morphometry and immunocytochemistry. Neuropathol Appl Neurobiol. 21, 41–9

Jeffrey, M., et al. 2000. Synapse loss associated with abnormal PrP precedes neuronal degeneration in the scrapie-infected murine hippocampus. Neuropathol Appl Neurobiol. 26, 41–54

Jesionek-Kupnicka, D., et al. 1997. Programmed cell death (apoptosis) in Alzheimer's disease and Creutzfeldt-Jakob disease. Folia Neuropathol. 35, 233–5

Johnston, A. R., et al. 1998. Synaptic plasticity in the CA1 area of the hippocampus of scrapie-infected mice. Neurobiol Dis. 5, 188–95

Jones, S., et al. 2005. Recombinant prion protein does not possess SOD-1 activity. Biochem J. 392, 309–12

Kanaani, J., et al. 2005. Recombinant prion protein induces rapid polarization and development of synapses in embryonic rat hippocampal neurons in vitro. J Neurochem. 95, 1373–86

Kaneko, K., et al. 1997. COOH-terminal sequence of the cellular prion protein directs subcellular trafficking and controls conversion into the scrapie isoform. Proc Natl Acad Sci USA. 94, 2333–8

Kapaki, E., et al. 2001. Highly increased CSF tau protein and decreased beta-amyloid (1–42) in sporadic CJD: a discrimination from Alzheimer's disease? J Neurol Neurosurg Psychiatr. 71, 401–3

Kardos, J., et al. 1989. Nerve endings from rat brain tissue release copper upon depolarization. A possible role in regulating neuronal excitability. Neurosci Lett. 103, 139–44

Keshet, G. I., et al. 2000. The cellular prion protein colocalizes with the dystroglycan complex in the brain. J Neurochem. 75, 1889–97

Khosravani, H., et al. 2008. Prion protein attenuates excitotoxicity by inhibiting NMDA receptors. J Cell Biol. 181, 551–65

Kim, E., Sheng, M., 2004. PDZ domain proteins of synapses. Nat Rev Neurosci. 5, 771–81
Kim, J. I., et al. 1999. Expression of cytokine genes and increased nuclear factor-kappa B activity in the brains of scrapie-infected mice. Brain Res Mol Brain Res. 73, 17–27
Kim, N. H., et al. 2000. Increased ferric iron content and iron-induced oxidative stress in the brains of scrapie-infected mice. Brain Res. 884, 98–103
Klamt, F., et al. 2001. Imbalance of antioxidant defense in mice lacking cellular prion protein. Free Radic Biol Med. 30, 1137–44
Klein, M. A., et al. 1997. A crucial role for B cells in neuroinvasive scrapie. Nature. 390, 687
Kleizen, B., Braakman, I., 2004. Protein folding and quality control in the endoplasmic reticulum. Curr Opin Cell Biol. 16, 343–9
Kunzi, V., et al. 2002. Unhampered prion neuroinvasion despite impaired fast axonal transport in transgenic mice overexpressing four-repeat tau. J Neurosci. 22, 7471–7
LaFerla, F. M., 2002. Calcium dyshomeostasis and intracellular signalling in Alzheimer's disease. Nat Rev Neurosci. 3, 862–72
Laine, J., et al. 2001. Cellular and subcellular morphological localization of normal prion protein in rodent cerebellum. Eur J Neurosci. 14, 47–56
Lashuel, H. A., et al. 2002. Alpha-synuclein, especially the Parkinson's disease-associated mutants, forms pore-like annular and tubular protofibrils. J Mol Biol. 322, 1089–102
Lasmezas, C. I., 2003. Putative functions of PrP(C). Br Med Bull. 66, 61–70
Lasmezas, C. I., et al. 1997. Transmission of the BSE agent to mice in the absence of detectable abnormal prion protein. Science. 275, 402–5
Lee, D. W., et al. 1999. Alteration of free radical metabolism in the brain of mice infected with scrapie agent. Free Radic Res. 30, 499–507
Lee, H. P., et al. 2005. Activation of mitogen-activated protein kinases in hamster brains infected with 263K scrapie agent. J Neurochem. 95, 584–93
Lee, K. J., et al. 2007. Cellular prion protein (PrPC) protects neuronal cells from the effect of huntingtin aggregation. J Cell Sci. 120, 2663–71
Leuba, G., et al. 2008. Differential changes in synaptic proteins in the Alzheimer frontal cortex with marked increase in PSD-95 postsynaptic protein. J Alzheimers Dis. 15, 139–51
Li, A., et al. 2007. N-terminally deleted forms of the prion protein activate both Bax-dependent and Bax-independent neurotoxic pathways. J Neurosci. 27, 852–9
Lledo, P. M., et al. 1996. Mice deficient for prion protein exhibit normal neuronal excitability and synaptic transmission in the hippocampus. Proc Natl Acad Sci USA. 93, 2403–7
Lopes, M. H., et al. 2005. Interaction of cellular prion and stress-inducible protein 1 promotes neuritogenesis and neuroprotection by distinct signaling pathways. J Neurosci. 25, 11330–9
Macreadie, I. G., 2008. Copper transport and Alzheimer's disease. Eur Biophys J. 37, 295–300
Magalhaes, A. C., et al. 2002. Endocytic intermediates involved with the intracellular trafficking of a fluorescent cellular prion protein. J Biol Chem. 277, 33311
Maglio, L. E., et al. 2004. Hippocampal synaptic plasticity in mice devoid of cellular prion protein. Brain Res Mol Brain Res. 131, 58–64
Makrinou, E., et al. 2002. Genomic characterization of the human prion protein (PrP) gene locus. Mamm Genome. 13, 696–703
Mallucci, G., et al. 2003. Depleting neuronal PrP in prion infection prevents disease and reverses spongiosis. Science. 302, 871–4
Mallucci, G. R., et al. 2002. Post-natal knockout of prion protein alters hippocampal CA1 properties, but does not result in neurodegeneration. EMBO J. 21, 202–10
Manson, J. C., et al. 1994. 129/Ola mice carrying a null mutation in PrP that abolishes mRNA production are developmentally normal. Mol Neurobiol. 8, 121–7
Manson, J. C., et al. 1999. A single amino acid alteration (101L) introduced into murine PrP dramatically alters incubation time of transmissible spongiform encephalopathy. EMBO J. 18, 6855–64
Marchut, A. J., Hall, C. K., 2006. Spontaneous formation of annular structures observed in molecular dynamics simulations of polyglutamine peptides. Comput Biol Chem. 30, 215–8

Martin, L. J., 2007. Transgenic mice with human mutant genes causing Parkinson's disease and amyotrophic lateral sclerosis provide common insight into mechanisms of motor neuron selective vulnerability to degeneration. Rev Neurosci. 18, 115–36

Mattson, M. P., et al. 2000. Calcium signaling in the ER: its role in neuronal plasticity and neurodegenerative disorders. Trends Neurosci. 23, 222–9

McKinley, M. P., et al. 1983. A protease-resistant protein is a structural component of the scrapie prion. Cell. 35, 57

McLennan, N. F., et al. 2004. Prion protein accumulation and neuroprotection in hypoxic brain damage. Am J Pathol. 165, 227–35

Miele, G., et al. 2003. Embryonic activation and developmental expression of the murine prion protein gene. Gene Expr 11, 1–12

Miele, G., et al. 2002. Ablation of cellular prion protein expression affects mitochondrial numbers and morphology. Biochem Biophys Res Commun. 291, 372–7

Minghetti, L., et al. 2000. Increased brain synthesis of prostaglandin E2 and F2-isoprostane in human and experimental transmissible spongiform encephalopathies. J Neuropathol Exp Neurol. 59, 866–71

Mironov, A., Jr., et al. 2003. Cytosolic prion protein in neurons. J Neurosci. 23, 7183–93

Mitterauer, B., 2004. Imbalance of glial-neuronal interaction in synapses: a possible mechanism of the pathophysiology of bipolar disorder. Neuroscientist. 10, 199–206

Montrasio, F., et al. 2000. Impaired prion replication in spleens of mice lacking functional follicular dendritic cells. Science. 288, 1257–9.

Moore, R. C., et al. 1999. Ataxia in prion protein (PrP)-deficient mice is associated with upregulation of the novel PrP-like protein doppel. J Mol Biol. 292, 797–817.

Moser, M., et al. 1995. Developmental expression of the prion protein gene in glial cells. Neuron. 14, 509–17.

Mouillet-Richard, S., et al. 2000. Signal transduction through prion protein. Science. 289, 1925–8.

Mouillet-Richard, S., et al. 2005. Modulation of serotonergic receptor signaling and cross-talk by prion protein. J Biol Chem. 280, 4592–601.

Moya, K. L., et al. 2000. Immunolocalization of the cellular prion protein in normal brain. Microsc Res Tech. 50, 58–65.

Negro, A., et al. 2000. Susceptibility of the prion protein to enzymic phosphorylation. Biochem Biophys Res Commun. 271, 337–41.

Nishikawa, S., et al. 2005. Roles of molecular chaperones in endoplasmic reticulum (ER) quality control and ER-associated degradation (ERAD). J Biochem. 137, 551–5.

O'Dell, B. L., 1976. Biochemistry of copper. Med Clin North Am. 60, 687–703.

Okamoto, P. M., et al. 2001. Dynamin isoform-specific interaction with the shank/ProSAP scaffolding proteins of the postsynaptic density and actin cytoskeleton. J Biol Chem. 276, 48458–65.

Pamplona, R., et al. 2008. Increased oxidation, glycoxidation, and lipoxidation of brain proteins in prion disease. Free Radic Biol Med. 45, 1159–66.

Pan, T., et al. 2002. Cell-surface prion protein interacts with glycosaminoglycans. Biochem J. 368, 81–90.

Petrakis, S., Sklaviadis, T., 2006. Identification of proteins with high affinity for refolded and native PrPC. Proteomics. 6, 6476–84.

Phillips, G. R., et al. 2001. The presynaptic particle web: ultrastructure, composition, dissolution, and reconstitution. Neuron. 32, 63–77.

Piccardo, P., et al. 2007. Accumulation of prion protein in the brain that is not associated with transmissible disease. Proc Natl Acad Sci USA. 104, 4712–7.

Pietri, M., et al. 2006. Overstimulation of PrPC signaling pathways by prion peptide 106–126 causes oxidative injury of bioaminergic neuronal cells. J Biol Chem. 281, 28470–9.

Politopoulou, G., et al. 2000. Age-related expression of the cellular prion protein in human peripheral blood leukocytes. Haematologica. 85, 580–7.

Prinz, M., et al. 2003. Oral prion infection requires normal numbers of Peyer's patches but not of enteric lymphocytes. Am J Pathol. 162, 1103–11.

Prusiner, S. B., 1991. Molecular biology of prions causing infectious and genetic encephalopathies of humans as well as scrapie of sheep and BSE of cattle. Dev Biol Stand. 75, 55–74.

Prusiner, S. B., 1997. Prion diseases and the BSE crisis. Science. 278, 245–51.

Prusiner, S. B., 1998. Prions. Proc Natl Acad Sci USA. 95, 13363–83.

Rapoport, M., et al. 2002. Tau is essential to beta -amyloid-induced neurotoxicity. Proc Natl Acad Sci USA. 99, 6364–9.

Rauk, A., 2008. Why is the amyloid beta peptide of Alzheimer's disease neurotoxic? Dalton Trans. 1273–82.

Re, L., et al. 2006. Prion protein potentiates acetylcholine release at the neuromuscular junction. Pharmacol Res. 53, 62–8.

Rhee, S. G., 1999. Redox signaling: hydrogen peroxide as intracellular messenger. Exp Mol Med. 31, 53–9.

Riek, R., et al. 1997. NMR characterization of the full-length recombinant murine prion protein, mPrP(23–231). FEBS Lett. 413, 282–8.

Rogers, M., et al. 1993. Conversion of truncated and elongated prion proteins into the scrapie isoform in cultured cells. Proc Natl Acad Sci USA. 90, 3182–6.

Roos, J., Kelly, R. B., 2000. Preassembly and transport of nerve terminals: a new concept of axonal transport. Nat Neurosci. 3, 415–7.

Roucou, X., et al. 2003. Cytosolic prion protein is not toxic and protects against Bax-mediated cell death in human primary neurons. J Biol Chem. 278, 40877.

Roucou, X., LeBlanc, A. C., 2005. Cellular prion protein neuroprotective function: implications in prion diseases. J Mol Med. 83, 3–11.

Roy, S., et al. 2005. Axonal transport defects: a common theme in neurodegenerative diseases. Acta Neuropathol. 109, 5–13.

Rudd, P. M., et al. 2001. Prion glycoprotein: structure, dynamics, and roles for the sugars. Biochemistry. 40, 3759–66.

Safar, J., et al. 1993. Thermal stability and conformational transitions of scrapie amyloid (prion) protein correlate with infectivity. Protein Sci. 2, 2206–16.

Sakaguchi, S., et al. 1996. Loss of cerebellar Purkinje cells in aged mice homozygous for a disrupted PrP gene. Nature. 380, 528–31.

Sales, N., et al. 1998. Cellular prion protein localization in rodent and primate brain. Eur J Neurosci. 10, 2464–71.

Sandberg, M. K., Low, P., 2005. Altered interaction and expression of proteins involved in neurosecretion in scrapie-infected GT1–1 cells. J Biol Chem. 280, 1264–71.

Sanes, J. R., Lichtman, J. W., 1999. Development of the vertebrate neuromuscular junction. Annu Rev Neurosci. 22, 389–442.

Santuccione, A., et al. 2005. Prion protein recruits its neuronal receptor NCAM to lipid rafts to activate p59fyn and to enhance neurite outgrowth. J Cell Biol. 169, 341–54.

Sarac, H., et al. 2008. Magnetic resonance spectroscopy and measurement of tau epitopes of autopsy proven sporadic Creutzfeldt-Jakob disease in a patient with non-specific initial EEG, MRI and negative 14–3–3 immunoblot. Coll Antropol. 32 Suppl 1, 199–204.

Scallet, A. C., Ye, X., 1997. Excitotoxic mechanisms of neurodegeneration in transmissible spongiform encephalopathies. Ann N Y Acad Sci. 825, 194–205.

Schmitt-Ulms, G., et al. 2001. Binding of neural cell adhesion molecules (N-CAMs) to the cellular prion protein. J Mol Biol. 314, 1209–25

Schneider, B., et al. 2003. NADPH oxidase and extracellular regulated kinases 1/2 are targets of prion protein signaling in neuronal and nonneuronal cells. Proc Natl Acad Sci USA. 100, 13326–31

Scott, J. R., et al. 1994. Unsuspected early neuronal loss in scrapie-infected mice revealed by morphometric analysis. Ann N Y Acad Sci. 724, 338–43

Shankar, G. M., et al. 2008. Amyloid-beta protein dimers isolated directly from Alzheimer's brains impair synaptic plasticity and memory. Nat Med. 14, 837–42

Shmerling, D., et al. 1998. Expression of amino-terminally truncated PrP in the mouse leading to ataxia and specific cerebellar lesions. Cell. 93, 203–14

Siskova, Z., et al. 2009. Degenerating synaptic boutons in prion disease: microglia activation without synaptic stripping. Am J Pathol. 175, 1610–21.
Siso, S., et al. 2002. Abnormal synaptic protein expression and cell death in murine scrapie. Acta Neuropathol (Berl). 103, 615–26.
Solforosi, L., et al. 2004. Cross-linking cellular prion protein triggers neuronal apoptosis in vivo. Science. 303, 1514.
Soto, C., 1999. Alzheimer's and prion disease as disorders of protein conformation: implications for the design of novel therapeutic approaches. J Mol Med. 77, 412–8.
Soto, C., Castilla, J., 2004. The controversial protein-only hypothesis of prion propagation. Nat Med. 10 Suppl, S63–7
Soto, C., Estrada, L. D., 2008. Protein misfolding and neurodegeneration. Arch Neurol. 65, 184–9
Spielhaupter, C., Schatzl, H. M., 2001. PrPC directly interacts with proteins involved in signaling pathways. J Biol Chem. 276, 44604–12
Stahl, N., et al. 1993. Structural studies of the scrapie prion protein using mass spectrometry and amino acid sequencing. Biochemistry. 32, 1991–2002
Steele, A. D., et al. 2006. Prion protein (PrPc) positively regulates neural precursor proliferation during developmental and adult mammalian neurogenesis. Proc Natl Acad Sci USA. 103, 3416–21
Stefani, M., Dobson, C. M., 2003. Protein aggregation and aggregate toxicity: new insights into protein folding, misfolding diseases and biological evolution. J Mol Med. 81, 678–99
Stockel, J., et al. 1998. Prion protein selectively binds copper(II) ions. Biochemistry. 37, 7185–93
Stutzmann, G. E., 2005. Calcium dysregulation, IP3 signaling, and Alzheimer's disease. Neuroscientist. 11, 110–5
Sudhof, T. C., 2004. The synaptic vesicle cycle. Annu Rev Neurosci. 27, 509–47
Swanson, D. A., et al. 1998. Identification and characterization of the human ortholog of rat STXBP1, a protein implicated in vesicle trafficking and neurotransmitter release. Genomics. 48, 373–6
Takahashi, R. H., et al. 2010. Co-occurrence of Alzheimer's disease beta-amyloid and tau pathologies at synapses. Neurobiol Aging. 31, 1145–52
Taylor, D. R., Hooper, N. M., 2007. The low-density lipoprotein receptor-related protein 1 (LRP1) mediates the endocytosis of the cellular prion protein. Biochem J. 402, 17–23
Taylor, D. R., et al. 2005. Assigning functions to distinct regions of the N-terminus of the prion protein that are involved in its copper-stimulated, clathrin-dependent endocytosis. J Cell Sci. 118, 5141–53
Telling, G. C., et al. 1995. Prion propagation in mice expressing human and chimeric PrP transgenes implicates the interaction of cellular PrP with another protein. Cell. 83, 79
Thackray, A. M., et al. 2002. Metal imbalance and compromised antioxidant function are early changes in prion disease. Biochem J. 362, 253–8.
Thannickal, V. J., Fanburg, B. L., 2000. Reactive oxygen species in cell signaling. Am J Physiol Lung Cell Mol Physiol. 279, L1005–28
Theiss, C., Meller, K., 2000. Taxol impairs anterograde axonal transport of microinjected horseradish peroxidase in dorsal root ganglia neurons in vitro. Cell Tissue Res. 299, 213–24
Tiruchinapalli, D. M., et al. 2003. Activity-dependent trafficking and dynamic localization of zipcode binding protein 1 and beta-actin mRNA in dendrites and spines of hippocampal neurons. J Neurosci. 23, 3251–61
Tobler, I., et al. 1996. Altered circadian activity rhythms and sleep in mice devoid of prion protein. Nature. 380, 639–42
Toggas, S. M., et al. 1996. Prevention of HIV-1 gp120-induced neuronal damage in the central nervous system of transgenic mice by the NMDA receptor antagonist memantine. Brain Res. 706, 303–7
Tu, J. C., et al. 1998. Homer binds a novel proline-rich motif and links group 1 metabotropic glutamate receptors with IP3 receptors. Neuron. 21, 717–26
Tuite, M. F., Cox, B. S., 2003. Propagation of yeast prions. Nat Rev Mol Cell Biol. 4, 878–90
Unterberger, U., et al. 2005. Pathogenesis of prion diseases. Acta Neuropathol. 109, 32–48

Ushkaryov, Y. A., et al. 1992. Neurexins: synaptic cell surface proteins related to the alpha-latrotoxin receptor and laminin. Science. 257, 50–6

Vassallo, N., Herms, J., 2003. Cellular prion protein function in copper homeostasis and redox signalling at the synapse. J Neurochem. 86, 538–44

Vey, M., et al. 1996. Subcellular colocalization of the cellular and scrapie prion proteins in caveolae-like membranous domains. Proc Natl Acad Sci USA. 93, 14945–9

Waites, C. L., et al. 2005. Mechanisms of vertebrate synaptogenesis. Annu Rev Neurosci. 28, 251–74

Wadia, J. S., et al. 2008. Pathologic prion protein infects cells by lipid-raft dependent macropinocytosis. PLoS One. 3, e3314

Wang, H., Tiedge, H., 2004. Translational control at the synapse. Neuroscientist. 10, 456–66

Ward, R. J., et al. 1998. Copper and iron homeostasis in mammalian cells and cell lines. Biochem Soc Trans. 26, S191

Watt, N. T., Hooper, N. M., 2003. The prion protein and neuronal zinc homeostasis. Trends Biochem Sci. 28, 406–10

Watt, N. T., et al. 2005. Reactive oxygen species-mediated beta-cleavage of the prion protein in the cellular response to oxidative stress. J Biol Chem. 280, 35914–21

Weise, J., et al. 2004. Upregulation of cellular prion protein (PrPc) after focal cerebral ischemia and influence of lesion severity. Neurosci Lett. 372, 146–50

White, A. R., et al. 1999. Prion protein-deficient neurons reveal lower glutathione reductase activity and increased susceptibility to hydrogen peroxide toxicity. Am J Pathol. 155, 1723–30

White, A. R., et al. 2001. Sublethal concentrations of prion peptide PrP106–126 or the amyloid beta peptide of Alzheimer's disease activates expression of proapoptotic markers in primary cortical neurons. Neurobiol Dis. 8, 299–316

Wong, B. S., et al. 2001. Aberrant metal binding by prion protein in human prion disease. J Neurochem. 78, 1400–8

Wong, B. S., et al. 2000. Prion disease: A loss of antioxidant function? Biochem Biophys Res Commun. 275, 249–52

Xia, C. H., et al. 2003. Abnormal neurofilament transport caused by targeted disruption of neuronal kinesin heavy chain KIF5A. J Cell Biol. 161, 55–66

Yu, J., Lyubchenko, Y. L., 2009. Early stages for Parkinson's development: alpha-synuclein misfolding and aggregation. J Neuroimmune Pharmacol. 4, 10–6

Yuan, J., Yankner, B. A., 2000. Apoptosis in the nervous system. Nature. 407, 802–9

Zakharov, V. V., et al. 2003. Natural N-terminal fragments of brain abundant myristoylated protein BASP1. Biochim Biophys Acta. 1622, 14–9

Zanata, S. M., et al. 2002. Stress-inducible protein 1 is a cell surface ligand for cellular prion that triggers neuroprotection. EMBO J. 21, 3307–16

Zhai, R. G., Bellen, H. J., 2004. The architecture of the active zone in the presynaptic nerve terminal. Physiology (Bethesda). 19, 262–70

Index

A

Aβ. *See* Amyloid β-protein
αB-crystallinopathy, 56
Aβ peptide, 273–275
AD. *See* Alzheimer's disease
Age-dependent neurodegenerative disorders, 206
Aggregation. *See* Protein aggregation
Alpha-synuclein (aSyn)
  gain or loss of function, 263
  intracellular trafficking, 264
  putative functions, 263
  Torpedo californica, 262
ALS8. *See* Amyotrophic lateral sclerosis type 8
Alzheimer's disease (AD)
  amyloid β-protein (Aβ)
    amyloid cascade hypothesis, 272–273
    genetics of, 273–275
    physiology of, 272
    synaptotoxicity of, 275–277
  APP proteolysis, 91
  cysteine-string protein, 167
  dementia
    neuropathological hallmarks, 270
    pathophysiology of, 271–272
  Hsp40 co-chaperones role, 91–92
  immunisation against Aβ
    animal models, 277–278
    patients, 279–282
  neuritic plaques, 90
  and prion diseases, 5
  sHSPs, 92
  tau effects, 91
Amyloid β-protein (Aβ)
  AD patients immunisation
    animal models, 277–278
    clinical course, 280–282
    pathology of, 279–280
  amyloid cascade hypothesis, 272–273
  genetics of
    peptide, 273–275
    tau protein, 275
  physiology of, 272
  synaptophysin staining *vs.* aβ42 staining, 280, 281
  synaptotoxicity of, 275–277
Amyloid cascade hypothesis, 272–273
Amyloid fibrils, aggregation
  generic feature, 16–17
  solubility and protofilaments, 16
  therapeutic potential of, 17
Amyloid precursor protein (APP), 5
Amyotrophic lateral sclerosis (ALS), 52
Amyotrophic lateral sclerosis type 8 (ALS8), 216
Anfinsen cage, 26
Apoptosis
  prion disease, 300–302
  small Hsp mutations, 63–64
  SUMOylation, 184
APP-like interacting protein 1 (APLIP1), 113–114
Assembly assistance chaperones, 30
aSyn. *See* Alpha-synuclein
α-synuclein mutant, Parkinson's disease, 89
Autonomous synapse bioactivity, 3
Autophagy
  chaperone-assisted degradation, 42–43
  small Hsp mutations, 63–64
Auxilin, clathrin-mediated endocytosis, 138–139
Axonal growth and degeneration, UPS, 204–206
Axonal transport
  anterograde delivery, 82
  cargo attachment regulation, 88
  components, 84

A. Wyttenbach and V. O'Connor (eds.), *Folding for the Synapse*,
DOI 10.1007/978-1-4419-7061-9, © Springer Science+Business Media, LLC 2011

Axonal transport (*cont.*)
  core motor complex assembly, 82–84
  dysfunctional disease
    Alzheimer's disease (AD), 90–92
    Huntington's disease (HD), 92–95, 243–244
    Parkinson's disease (PD), 89
    synaptic dysfunction, 88
    torsional dystonia, 90
  kinesin and dynein, 84
  neuronal cytoskeleton, 84–85
  vesicle formation and trafficking, 85–87

B

BAG-1 co-chaperone, 39–41
Basal ganglia circuitry, 259–261
Brain-derived neurotrophic factor (BDNF), 243
β structure, amyloid fibrils, 16–17

C

Ca^{2+} homeostasis, 165–166
Cell differentiation, small Hsp, 53
Cellular adhesion, 297
Cellular prion protein (PrP^{C})
  biochemical properties, 290
  conformational changes, 298
  genetic background, 290
  localization of, 300, 301
  pathophysiological implications
    mitochondria, apoptosis, and oxidative stress, 300–302
    prion protein and nerve terminal, 299–300
    synapse and under-explored mechanisms, 302–303
  physiological functions
    cellular adhesion, 297
    cellular signaling, 296–297
    copper, 297
    differentiation, 293–294
    endocytosis, 297
    functional implications, 294–297
    neurite outgrowth, 293–294
    neuronal development, 293–294
    neuroprotective functions, 292–293
    protection against oxidative stress, 293
    synaptic protein interactions, 295–296
  structure of, 290, 292
Chaperone-assisted degradation
  autophagy, 42–43
  Hsp70 proteins, 36–37
  macroautophagy, 43–44
  regulation, 41–42
  ubiquitin/proteasome system
    BAG-1 co-chaperone, 39–41
    CHIP-mediated ubiquitylation, 38–39
    initial substrate recognition, 37–38
Chaperone-mediated autophagy (CMA), 42–43
Chaperones
  assembly assistance, 30
  biochemical organization
    ATP binding, 131
    clathrin coats and Hsc70 chaperones, 137–139
    folding activity, 131–133
    NSF, fusion protein, 133–137
    selective targeting, 139
  co-chaperones, 14
  and disease, 31
  distinct families, 12–14
  folding assistance
    bacteria, archaea and eukarya, 25
    large chaperones, 28–30
    small chaperones, 23–28
  function, 11–13
  non-covalent binding, 12
  nucleoplasmin, 10
  parsimony principle, 10
  presence and distribution in
    cell compartments to brain regions, 124–125, 130
    Hsp60, Hsp70, Hsc70 and Hsp90, 127–128
    small heat shock proteins, 126–127
    synaptic distribution, 128–129
  protein self-assembly, 10
  rubisco, 10–11
Chaperonopathies, 64–65
Charcot-Marie-Tooth (CMT) disease, 58–59
CHIP-mediated ubiquitylation, 38–39
Clathrin coats, 137–139
Co-chaperones
  auxilin, 85
  Bag-1, 27, 39–40
  Bag-3, 43
  CHIP, 38
  definition, 23
  domains, 146–148
  GrpE, 27
  HSJ1, 39
Complexin (Cplx) expression
  behavioural abnormalities in, 239–240
  Ca^{2+}-triggered exocytosis, 237–239
  knockout mice studies, 240–241
  levels in HD brains, 239

Compound muscle action potentials (CMAPs), 154
Conformational breathing, 19–20
Copper and prion protein, 297
CSP/Hsc70/SGT complex
  biochemical features, 160–161
  client protein interaction, 161
  SGT proteins, 159–160
Cysteine-string protein (CSP)
  AD, PD, and polyQ diseases, 167
  cysteine-string domain, 157–158
  evolutionary conservation, 148–149
  functions
    Ca^{2+} homeostasis, 165–166
    Ca^{2+}-triggered vesicle fusion, 163
    synaptic vesicle exocytosis, 164–165
  Hsp70 protein
    exocytosis and endocytosis, 148
    functional versatility, 146–148
    J domain-containing proteins, 148
    pleiotropy, 146
  J domain, 156–157
  linker domain, 158–159
  neuronal and non-neuronal cell expression, 149–150
  neuroprotective role, 161–163
  presynaptic terminals, 145–146
  significance
    neurodegenerative defects, 150–151
    synaptic defects, 151–156
  trimeric CSP/Hsc70/SGT complex
    biochemical features, 160–161
    client protein interaction, 161
    SGT proteins, 159–160

D
Degradation pathway
  chaperone-mediated autophagy (CMA), 42–43
  Hsp70 proteins, 36–37
  macroautophagy, 43–44
  regulation, 41–42
  ubiquitin/proteasome system, 37–41
Dementia
  neuropathological hallmarks, 270
  pathophysiology of, 271–272
DnaK, 27–28
Docking mechanisms, mitochondria, 114–115
Down's syndrome (DS), 273
Dyneins, 3

E
Electron microscopy (EM) studies, 2
Endocytosis
  Huntington's disease (HD), 244–245
  prion diseases, 297
  PrP^C trafficking, 297
ER docking proteins, 218–219
Exocytosis, Huntington's disease (HD), 237–239

F
Folding assistance chaperones
  bacteria, archaea and eukarya, 25
  large chaperones, 28–30
  small chaperones, 23–28

G
Gammaaminobutyricacid decarboxylase (GAD), 296
GFP-VAPB fusion proteins, 221
Glial cells, 303–304
GroEL/ES chaperone system, 28–30

H
HD. *See* Huntington's disease; Huntington's disease, synaptic dysfunction
Heriditary sensory neuropathy (HSN), 65
Hsc70, clathrin-mediated endocytosis, 137–139
HSJ1 co-chaperone, 39
Hsp70 protein
  chaperone-assisted degradation, 36–37
  cysteine-string protein (CSP)
    exocytosis and endocytosis, 148
    functional versatility, 146–148
    J domain-containing proteins, 148
    pleiotropy, 146
  DnaK, 27–28
Huntington's disease (HD)
  Hsp70 role, 92
  Htt aggregation, 92–94
  polyQ aggregates, 93–94
  ROS and toxicity, 95
  synaptic dysfunction
    axonal transport disruptions, 243–244
    complexin levels, 239–240
    Cplx I and Cplx II expression, 239–241
    definition of, 241–242
    endocytosis disruptions, 244–245
    exocytosis, 237–239
    mouse models, 235

Huntington's disease (HD) (*cont.*)
mutant huntingtin role, 241–242
neuropathology of, 234
neurotransmitter, 234–235
protein aggregation, 242–243
proteins distribution, 239
receptor abnormalities, 234–235
R6/2 mouse, abnormality and neurochemistry, 236–237
SNARE proteins role, 237–239
synaptic plasticity changes, 245–246
ubiquitin-proteasome system (UPS), 242–243
ubiquitin-proteasome system, 208–209
Hypoxia-inducible transcription factor (HIF1α) activity, 52

I
Immunisation, amyloid β-protein
animal models
disease progression, 278
pathology attenuation, 277–278
clinical course, 280–282
pathology of, 279–280
synaptophysin staining *vs.* Aβ42 staining, 281
Inherited degenerative diseases, small Hsp
aggregates formation, 60–61
altered protein-protein interaction, 62–63
apoptosis and autophagy, 63–64
cytoskeletal disorganization, 61–62
inefficient chaperoning activity, 59–60
localization, 57
motor neuropathies, 58–59
myofibrillar myopathies, 56–58
Intracellular trafficking, 263–264

K
Kinesin
KIF5 genes, 108
motor proteins, 3
TRAK family, mitochondrial transport
aggregation, 109–110
domain structures, 111
milton gene, 108
Miro1 and Miro2, 110
N-acetylglucosamine transferase interaction, 111
trafficking complex, 112

L
LBs. *See* Lewy bodies
Lewy bodies (LBs), 89, 258–259
Long-term depression (LTD), 276
Long term potentiation (LTP), 276

M
Macroautophagy, 43–44
Macromolecular crowding, 20–23
Major sperm protein (MSP) domain
ATF6α interaction, 224
properties, 222, 223
P56S mutation consequence, 226
UPR-regulated promoter, 224–226
McKusick-Kaufman syndrome (MKKS), 64–65
Milton gene, mitochondrial transport, 108–109
Missense mutations, small Hsp, 66–67
Mitochondrial transport
actin filaments based movement, 108
APP-like interacting protein 1 (APLIP1), 113–114
distribution and mobility
anterograde transport, 106–107
enriched sites, 106
fusion and fission, 107
mitotracker dyes, 106, 107
docking mechanisms
stationary mitochondria, 114
syntaphilin role, 114
TRAK/kinesin/miro complex, 114–115
functions of, 105–106
impaired transport, 108
microtubule based movement, 108
prion disease, 300–302
Ran-binding protein 2 (RanBP2), 113
regulation
Ca^{2+} influx, 115
GTPase cycle inhibition, 116
signalling pathways, 116
syntabulin, syntaxin-binding protein, 112–113
TRAK family, kinesin adaptor proteins
aggregation, 109–110
domain structures, 111
milton gene, 108
Miro1 and Miro2, 110
N-acetylglucosamine transferase interaction, 111
trafficking complex, 112
Mitotracker dyes, 106, 107
MND. *See* Motor neuron disease
Motor neuron disease (MND), 215–216

Motor proteins, 2–3
MSP domain. *See* Major sperm protein domain
Multilameller bodies induction, 221
Myofibrillar myopathy, small Hsp, 56–58

N
N-ethylmaleimide sensitive fusion (NSF) protein
 postsynaptic cell, 136
 SNAP receptors (SNAREs), 133–136
 synapse selectivity mechanism, 136–137
Neurite outgrowth, 293–294
Neurodegenerative disorders, 257–258
Neurofibrillary tangles (NFTs), 270
Neurofilament L (NF-L), point mutations, 62
Neuronal development, 293–294
Nucleoplasmin, 10
Nucleosomes, 10

O
Oxidative stress, prion disease, 300–302

P
Parkin, 44
Parkinson's disease (PD), 4
 cysteine-string protein, 167
 MPTP neurotoxicity, 89
 synaptic dysfunction
  alpha-synuclein (aSyn), 262–264
  basal ganglia circuitry, 259–261
  endocytosis, 264
  folding/unfolding/misfolding reactions, 259
  functional alterations, 261–262
  lewy bodies, 258–259
  protein misfolding diseases, 257–258
 α-synuclein mutant, 89
Point mutations, neurofilament L (NF-L), 62
Polyglutamine (polyQ) diseases
 cysteine-string protein, 167–168
 Huntington's disease, 93–94
 ubiquitin-proteasome system (UPS)
  age-dependent neurodegenerative disorders, 206
  in vitro systems, cell models, 206–207
  synaptic dysfunction, Huntington's disease (HD), 208–209
  transgenic mouse models, 207–208
Prion diseases
 cellular prion protein, 290
 conformational changes, native PrP^C, 298
 overview of, 290, 291
 physiological functions PrP^C
  cellular adhesion, 297
  cellular signaling, 296–297
  copper, 297
  differentiation, 293–294
  endocytosis, 297
  functional implications, 294–297
  neurite outgrowth, 293–294
  neuronal development, 293–294
  neuroprotective functions, 292–293
  protection against oxidative stress, 293
  synaptic protein interactions, 295–296
 prion and glial cells, 303–304
 PrP^C pathophysiological implications
  mitochondria, apoptosis, and oxidative stress, 300–302
  prion protein and nerve terminal, 299–300
  synapse and under-explored mechanisms, 302–303
Protein aggregation
 chaperone
  assembly assistance, 30
  and disease, 31
  folding assistance, 23–30
  function, 11–13
  nucleoplasmin
  rubisco
 complications of, 14–15
 definition, 15–16
 in vivo folding and assembly, 19–20
 macromolecular crowding, 20–23
 mechanisms, 18–19
 structure, 16–17
 ubiquitin-proteasome system, 242–243
Proteinopathies, 5
PrP^C. *See* Cellular prion protein
P56S mutation, 228–229

R
Ran-binding protein 2 (RanBP2), 113
Reactive oxygen species (ROS), prion disease, 301–302
R6/2 mouse
 behavioural abnormalities, 236
 neurochemistry of, 236–237
Rubisco, 10–11

S
Seizures, small Hsp, 55–56

Small heat shock proteins (small Hsp's)
chaperonopathies, 64–65
functions
anti-apoptotic activity, 53–54
cell differentiation, 53
cellular homeostasis, 52–53
heat-shock factors (HSFs), 52
missense mutations, 66–67
mutations, inherited degenerative diseases
aggregates formation, 60–61
altered protein-protein interaction, 62–63
apoptosis and autophagy, 63–64
cytoskeletal disorganization, 61–62
inefficient chaperoning activity, 59–60
localization, 57
motor neuropathies, 58–59
myofibrillar myopathies, 56–58
in neurological disease
PD and HD, 55
protein aggregation, 54
seizures, 55–56
synapse regulation, 54
α-synuclein toxicity, 55
protein misfolding stress induction, 126–127
structure, 50–52
Soluble N-ethylmaleimide-sensitive factor attachment protein receptor (SNARE)
fundamental role, 135–136
Huntington's Disease, 237–239
NSF chaperoning, 136
synaptobrevin, 133–134
Spastic paraplegia, 64
SUMOylation, 4
Synaptic origami, 2
Synaptic protein interactions, 295–296
Synaptobrevin, 134
Synaptophysin, 299–300
Syntabulin, mitochondrial transport, 112–113
Syntaphilin, 114
Syntaxin-binding protein. *See* Syntabulin, mitochondrial transport

T

Tau protein
dementia, 91
mutations, 275
pathological characteristics in, 270
and SUMOylation, 191–192
Torsional dystonia, 90
TRAK family, kinesin adaptor proteins
aggregation, 109–110
domain structures, 111
milton gene, 108
Miro1 and Miro2, 110
N-acetylglucosamine transferase interaction, 111
trafficking complex, 112

U

Ubiquitin interaction motifs (UIMs), 39
Ubiquitin-proteasome system (UPS)
axon, growth and degeneration, 204–206
BAG-1 co-chaperone, 39–41
CHIP-mediated ubiquitylation, 38–39
degradation reaction, 202
Huntington's disease, 242–243
initial substrate recognition, 37–38
polyglutamine (polyQ) diseases
age-dependent neurodegenerative disorders, 206
in vitro systems, cell models, 206–207
synaptic dysfunction, Huntington's disease (HD), 208–209
transgenic mouse models, 207–208
regulatory role
postsynaptic terminal, 203–204
presynaptic terminal, 202–203
ubiquitination reaction, 201–202

V

VAPB aggregates and neurodegeneration
ALS, 219–220
ALS8 mutant VAPB^{P56S} aggregation, 220
amyotrophic lateral sclerosis type 8 (ALS8), 216
Aplysia californica, 216
domain structure, VAP proteins, 216, 218
ER docking proteins, 218–219
GFP-VAPB fusion proteins, 221
intracellular aggregates, 227
major sperm protein (MSP) domain
ATF6α interaction, 224
properties, 222, 223
P56S mutation consequence, 226
UPR-regulated promoter, 224–226
membrane proteins, luminal domains, 227–229
microtubules, 218
motor neuron disease (MND), 215–216
multilameller bodies induction, 221
VAP gene, 216

W

Wallerian degeneration (Wld^{s}), 205